Scotland After the Ice Age

Scotland After the Ice Age

Environment, Archaeology and History,
8000 BC – AD 1000

Edited by Kevin J. Edwards
and Ian B. M. Ralston

Edinburgh University Press

Scotland: Environment and Archaeology, 8000 BC – AD 1000
was first published in 1997 by John Wiley & Sons.

Reprinted 2005

Edinburgh University Press Ltd
22 George Square, Edinburgh

Printed and bound in Great Britain by
The Bath Press, Bath

A CIP record for this book is available from the British Library

ISBN 0 7486 1736 1 (paperback)

Contents

List of Figures vii

List of Plates ix

List of Tables xi

List of Authors xiii

Preface xvii

1 Environment and People in Prehistoric and Early Historical Times:
   Preliminary Considerations 1
   Kevin J. Edwards and Ian B. M. Ralston

2 Climate Change 11
   Graeme Whittington and Kevin J. Edwards

3 Geomorphology and Landscape Change 23
   Colin K. Ballantyne and Alastair G. Dawson

4 Soils and Their Evolution 45
   Donald A. Davidson and Stephen P. Carter

5 Vegetation Change 63
   Kevin J. Edwards and Graeme Whittington

6 Faunal Change
   The Vertebrate Fauna: *Finbar McCormick and Paul C. Buckland* 83
   Land Snails: *Stephen P. Carter*   104
   Insects: *Paul C. Buckland and Jon P. Sadler*   105

7 The Mesolithic 109
   Bill Finlayson and Kevin J. Edwards

8 The Neolithic 127
   Gordon J. Barclay

9 The Bronze Age 151
   Trevor G. Cowie and Ian A. G. Shepherd

10   The Iron Age   169
     Ian Armit and Ian B. M. Ralston

11   The Roman Presence: Brief Interludes   195
     William S. Hanson

12   The Early Historic Period: An Archaeological Perspective   217
     Ian B. M. Ralston and Ian Armit

13   The Early Norse Period   241
     John R. Hunter

14   Environment and Archaeology in Scotland: Some Observations   255
     Kevin J. Edwards and Ian B. M. Ralston

References   267

A Guide to the Literature since 1996   311

Index   321

List of Figures

1.1   A chronological guide showing archaeological and environmental subdivisions   9
1.2   The counties of Scotland prior to local government reorganization in 1975   10
2.1   Locations of the Polar Front through time   12
3.1   The structural provinces of Scotland, showing the distribution of major rock types   24
3.2   Limits of Late Devensian glaciation   28
3.3   Generalized distribution of drift deposits in Scotland   31
3.4   Quadratic trend surface maps for (a) the Main Lateglacial Shoreline and (b) the Main Postglacial shoreline in Scotland   34
3.5   Time–altitude graph depicting relative sea-level curves   38
3.6   Extent of inundation of east-central Scotland during the Main Postglacial Transgression   40
3.7   Sediment deposition rate at Braeroddach Loch   43
4.1   Schematic distribution of soils for a typical area in the Southern Uplands   50
4.2   Distribution of land capability classes for agriculture   51
4.3   Map of archaeological sites with buried soils   55
5.1   Woodland in Scotland *c.* 5000 BP (3780 cal BC)   65
5.2   Isochrone maps of *Corylus* and *Ulmus* pollen for Scotland   66
5.3   Location of sites mentioned in Chapter 5   68
5.4   Pollen diagram for Black Loch II, Fife   69
5.5   Mesolithic age pollen spectra from Loch an t-Sìl, South Uist   71
5.6   Pollen diagram for Black Loch I, Fife   77
5.7   Pollen diagram for the period *c.* 6330 BP (5270 cal BC) to the present from Lochan na Cartach, Barra   78
5.8   Pollen diagram for Saxa Vord, Unst, Shetland   80
6.1   Lateglacial to mid Holocene sites referred to in Chapter 6   84
6.2   Late Holocene sites referred to in Chapter 6   85
7.1   Sites and key areas mentioned in Chapter 7   111
7.2   Mesolithic artefacts   114
7.3   Radiocarbon dates for the Mesolithic   118
8.1   Map showing sites mentioned in Chapter 8   130
8.2   Distribution map of certain and possible henges and small hengiform enclosures (less than 20 m in diameter); cursus monuments; Clava cairns and recumbent stone circles   136
8.3   Pit-defined enclosures and cursus monuments in Scotland   137
8.4   Orkney buildings   145

8.5	Structures on the mainland and the Western Isles	146
9.1	Map showing sites mentioned in Chapter 9	152
9.2	Balnabroich, Strathardle, Perthshire archaeological landscape	160
9.3	Reconstruction of the roundhouse shown in Plate 9.6	162
10.1	The conventional scheme for the subdivision of the Scottish Iron Age	171
10.2	Simplified plans of Scottish Iron Age sites	172
10.3	Sites mentioned in Chapter 10	173
10.4	Schematic representation of the principal phases of enclosure represented at Hownam Rings, Roxburghshire and Broxmouth, East Lothian	177
10.5	Distribution of Scottish hillforts and duns	181
10.6	Plans, and access maps indicating differential complexity, of complex Atlantic roundhouses and wheelhouses	186
10.7	Landscape subdivisions in the cropmark record around Castlesteads and Newton in Midlothian	190
11.1	The Roman occupation of Scotland in the early Antonine period (*c.* AD 142–158)	196
11.2	The Roman occupation of Scotland at its furthest extent in the Flavian period (*c.* AD 84–87)	199
11.3	Hadrian's Wall, third-century Roman forts and non-Roman sites mentioned in Chapter 11	200
12.1	Distribution of 'pit' names, Pictish symbol stones and Pictish silver chains	220
12.2	Sites mentioned in Chapter 12	223
12.3	Pictish buildings in a variety of architectural styles	227
12.4	British and Anglian buildings in East Lothian	228
12.5	The fauna of Pictland as depicted on symbol stones	235
13.1	General area of Norse influence in northern and western Scotland	242
14.1	The number of excavations assigned to archaeological periods, showing the proportions of Scottish excavations featuring environmental and non-environmental investigations	258
14.2	The major types of environmental data obtained from Scottish excavations expressed as percentages of all environmental investigations	260
14.3	The percentages of Scottish excavation reports featuring environmental investigations over the review period	261

List of Plates

2.1 Pine stumps appearing from beneath peat at Clatteringshaws Loch, Stewartry of Kircudbright   16

3.1 Quartzite mountains rising above a platform of glacially scoured gneiss bedrock, Arkle, Sutherland   25

3.2 Glaciated landscape of the Loch Quoich area, north-western Highlands   29

3.3 Fertile deltaic and outwash deposits amid the mountains of the north-western Highlands   29

3.4 Chains of hummocky recessional moraines deposited at the end of the Loch Lomond Stadial, Luib, Isle of Skye   32

3.5 Relict cliffline in western Jura uplifted as a result of glacio-isostatic processes   35

3.6 Fossil clifflines with Lateglacial unvegetated raised beach ridges in western Jura   36

4.1 Braeroddach Loch, Aberdeenshire   60

5.1 Coppiced hazel on the Isle of Mull   70

5.2 Black Loch, Fife   74

5.3 Lochan na Cartach, Barra   79

5.4 Peat cutting in blanket peat on slopes of Reineval, South Uist   81

7.1 Mesolithic shell middens on Oronsay   110

7.2 The machair at Bàgh Siar, Vatersay   113

7.3 Kinloch, Isle of Rhum, with the Cuillins of Skye in the background   116

7.4 Terraces of the River Dee, near Banchory, Kincardineshire   121

8.1 The chambered cairn at Cairnholy, Dumfriesshire   131

8.2 The long barrow known as Herald Hill, Perthshire   132

8.3 Excavations on the north-east cairn at Balnuaran of Clava   133

8.4 The excavated henge monument at North Mains, Strathallan, Perthshire   134

8.5 The long cairn at Auchenlaich near Callendar   135

8.6 An aerial view of the Neolithic enclosure at Douglasmuir, Angus   138

8.7 Recumbent stone circle at Loanhead of Daviot, Aberdeenshire   138

8.8 A sherd of Grooved Ware with residues of its contents   144

8.9 The building at Balbridie, Kincardineshire, under excavation   147

9.1 Jet necklace from Pitkennedy, Angus   154

9.2 Bronze Age burial from Cnip Headland, Uig, Lewis   155

9.3 Reconstruction of the facial features of the Bronze Age adult male from Cnip   156

9.4 The remains of the wooden disc wheel found at Blair Drummond Moss   157

9.5   Balnabroich, Strathardle, Perthshire. Hut-circles, field systems and
       small cairns showing from the air   159
9.6   Large roundhouse in the course of excavation, Lairg, Sutherland   161
9.7   Ard marks at Rosinish, Benbecula   163
9.8   Ox yoke found in a bog at Loch Nell, Argyll   164
9.9   Traces of a relict prehistoric landscape at Drumturn Burn, Alyth,
       Perthshire   167
10.1 Hillfort, Hownam Law, Roxburghshire, with hut platforms   174
10.2 Cropmarks of an unenclosed settlement in the North Esk valley,
       Angus   175
10.3 Remnants of an experimental timber-laced wall   178
10.4 Traprain Law, East Lothian   180
10.5 Broch at Dun Carloway, Lewis   184
10.6 Hut Knowe, Roxburghshire: an enclosed settlement with an external
       trackway and bounded plots of cord rig agriculture   191
11.1 The line of the Antonine Wall ditch and upcast mound across Croy
       Hill from the air   197
11.2 The line of the road, picked out by its quarry pits, and an adjacent
       timber watchtower on the Gask Ridge at Westerton   201
11.3 The Antonine auxiliary fort and parts of two adjacent temporary
       camps at Glenlochar   202
11.4 The 63 acre (25 ha) Severan camp at Kirkbuddo, Angus, showing as
       cropmarks   210
11.5 Ditched field systems outside the *vicus* at Inveresk showing as
       cropmarks   213
11.6 The broch, unenclosed settlement and earlier hillfort at Edin's Hall   215
12.1 The hillfort at Dundurn, Perthshire   224
12.2 The timberlaced rampart (built of reused oak) at Green Castle,
       Portknockie   225
12.3 The tidal islet of the Brough of Birsay and the site of Buckquoy   232
12.4 A battle scene on the Class II slab at Aberlemno Kirkyard, Angus   236
12.5 The Dupplin Cross, Perthshire   237
12.6 Aerial view of cropmarks near Boysack in the Lunan Valley, Angus   238
13.1 The eroded boat burial at Scar, Sanday, Orkney during excavation   246
13.2 Sorisdale, Coll, featuring the fertile pocket of land and sheltered bay
       – characteristics favoured by early Norse settlers   247
13.3 The outline of House 1 on site 2 at Skaill, Deerness, Orkney   250
13.4 Increasing use of cattle for traction at Pool, Orkney illustrated by
       metapodial showing extension of articular end and infection,
       probably resulting from arthritis   252
13.5 Scapula from a red deer at Pool showing new bone formation
       resulting from wound to tissue which seems to have been caused by a
       projectile   253
14.1 Deflation of machair sands by wind causing damage to a Bronze Age
       settlement at Cladh Hallan, South Uist   262
14.2 These rat holes have pierced the sandy soils associated with a
       possible Iron Age settlement on Sandray   263

List of Tables

4.1	Classification of Scottish soils	47
4.2	Main soil associations, their extent and parent materials in Scotland	48
4.3	Occurrence in rank order of major soil groups or subgroups in Scotland	49
4.4	Extent of land capability classes for agriculture in Scotland	52
4.5	Sites for which there is published information on buried soils in Scotland	54
6.1	Animal bone from mesolithic sites at Oronsay, Morton and Carding Mill Bay	89
6.2	Distribution of bird fragments from a selection of Scottish sites	92–6
6.3	Fish from a selection of sites of varying date	97–8
11.1	Estimated Roman garrisons in Scotland	204
11.2	Dated pollen diagrams and the onset of major forest clearance	209
11.3	Tree species from Roman forts represented by macrofossil evidence	211

List of Authors

Ian Armit   MA, PhD (Edinburgh) is Senior Lecturer in Archaeology at The Queen's University of Belfast and was formerly an Inspector of Ancient Monuments with Historic Scotland. He has written numerous books and articles on Scottish and north-west European archaeology and is currently co-directing a programme of excavation on the major Scottish hillfort of Traprain Law. His most recent books are *Scotland's Hidden History*, *Celtic Scotland* and *The Archaeology of Skye and the Western Isles*.

Colin K. Ballantyne   MA (Glasgow), MSc (McMaster), PhD (Edinburgh), DSc (St Andrews) is Professor of Physical Geography at the University of St Andrews. His principal research interests are in the fields of glacial, periglacial, paraglacial and hillslope geomorphology and Late Quaternary landscape evolution, particularly in Scotland. He is co-author of *The Periglaciation of Great Britain*.

Gordon J. Barclay   MA, PhD (Edinburgh) is editor of the *Proceedings of the Society of Antiquaries of Scotland* and Principal Inspector of Ancient Monuments for Historic Scotland. He has excavated and written extensively on the prehistory of east-central Scotland and he is particularly interested in the historiography of prehistory. He has published a monograph (with G. S. Maxwell) on the excavations of the Cleaven Dyke, and *Farmers, Temples and Tombs*, a study of Neolithic Scotland.

Paul C. Buckland   BSc, PhD (Birmingham) is Professor in the Department of Archaeology and Prehistory, University of Sheffield. He formerly held posts with the York Archaeological Trust and Doncaster Museum before becoming Lecturer in Geography at the University of Birmingham. His principal interests lie in the insect biogeography of Atlantic islands and the conservation of wetlands.

Stephen P. Carter   BSc (Bristol), PhD (London) is a Director of Headland Archaeology Ltd in Edinburgh. He has worked as a consultant soil scientist and holds an honorary position in the Department of Environmental Science, University of Stirling. He is particularly interested in sediments from archaeological contexts in Scotland, including the use of soil micromorphology.

Trevor G. Cowie   MA (Edinburgh) is Curator of the Bronze Age collections in the Department of Archaeology, National Museums of Scotland and was a field archaeologist with the former Central Excavation Unit, Scottish Development Department. His research interests include the Neolithic and Bronze Age periods and he is a co-author of *Symbols of Power at the Time of Stonehenge*.

Donald A. Davidson   BSc (Aberdeen), PhD (Sheffield) is Professor in the Department of Environmental Science, University of Stirling, having held posts in the Universities of Sheffield and Wales, and a Readership in Geography at the University of Strathclyde. He has worked on geoarchaeological aspects of projects in Greece and Scotland. His current research interests include applications of soil micromorphology and the spatial variability of soil properties. He is the author or co-editor of numerous books including *Principles and Applications of Soil Geography* and *Geoarchaeology: Earth Science and the Past*.

Alastair G. Dawson   MA (Aberdeen), MS (Louisiana), PhD (Edinburgh) is Professor in Quaternary Science in the School of Science and the Environment at Coventry University. His principal research interests lie in Quaternary palaeoclimatology, in particular Late Holocene climate changes in the North Atlantic region. He has also published widely on aspects of Quaternary sea-level changes and tsunami research. He is author of *Ice Age Earth*.

Kevin J. Edwards   MA (St Andrews), PhD (Aberdeen) is Professor of Physical Geography in the Department of Geography and Environment, University of Aberdeen and Adjunct Professor in the Graduate School, The City University of New York. He was on the staff of the Departments of Geography at the Universities of Belfast and Birmingham before becoming Professor and Head of the Department of Archaeology and Prehistory, University of Sheffield. His research interests focus on Scotland and the North Atlantic region, and include applications of palynology, sedimentology and tephra studies in archaeology. For ten years he was co-editor of the *Journal of Archaeological Science* and he is currently Associate Editor of *Environmental Archaeology*.

Bill Finlayson   MA, PhD (Edinburgh) was Manager of the Centre for Field Archaeology at the University of Edinburgh until his appointment as Director of the Council for British Research in the Levant, based in Amman. He has been responsible for a wide variety of archaeological projects in Scotland and as far afield as Jordan. His research interests lie primarily within the Mesolithic, and focus especially on the use of lithic materials. He is the author of *Wild Harvesters*.

William S. Hanson   BA, PhD (Manchester) is Professor of Roman Archaeology in the Department of Archaeology, University of Glasgow. His research interests concentrate on the northern provinces of the Roman Empire, particularly the interaction between Rome and the indigenous population. His publications include *Agricola and the Conquest of the North* and *Rome's Northwest Frontier: The Antonine Wall*. He co-edited *Scottish Archaeology: New Perceptions*.

John R. Hunter   BA (Durham) PhD (Durham and Lund) is Professor of Ancient History and Archaeology at the University of Birmingham and was formerly Reader in Archaeology at Bradford University. He is experienced in multi-period survey and excavation in Scotland, particularly in the Northern and Western Isles. His books include *Rescue Excavations at the Brough of Birsay* and *Fair Isle: the Archaeology of an Island Community*. He is also a co-editor of *Studies in Crime: an Introduction to*

Forensic Archaeology, The Archaeology of Britain and *Archaeological Resource Management in the UK.*

Finbar McCormick   MA (Cork), PhD (Belfast) is Lecturer in Scientific Archaeology and Head of the School of Archaeology and Palaeoecology in The Queen's University of Belfast. He was formerly an archaeologist with AOC (Scotland) Ltd. His research interests are primarily on the evolution of the relationship between people and animals in Ireland and Scotland.

Ian B. M. Ralston   MA, PhD (Edinburgh) is Professor of Later European Prehistory in the Department of Archaeology, University of Edinburgh, having been a lecturer in the Department of Geography, University of Aberdeen. He maintains research interests in the later prehistory of France and in aspects of Scottish archaeology. He authored *Les Enceintes fortifiées du Limousin*, and (with O. Buchsenschutz) *Les remparts de Bibracte* and a study of Iron Age Bourges. He co-edited *Archaeological Resource Management in the UK* and *The Archaeology of Britain.*

Jonathon P. Sadler   MSc (Birmingham), PhD (Sheffield) is Senior Lecturer in Biogeography in the School of Geography, Earth and Environmental Sciences, University of Birmingham and was formerly a Research Fellow in the Department of Archaeology and Prehistory at the University of Sheffield. He has carried out extensive palaeoentomological research associated with archaeological projects in Scotland, Iceland and Greenland.

Ian A. G. Shepherd   MA (Edinburgh) is Principal Archaeologist for Aberdeenshire Council and held a similar post with the former Grampian Regional Council. His research interests centre on the earlier Bronze Age of western Europe. He is a former editor of the *Proceedings of the Society of Antiquaries of Scotland* and the author of *Powerful Pots; Beakers in North-East Prehistory* and *Aberdeen and North-East Scotland.*

Graeme Whittington   BA, PhD (Reading) is Emeritus Professor of Geography in the School of Geography and Geosciences, University of St Andrews. His research interests lie in Scotland and are concerned with landscape and environmental change, particularly palynological and sedimentological approaches. His publications include *An Historical Geography of Scotland* and *Fragile Environments.*

Preface

This paperback represents an updated edition of *Scotland: Environment and Archaeology, 8000 BC – AD 1000*, published by Wiley in 1997. Our aim then was to provide an overview of Scotland's natural environment and the human communities which inhabited it from the end of the last Ice Age until about a thousand years ago. That book represented a successful collaborative venture involving numerous colleagues working across a range of disciplines. We are indebted to our original publisher, Iain Stevenson, for his initial guidance and to John Davey and his colleagues at Edinburgh University Press for their subsequent support.

The new feature of the present volume is a selective guide to the literature that has emerged over the last few years. The quantity of research published between 1996 and early 2002 has been impressive. We trust that, with the assistance of the contributing authors, we have identified many of the key items and have conveyed a flavour of some newer developments. Much has had to be omitted and the full significance of some such contributions will undoubtedly only become apparent in the future. Final responsibility for this selection rests with us.

In this more accessible format, *Scotland After the Ice Age* should continue to serve those interested in the intertwined natural and cultural dimensions of the country's past. We hope that readers will be able to appreciate the advances that have been made over recent decades and that these essays will act as a spur to increasing integration among the various research fields.

Like its predecessor, this volume owes much to our wives, Rachel and Sandra, and to our children, Calum, Fraser, Natalie and Tom, who again have put up with playing second fiddle to computer screens and keyboards.

Kevin Edwards and Ian Ralston,
Maryculter and Kinross, June 2002

1 Environment and People in Prehistoric and Early Historical Times: Preliminary Considerations

KEVIN J. EDWARDS AND IAN B. M. RALSTON

INTRODUCTION

During and since the final stages of deglaciation some 10 000 years ago, the Scottish landscape has been dramatically fashioned by natural and human forces. Climate and geology have influenced the physical make-up of the territory and, to an extent, have dictated the biogeographical patterns revealed by records of flora and fauna. The British Isles, initially forming part of a more substantial landmass and linked by a landbridge to continental Europe at the outset of the period, became fully detached by *c.* 7000 BP.

Taking the long view, Scotland's location on the Atlantic edge of north-western Europe, and its often inhospitable terrain, have not noticeably inhibited human settlement. That said, when the record is examined closely, instances of land formerly occupied and now abandoned, or now used much less intensively than in previous centuries, are numerous. In the relatively recent past, processes such as the Highland Clearances and the Improving Movement have profoundly modified the appearance of substantial tracts of the countryside, as well as having had a serious impact on the distribution and socio-economic life of the human population. Whilst these large-scale changes within the last few centuries have generated a substantial literature, the earlier changes that affected the countryside and its inhabitants have tended in the main to be considered in specialist publications.

AIM

The principal aim of this account is to make information on the environmental and archaeological records for Scotland for the period from the first clear human

Scotland: Environment and Archaeology, 8000 BC – AD 1000. Edited by Kevin J. Edwards and Ian B. M. Ralston.
© 1997 The editors and contributors. Published in 1997 by John Wiley & Sons Ltd.

presence until approximately the emergence of the medieval state, more accessible. The intention is to bring together information and ideas that have been put forward in a wide range of academic literature, more particularly since the 1970s, in a readable overview. This is not the place to indulge in historiographical discussions of the development of research in these fields. It would be inappropriate, however, not to acknowledge present-day indebtedness to pioneering work, and contributors have in some cases included reference to key early studies within their fields. The primary focus of this volume also differs from previous general studies of Scottish prehistory (Childe 1935, 1946; Piggott 1962, 1982; Ritchie and Ritchie 1981, 1991). Tools, ornaments and funerary practices, for instance, are less fully examined, and other categories of archaeological data which can be juxtaposed more readily with environmental information, such as settlement sites and evidence for agriculture, are given proportionately greater prominence.

Scotland offers many circumstances for the excellent preservation of data for past environmental conditions, and to some extent its archaeological sites provide similar opportunities for considering environmental dimensions. In this treatment, these environmental perspectives have been highlighted where this is appropriate, but neither to the total exclusion of other approaches nor, in cases where available information is very weak, by overplaying the significance of the data. What follows is not a conventional archaeological account of Scotland's past, although it deliberately retains enough of the familiar (e.g. the use of the standard system of Stone, Bronze and Iron Ages) to enable the reader to navigate a course. Neither is this a book about environmental archaeology as such. That would require the practice of multidisciplinary approaches to extract and evaluate information on the environment from archaeological contexts or their wider settings. Many of the approaches that underpin the present study are indeed identical, but their deployment has been to furnish perspectives for wider interpretation. Less attention is paid to the strengths and weaknesses of the various suites of environmental data than would be the case in a work where taphonomic and methodological considerations were a major emphasis.

The focus here is upon the nature and extent of human–environment interactions since the final melting of the Lateglacial ice sheets. The early Holocene (Postglacial) was a time when natural environmental changes were already transforming the appearance and resources of the land: it is therefore appropriate that the immediately antecedent conditions are examined in a number of chapters in order to set the scene. It is shown that indications of human exploitation of, and by extension, impact on the environments of Scotland are traceable as far back as human communities are demonstrably present. At times when naturally occurring changes were particularly dynamic, a human presence is sometimes difficult to detect; in later periods, natural changes seem to be overwhelmed by those attributable to human action.

Environment, as considered here, is not simply a mute context for human endeavour – rather, it formed an integral part of life in the past. Until very recently – and certainly well within the present millennium – Scottish communities were exclusively rural and their natural environments formed, consciously or unconsciously, part of the fabric of their lives. To say this is not to imply a deterministic stance. Environmental factors may restrict, but do not eliminate, human choices.

Other factors – social, economic and religious, for instance – were also of importance in contributing to change and stability, and the chapters which follow, whilst
keeping environmental circumstances in mind, also outline some of these influences.
The opposition between environmental and cultural approaches to the study of the
archaeological record found in some recent publications (e.g. J. Thomas 1990), is
inappropriate (cf. rebuttals by O'Connor [1991] and Wilson [1995]). Our position
remains that the environment should not be relegated to a benign neutrality and
disregarded as 'noise' in the creation of a social archaeology, any more than it
should be seen as a series of backcloths with varied designs, in front of which the
actors perform with insouciance.

Studies of the environment and of earlier human communities are both necessarily multidisciplinary, the range of approaches now available exceeding the
capacity of any individual. Whilst the contributors have collectively addressed many
research fields, a work of this kind makes no pretence at being fully comprehensive.
Amongst resources, plant macrofossils and exploitable minerals and ores are some
of the topics which might be considered to merit more extended treatment. The
impact of external trade and contact and a consideration of human population
dynamics might also have enjoyed greater prominence. None the less, the major
active research areas are all included. The levels of integration found in published
research associated with particular topics or cultural periods is variable and the
following contributions reflect this. Thus, the discussion of climate is unidirectional
in that changing climate may influence landscape, soil, vegetational and agricultural
development, but these variables are not shown to have a demonstrable impact
upon climate. During the Mesolithic, a strong interrelationship between environmental constraints and opportunities and human life is assumed, not least because
of the characteristics of the archaeological record. Inferences focusing upon social
behaviour at this time have to allow for the sparseness of exploitable data within
Scotland.

Paradoxically, as more recent prehistoric and historical times are approached,
and knowledge of environmental change might be thought likely to be no less
than in earlier periods, consideration of the influence of environment upon the
human communities seems to be less prominently developed in the literature.
There are a number of reasons underpinning this assertion, which, if overstated, is
not grossly so. On the one hand, certain of the environmental indicators, potentially of great sensitivity in the early periods, become less useful as measures of
change. Woodland, once reduced, is a case in point. Other indicators remain
potentially useful (e.g. insects or stable isotopes), but are little-studied because of
the difficulties involved in obtaining suitable deposits, while the environmental
signals derived from lake sediments could result from a complex mixture of
natural and anthropogenic processes. On the other hand, this reduced emphasis
may be a function of less responsive environmental indicators and hence a relative
lack of interest on the part of environmental researchers. It may be that the
record is overwhelmed by the weight of physical archaeological evidence and
historical information that is available, or indeed, other than when natural
disasters occurred, perhaps there was less environmental influence on the affairs of
people. At the very least, the environment had to be permissive in terms of the
ecological requirements of biota and agricultural practices.

THE SURVIVAL AND DETECTION OF THE ARCHAEOLOGICAL RECORD

Authors were not asked to address in detail the matter of the survival and detection of sites attributable to earlier human communities in the landscapes of Scotland. This question, well rehearsed from different perspectives by Stevenson (1975) and Barclay (1992), is however intimately related to environmental concerns, as the locations of the sites of prehistoric and early historic activities have a greater or lesser chance of being identified substantially as a result of the materials of which they were made, their positions within the varied topography of the country, and subsequent land-use histories.

Current evidence suggests that few major sites occur above the 500 m contour, although there are exceptions, such as Ben Griam Beg in Sutherland (Chapter 9). In the altitudinal band below this, and above the intensively cultivated sectors of the countryside, upstanding monuments survive in substantial numbers and over considerable areas, in permanent grassland, heathland, and in some afforested settings. Alongside clearly visible stone structures, other sites, less immediately apparent, can be noted in quantity. The latter include banks of stones delineating field systems and, in some circumstances, the traces of early cultivation practices within them. Leaving aside the more massive drystone monuments, perhaps too remote to have served as convenient quarries for subsequent reuse, and some land boundaries which may have demarcated parcels of land for extended periods, most of these sites have survived on the surface because subsequent episodes of land use have been insufficiently intensive to prompt their demise. This zone of likely survival – particularly in areas like the short grass moorlands, clipped by large flocks of sheep – represents one of the great resources of the Scottish archaeological record. For here, soils, vegetation and land use have interacted to enable the survival of extensive surface traces attributable to some of the early periods. It has long been appreciated that some of these archaeological landscapes include amongst their visible remains not only the traces of stone-built structures, but also the former stances of those built of wood. Improved fieldworking techniques, built on familiarity with these landscapes and supported in particular by oblique aerial photography, continue to reveal extensive and finely detailed traces, on occasion disappearing below peatland and thus indicative of how that substance now obscures areas previously used by human communities.

In contrast, the lowlands, whether improved substantially for pasture as in the South-West, or given over to arable crops as is more frequently the case in eastern Scotland, have a much reduced tally of monuments surviving from earlier periods. Here, we have entered the zone of likely destruction where, as a result of cultivation since Medieval times, and perhaps most importantly during the periods of intense agricultural improvement of recent centuries, surface indications of many archaeological sites have been wholly eliminated. Artefact collection through field-walking, whilst it can be successfully employed in Scotland, is less used than further south in Britain. Such collection (sometimes supplemented by test-pitting) is of considerable significance for earlier prehistory, notably the Mesolithic period, where, leaving aside midden accumulations, sites are difficult to identify. In favoured lowland zones, the former extent of human activity for all subsequent periods is most extensively displayed through aerial photography of cropmarks. Although pioneering flights were made over Scotland in the 1920s, the scale and regularity of overflying for archaeological reconnaissance has increased dramatically only since the mid 1970s, i.e. relatively recent in British terms. In areas where free-draining soils, especially located over glaciofluvial sands and gravels, coincide with reduced rainfall, extensive, multi-period suites of cropmarks can be recorded. These cropmarks include categories of sites not readily comparable with those in the corpus of upstanding monuments. Archaeologists are only now coming to grips with this wealth of new information. Cropmark records also furnish numerous indications that lower-altitude areas, more favourable for settlement on environmental grounds, were extensively and repeatedly employed in the past.

This brief delineation of zones of likely survival and destruction is necessarily a simplification. In reality, complex patterns result from the interplay of soil, climatic, land-use and other factors. The insertion of modern infrastructure, including motorways and urban sprawl, has removed archaeological evidence wholesale. The long coastline sees the dovetailing of zones of survival on hard rock areas and sometimes within sand-dune systems, with other sectors where natural and human destructive processes have long been at work. Areas masked by peat growth have been mentioned for their potential blanketing of archaeological features, thereby often ensuring good preservation. The numerous bodies of water (and their immediate littorals), used for settlement at least intermittently from the Neolithic, and most famously in the case of the crannogs of late Bronze Age and more recent dates, also provide excellent milieux for the survival of organic artefactual and environmental data. The diver with drysuit and aqualung, and the airborne photographer, are perhaps the two most obvious symbols of the twentieth century search for archaeological remains in Scotland. They here serve to emphasize the incompleteness of the Scottish archaeological record, and the direct relationships between survival and detection of archaeological remains, and environmental and land-use factors.

POPULATION: THE UNQUANTIFIED PARAMETER

In almost all cases, assessing the scale of the impact that is detectable at a particular point is fraught with difficulties, compounded by the fact that archaeology has few, if any, methods at its disposal to estimate previous human population sizes. Compared to the perspectives of researchers of a generation ago, there is a readiness to accept higher figures for earlier populations, although absolute numbers remain elusive. Thus the suggestion of about 70 as the Mesolithic population of Scotland (Atkinson 1962, 7) is considered a serious underestimate (Chapter 7), not only in terms of the carrying capacity of the environment, but also with regard to reproductive viability. Burgess (1989) has proposed a population graph for Britain from 5000 BC until modern statistics become available. Necessarily speculative in terms of absolute numbers, his diagram makes explicit the likelihood of oscillations in human numbers in early times, as well as suggesting, though in the absence of secure evidence, environmental factors, more particularly volcanic events, as a control (cf. Chapter 9). It is a frequent assumption that the development of agricultural regimes may have permitted increasing populations, as milk-and-mush diets made earlier weaning and adequate nutrition of more children a feasibility, and arguably, made the availability of more hands a distinct advantage in undertaking the tasks associated with agricultural production. The literature contains frequent references to demographic expansion (e.g. Atkinson 1968; though see Bradley 1978), but within Scotland this is usually wholly unquantified, other than at the individual monument scale, where speculative forays have been made, as in the case of Isbister and other chambered cairns in Orkney (Hedges 1982; Fraser 1983). For individual archaeological sites, population estimates have either been derived from sets of assumptions about labour requirements for their construction, or, as in the Isbister case, the characteristics of the human skeletal material they contain. This latter approach is predicated on a consideration of whether all, or part of the population of the locality is comprised within the bone assemblage. Even less secure are guesses about population sizes based on the likely number of inhabitants within buildings. At a wider scale, estimates have been made about likely population figures in relation to carrying capacities where near-contemporaneity of extensive sets of sites can be assumed (Fojut 1982). A special case of the impact of a particular population is offered by the Roman army (as discussed in Chapter 11), and even here, although this perspective is underpinned by knowledge from written sources of the size of military units at the time, there remains a considerable measure of uncertainty.

From an environmental perspective, Rackham (1980) has estimated that a minimum population of about half a million adults would have been needed in Britain to reduce elm pollen productivity sufficiently for it to be registered as the widespread elm decline. Whilst this provides us with a 'national statistic', it is, as the author makes plain, one fraught with uncertainty to the point of improbability.

All such approaches to former populations are useful heuristically as an aid to envisaging the possible structures of previous social groups and the potential scale of impact of communities on their environments. The foregoing examples indicate how tentative such attempts necessarily are, and thus how unsure we remain about the size of previous human populations.

PRACTICAL CONSIDERATIONS

The cultural framework and its chronology

The contributors have adopted a relatively traditional perspective on the classification systems employed here. An initial note on the use of these is provided by way of explanation.

Radiocarbon dating

Current understanding both of the chronologies of environmental change and of the duration of human settlement in Scotland is underpinned by the radiocarbon timescale. Early confidence that this was straightforwardly comparable with calendar years has long proved to be misfounded. A complication is introduced by the fact that the radiocarbon chronology is subject to a number of systematic errors, not least because of long-term fluctuations in the ^{14}C content of the earth's atmosphere. Fortunately, it is now possible to calibrate the radiocarbon dates to accommodate some of these effects. Tree-ring calibration (dendrochronology) of radiocarbon dates demonstrated the variable discrepancies between historical years and those derived from the radiocarbon method (e.g. the radiocarbon date for the elm decline of c. 5100 radiocarbon years ago becomes, upon correction, c. 5780 calendar years ago).

It has been the convention in the environmental sciences to work with radiocarbon (^{14}C) dates expressed as years before present (BP; where 'present' equals AD 1950). Thus, for the example of the elm decline, the date would be presented as 5100 BP. This is perfectly satisfactory where the uncorrected radiocarbon time-scale is used solely, but occasions particular difficulties when other sources of dating are employed, or when the true duration of an event or process is of significance. These circumstances apply in many archaeological contexts. Moreover, since the development of the radiocarbon technique, archaeologists have generally expressed uncalibrated ^{14}C dates by reference to the BC/AD convention of historians.

As confidence in calibration procedures has increased (most notably after the publication of a special issue of the journal *Radiocarbon* in 1986), Quaternary scientists are increasingly adopting calibrated time-scales. This trend has been accelerated by the availability of computer calibration programs (e.g. Stuiver and Reimer 1993). To accommodate these developments, the following conventions have been adopted throughout this work:

(i)   Uncalibrated radiocarbon dates will be cited in years BP.
(ii)   Calibrated radiocarbon dates will be cited as years cal BC or cal AD, using the correction software of Stuiver and Reimer (1993), but only for dates more recent than 10 000 BP. Figures following BP dates, and labelled cal BC or cal AD as appropriate, identify the corresponding calibrated chronological ranges at one standard deviation. The term 'cal BC' will be applied to longer time-spans (e.g. millennia), where the period concerned has been primarily defined by use of calibrated radiocarbon dates.
(iii)   The use of BC or AD in relation to a specific date will mean that this has not been derived from radiocarbon, but normally from historical sources.
(iv)   The calibrated radiocarbon date derived from a raw radiocarbon determination (i.e. including the ±1 standard deviation associated with it) will be expressed, rounded to the nearest decade, as a one standard deviation corrected range or the extremes of corrected ranges where these are produced (e.g. 3880±60 BP becomes, upon conversion and rounding 2460–2200 cal BC).

Impact of calibration on the cultural framework

In general terms, the availability of radiocarbon dating has had the effect of lengthening the chronological spans of the earlier prehistoric periods from those envisaged by mid-century scholars, who were largely reliant on cross-dating with the historical civilizations of the Mediterranean Basin, using material comparable to that found in Britain from parts of continental Europe. For example, the likely start date, and the duration, of the New Stone Age (Neolithic) of Scotland was recognized, as radiocarbon dates became available, to lie centuries before Piggott's (1954) estimate of *c.* 2000 BC; the impact of calibration has been to extend the timespan, and to push back the initial appearance of Neolithic traits yet further. It follows that the lengthening chronology for human settlement of Scotland extends the time-scale over which human–environment interactions can have taken place.

The traditional archaeological subdivisions of cultural periods have also been retained (Figure 1.1). These are still firmly embedded in the literature on which this book is based, and despite the shortcomings of divisions essentially founded on technological developments, they serve as useful labels to enable correlations to be achieved. The Mesolithic period, associated with hunter–gatherer–fisher activity, occupies the span up to around the beginning of the fourth millennium cal BC, with the Neolithic continuing for about one and a half millennia thereafter. The first appearance of copper and copper-alloy equipment heralds the Bronze Age from perhaps a little before *c.* 2500 cal BC. Iron's appearance as the preferred material for edge tools, marking the Iron Age, may be estimated to have occurred around the seventh century BC, but environmental conditions within Scotland mean that ferrous material does not survive well. Use of these labels should not prejudice the possibility of considerable overlaps between some of the technological and subsistence strategies implied by these terms, most notably those conventionally understood by Mesolithic and Neolithic (Armit and Finlayson 1992).

Geographical units

A further complication is produced by the fact that twice in the last quarter of a century, the map of local authority areas in Scotland has been substantially redrawn. The local government reorganization effected in Scotland in 1996 disposed of the two-tier system of Regions and Districts that had been in place since 1975. The new local authority areas, a system set in place whilst this book was in preparation, are unfamiliar, and include areas both too large (e.g. Highland) and too small (e.g. East Renfrewshire) to be of great service; equally, other names (e.g. Aberdeenshire) are now attached to different areas than those traditionally associated with them. The two-tier system of Regions and Districts that preceded them, in use for only two decades, has too rapidly been consigned to history for it to have been adopted here. The most useful course of action seemed to us to be to revert to the pre-1974 counties as the best-established convenient geographical units, and any reference to counties in the following chapters thus refers to these (Figure 1.2).

On a broader scale, the restriction of the scope of the present study to the modern political boundaries of Scotland, whilst conforming to recent traditions of archaeological heritage management, is culturally and environmentally meaningless for earlier times. It should thus be regarded largely as a convenient means of providing recognizable geographical limits to the present study, beyond which contributors have been permitted to stray as the nature of their subject material warranted.

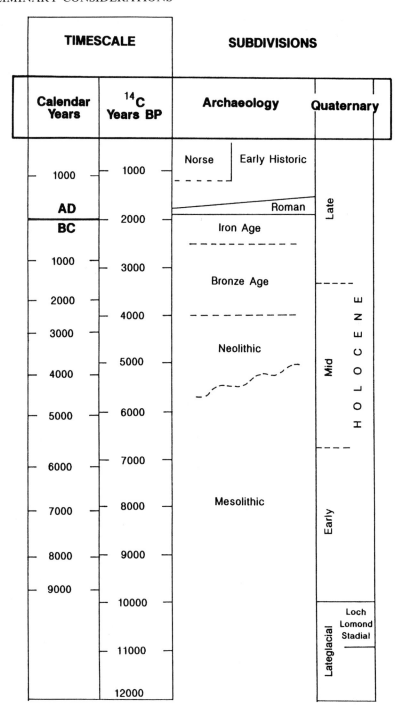

Figure 1.1 A chronological guide showing archaeological and environmental (Quaternary period) subdivisions

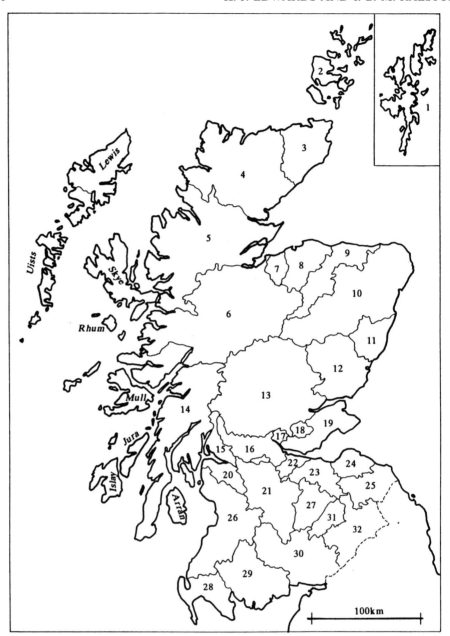

Figure 1.2 The counties of Scotland prior to local government reorganization in 1975.
1: Shetland; 2: Orkney; 3: Caithness; 4: Sutherland; 5: Ross and Cromarty; 6: Inverness;
7: Nairn; 8: Moray; 9: Banff; 10: Aberdeen; 11: Kincardine; 12: Angus; 13: Perth; 14: Argyll;
15: Dunbarton; 16: Stirling; 17: Clackmannan; 18: Kinross; 19: Fife; 20: Renfrew; 21: Lanark;
22: West Lothian; 23: Midlothian; 24: East Lothian; 25: Berwick; 26: Ayr; 27: Peebles;
28: Wigtown; 29: Stewartry of Kirkcudbright; 30: Dumfries; 31: Selkirk; 32: Roxburgh

2 Climate Change

GRAEME WHITTINGTON AND KEVIN J. EDWARDS

INTRODUCTION

In considering the activities of past populations in relation to the weather they experienced, it is necessary to prevent current perceptions of Scotland's climate from intruding. Current perceptions are derived from those conditions which operate at the present, and yet it can be demonstrated that weather conditions, at the global level at least, have been far from stable over time. There have been large-scale and long-term variations in climate during the Quaternary period. Such changes occurred during the Ice Age and its aftermath, and thus it might be expected that Scotland, owing to its intimate relationship with the accumulation and wasting of the vast ice sheets (Price 1983), has also experienced climatic disturbance. That could well have continued into the present interglacial period due to the time needed for recovery from the severe conditions of the Ice Age, especially as Scotland is located in a particularly sensitive climatic zone. This chapter will explore the possibility of climate change during the nine millennia following the removal of the last permanent ice from Scotland. A first consideration will be the interplay between the country's geographical location and the effect of patterns of atmospheric circulation. This knowledge must then be mediated by acknowledging the difficulties involved in extrapolating back many thousands of years. Such extrapolation requires the use of proxy evidence which itself may be incomplete and difficult to exploit. Thus a second focus of enquiry will be the sources of evidence which are available in relation to their potential and to their inherent problems. Climate change may also operate over different time-scales and so it will be considered from long-term and short-term viewpoints.

THE GEOGRAPHICAL LOCATION OF SCOTLAND AND ITS CLIMATIC IMPLICATIONS

Scotland's location at the western extremity of the Eurasian land mass and its proximity to the northern Atlantic Ocean, has profound significance for both its short- and long-term climate. Furthermore, it is important to emphasize the country's northerly position which gives it a latitudinal equivalence with Labrador.

Scotland: Environment and Archaeology, 8000 BC – AD 1000. Edited by Kevin J. Edwards and Ian B. M. Ralston.
© 1997 The editors and contributors. Published in 1997 by John Wiley & Sons Ltd.

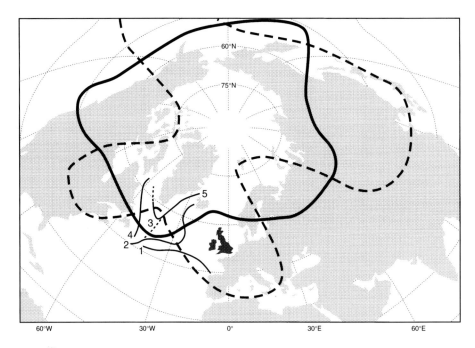

Figure 2.1   The solid line surrounding the North Pole represents a flow line of the strongest winds in the circumpolar vortex. The dashed line is an example of the deformation which that flow can undergo. The numbered lines represent positions of the Polar Front at different times. 1: refers to the position from 11 000–10 000 BP; 2: 13 000–11 000 BP; 3: 10 000–9000 BP; 4: 9000–6000 BP; 5 from 6000 BP onwards

Taken together, these features mean that there is not only exposure to both oceanic and continentally generated climatic factors, but that the elements associated with one of the earth's most important climatic mechanisms, the circumpolar vortex, are of immediate importance. As part of this system, air flow, in the northern hemisphere, occurs in a series of waves which migrate around the pole, moving from west to east in an irregular pattern (Figure 2.1). The waves generate low-pressure areas, where polar air extrudes southwards towards the equator, and ridges of high pressure which allow tropical air to intrude towards the pole.

The contrast between the temperature and humidity properties of air that originates in the polar regions and air that develops in the subtropics, means that their contact generates a further important major climatic feature – the Polar Front. The location of the Front varies, being further south in winter than in summer, but it has also altered its position many times in relation to long-term changes in atmospheric circulation (Ruddiman and McIntyre 1981; Figure 2.1). The importance of the Polar Front to Scotland's climate lies in the fact that it creates mid-latitude travelling depressions; these act as a mechanism for transferring heat from low latitudes and cold air from the opposite direction. The structure of these depressions ensures an alternation of warmer and colder air, with considerable levels of precipitation and strong winds, for the areas that lie in their track. Thus the location of the Polar Front, in any long-term perspective, determines to a large extent the climate experienced by the Scottish land mass.

Whether one type of climatic regime prevails over another depends upon the length of time that atmospheric circulation remains constant. Over decades or even centuries, when the movement of depressions along the Polar Front is the dominant climatic process, Scotland will experience mild, wet winters and cool, damp summers. With anticyclones prevailing, winters will be cold, with precipitation types and amounts depending upon the location of the anticyclone's centre, and summers will be warm or even hot.

SCOTLAND'S HOLOCENE CLIMATE

The disappearance of permanent ice from Scotland was inevitably followed by an amelioration of climate. Much has been written on this topic and a general consensus had been reached which allowed a synthesis of the nature of the Holocene climate in north-west Europe as a whole (Lamb 1977). This view has been challenged in recent years and a re-evaluation of Scotland's climate in the Postglacial period is necessary.

The traditional view

The work in the early twentieth century of two Scandinavian botanists, A. Blytt and R. Sernander, led to the widespread acceptance of four distinct climatic periods for north-west Europe since 10 000 BP. During a Boreal phase, in which anticyclonic conditions dominated, temperatures rose and summers became warmer than those of today, while winters, although milder, still witnessed frosts. The following Atlantic period (7500–5000 BP; 6270–3780 cal BC) was considered to provide the Postglacial Climatic Optimum during which the summer temperatures, already at their maxima, were some 2–3 °C greater than those of today (cf. Atkinson *et al.* 1987), while the 'cold' seasons were mild and rainfall was plentiful. Between about 5000–2500 BP (3780–630 cal BC) (the sub-Boreal), conditions akin to those of the Boreal were thought to prevail, but the climate was supposedly less dry due to oscillating patterns of latitudinal and meridional atmospheric circulation. Finally came the sub-Atlantic period (*c.* 2500 BP to the present) during which summer temperatures fell by as much as 2 °C, rainfall totals increased and mild winters prevailed, following the establishment of the Polar Front as a predominant factor in Scotland's climate. This climatic division has tended to remain in currency, especially in the archaeological literature (Magny 1982; Vermeersch and Van Peer 1990), but also in the writings of other disciplines (Serebryanny and Orlov 1993), although like the cultural divisions of Neolithic, Bronze and Iron, more through convention than proven scientific accuracy.

A revised view

A modification of the traditional position appeared as early as 1949, when Firbas proposed a simpler tripartite division of the postglacial period into Vorwärmezeit (the time of warming), Wärmezeit (the warmest times) and Nachwärmezeit (the time of cooling). By 1964, McVean was commenting on how little change took place in the Scottish forests during Postglacial times.

Recently, a more radical reappraisal has occurred. This has been stimulated by the recognition that the inferences about climate change have been based overwhelmingly on assumptions derived from the pollen stratigraphical record, and these may be inherently flawed (Birks 1982; Davis and Botkin 1985). Furthermore, the use of climatic modelling (Kutzbach and Guetter 1986; COHMAP 1988), the development of biological databases (e.g. Huntley and Birks 1983) and the statistical treatment of pollen data (e.g. Guiot et al. 1989; Pons et al. 1992) have led to sharper insights into climatic behaviour. Current thinking confirms earlier ideas both on the fast rate of temperature and precipitation increase between 10 000 and 7000 BP (5840 cal BC), and also that by the latter date Scotland would have been drier and warmer than at present (Birks 1990, Figure 1). Since that time, however, it is argued that changes in climatic regime have been only minor.

LONG-TERM CHANGES

If the pattern of climate change for north-west Europe also applies, as seems probable, to Scotland, an examination of the Scottish proxy evidence might be expected to provide corroborative evidence. Three such sources will be examined in this section and, because of the extremely important part played by vegetational history in the exploration and defining of past climate change, evidence drawn from that source will be considered first.

The evidence from vegetational change

The nature of the evidence

The temperature and moisture requirements of plants make them potential indicators of climate change. While some plants have adapted to a wide range of conditions, others are extremely sensitive to climatic fluctuations, especially if they are near the limits of their climatic range. Most of Scotland was glaciated until about 18 000 years ago (Chapter 3) and since that time a generally ameliorating climate has allowed colonization by plants from the east and south, especially up to c. 7000 BP (5840 cal BC) when Britain was finally severed from the European mainland (Funnel 1995). If major climatic oscillations have occurred in Postglacial times, it might be expected that changes in Scotland's vegetation patterns would reveal them. There are two main routes to the investigation of vegetational change, one at a microscopic and the other at a macroscopic scale.

The microscopic scale lies in the study of biostratigraphical evidence provided by pollen analysis. Plants release pollen and spores which are incorporated into accumulating peat and lake deposits, thus providing a chronostratigraphical record of the vegetation of an area. The speed at which and the length of time over which the pollen-preserving matrix accumulates determines the detail to which the vegetation reconstruction can be taken. In optimum conditions, such as where annually varved sediments are available, such reconstruction can be very sensitive indeed (Peglar 1993).

This line of enquiry into climate change does, however, present problems. Depositional hiatuses may occur, leading to breaks in sedimentary sequences. More fundamental perhaps is the lag effect in plant colonization. Climatic conditions may well be suitable for particular plant species to thrive but this will depend upon many factors. It is necessary for seeds to arrive in an area and also for soil conditions to be suitable for their success. Furthermore, trees, for example, not only need to migrate into an area but require time before they reach the flowering stage; this will be accomplished only tardily in areas where the climate is ameliorating slowly, but more speedily where conditions change rapidly. Different trees and shrubs, even in their optimum environment, may also achieve reproductive maturity at varying rates: *Quercus* (oak) achieves this state more slowly than do either *Betula* (birch) or *Corylus avellana* (hazel) and thus, although present in an area, it may be temporarily palynologically silent.

This problem of the sensitivity of the biostratigraphical record is illustrated very clearly by the Postglacial record of *Alnus glutinosa* (alder). Birks (1989) mapped the recolonization of Scotland by alder and showed that it had arrived in much of the country by about 6500 BP (5440 cal BC), but subsequent work (e.g. Whittington *et al.* 1991a) indicated that in one area of eastern Scotland it was present about a thousand years earlier. It is now realized that this species, previously accepted as an indicator of increasing wetness, has an extremely complicated history which may not be directly related to climate change (Chambers and Elliott 1989; Bennett and Birks 1990; Edwards 1990; *pace* Tallantire 1992).

At the macroscopic level, the remains of trees, preserved in peat (Plate 2.1) or lacustrine sediments, also serve as evidence of previous climatic regimes. The current treeless nature of the Outer Hebrides contrasts with the pine stumps still to be found buried there in peat (Wilkins 1984) and the recording, by Beveridge (1911), of birch and hazel wood in intertidal deposits (see also Fossitt 1996). Similarly, the higher areas of Scotland are today treeless although, well above the current altitudinal limit of pine growth, there are root remains in blanket peat (Pears 1968, 1970). Attempts have been made to use this evidence to show that greater warmth existed in the earlier Holocene, but the dates of these macrofossils are not well established and there is a danger of assuming that they are all contemporary rather than representing different phases of adjustment to climatic amelioration (Lowe 1993).

The exploitation of the evidence

With this awareness that past vegetation patterns can be difficult to use as proxy evidence for climate change, it is now possible to attempt to exploit it by addressing some of the main features of the until-recently accepted climatic variability that would have affected Scotland.

The first point which might be examined is that of a rapid climatic amelioration by about 10 000 BP, since when conditions have remained more or less constant at the macro-scale. If this is the case, the broad-leaved arboreal taxa, which demand thermophilous conditions, should not only have spread (*sensu* Birks 1989) into

Plate 2.1   Pine stumps appearing from beneath peat at Clatteringshaws Loch, Stewartry of Kircudbright. Copyright: K. J. Edwards

Scotland by the time of the supposed Climatic Optimum, but maintained their presence and distribution thereafter. *Betula* was established in the central and eastern areas by 10 000 BP and in the north and west by 9500 BP (8560 cal BC). *Corylus avellana* was present in western Scotland at 9500 BP and by 9000 BP (8030 cal BC) had spread to the south and east. *Ulmus* (elm) had advanced to cover virtually the whole of the mainland by 8500 BP (7530 cal BC) and *Quercus* was north of the Forth–Clyde lowlands by the same date. The later colonizing *Tilia* (lime) and *Fraxinus excelsior* (ash) had a restricted spread into Scotland (Birks 1990). This may well serve to emphasize that it is not climatic conditions alone which control the establishment of vegetation, but that conditions of soil, seed spread and competition are as important in many cases. This is also an example of where the proxy evidence may always be incomplete and therefore misleading. It is noticeable that Birks's maps of spread for *Tilia* and *Fraxinus* reveal few sites from the eastern lowlands. Two sites in Scotland, investigated after Birks's data collection, have *Tilia* present by 7000 BP and 6200 BP (5840 and 5140 cal BC) and *Fraxinus* by 8400 BP and 7000 BP (7460 and 5840 cal BC) (Whittington *et al.* 1991a,b). It therefore appears that temperatures had achieved the threshold demanded by the warmth-loving trees for them to become established at an early date in the Holocene. This lends support to the contention that the rise in temperature had indeed been rapid. Furthermore, an examination of pollen diagrams from across Scotland reveals stability in the nature of the woodland until about 5000 BP (3780 cal BC). At that time, the effects of human activity in the landscape appear to undermine the worth of vegetation patterns as proxy climate data. It must be pointed out, however, that even after that date there is no fundamental change in the distribution and composition of the arboreal and shrub taxa which survived.

The foregoing evidence suggests that the main features of Scotland's Postglacial climate were not only established at an early date but that they have shown very little alteration since then. That such a statement is largely derived from pollen and associated data makes it vital to remember the problems created by vegetation inertia (Smith 1985), and the debate about vegetation spread as against vegetation and climate equilibrium (Davis and Botkin 1985; Webb 1986), and atmospheric carbon variability (Isdo 1989), as well as those inherent in the method of pollen analysis.

Pertinent to this topic of Holocene climate change, however, remains the problem of such a long-accepted division as the Atlantic period and its wetness. Can it be dismissed in line with the new climatic framework that is emerging? Two main sources of proxy data are useful in this matter; one will be discussed here and the second in the following section.

The spread and rapid expansion of *Alnus*, because of its affinity with wet sites, has been used as an indicator of a major change in climate brought about by a marked increase in rainfall. As already stated, however, the initiation and spread of *Alnus* were certainly not as synchronous as might be expected if climate change over Scotland as a whole, associated with a long-term pluvial regime, was the main cause. The need to consider carefully the nature of the proxy data is again evident. Pollen survival requires the existence of anaerobic conditions. These exist in bogs and lacustrine sediments, the very areas in which *Alnus* flourishes. Thus the alder appearance and rise, although real, may well be no more than a response to hydrological changes of a local nature and not related to a fundamental shift in climate.

The danger of modern climatic perceptions clouding the interpretation of past climatic periods was emphasized above. If pollen-analytical data are used circumspectly, they can function as an important correction factor. Contrary to current perceptions concerning the length of the growing season and exposure to high winds there (Lowe 1993), the pollen record shows that the Outer Hebrides and Shetland were wooded (Bennett *et al.* 1990, 1992; Edwards 1996a; Fossitt 1996), even down to the western shoreline (Whittington and Edwards, unpublished), from 8000 BP (6840 cal BC), if not earlier, and remained so until at least 5000 BP (3780 cal BC) – the time at which human interference in the landscape became much more intense.

The evidence of peat development

The nature of the evidence

A distinctive feature of Scotland's landscape is the frequent occurrence of soligenous wetlands (valley mires) and bogs. Of these, bogs should be exploitable as a source of information on climatic variation. They occur in two forms: the infrequent, raised (ombrogenous) bogs are often large, as in the wetter west, e.g. Claish Moss (Argyllshire) and Moss of Cree (Wigtownshire), but smaller and less frequent in the drier east, e.g. Bankhead Moss (Fife); the frequently occurring blanket bogs are widespread (e.g. in the Outer Isles and the Cairngorm and Monadhliath

Mountains), but can also appear as deep unbroken stretches, as on Rannoch Moor. Unlike the valley mires, bogs rely on atmospheric conditions, particularly high rainfall and low evaporation rates, for their initiation and continued existence. An examination of the vertical stratigraphy of bogs, particularly of the ombrogenous type, can often reveal considerable differences in its constituent vegetational components. It also exhibits differences in the humification and compaction which have taken place. Blanket bogs also cover large areas which peat digging (for domestic and agricultural purposes), natural erosion and air photography reveal as formerly having been exploited for settlement and agriculture. As the bogs are a response to moisture availability, it might be expected that all of the above features of bog stratigraphy would be related to variability in precipitation amounts and humidity levels (cf. Aaby 1978). In theory, therefore, the recovery of dates of initiation should indicate the time at which the climate became significantly wetter. Furthermore, drier periods should cause a slowing down or even cessation in peat development, marked by a change in the vegetation succession, which should be apparent in the peat stratigraphy and should therefore be datable. After such a dry period, an increase in wetness should bring about regrowth at the peat surface, especially with regard to a resurgence of *Sphagnum*, thus leading to what is known as a recurrence surface (Gore 1993).

The exploitation of the evidence

Exploitation of these features for the deriving of proxy climatic data can be difficult. The credibility of the Atlantic and sub-Atlantic phases as contributors to climate change in Scotland has depended to a large extent upon the growth of bogs, both ombrogenous and blanket. There are major problems in this equation because peat growth can occur as a result of the interplay of factors other than those of a direct climatic origin. Peat may be a response to a sudden increase in rainfall, but might also represent the climax stage of soil development as part of the long-term operation of pedogenic processes in which leaching led to acidic and anaerobic conditions becoming dominant (Ball 1975). If that is the case, there is no need to posit a sudden deterioration in climate for the onset of peat. It could develop at any time when the appropriate critical conditions had been reached. That would also help to explain a further factor which tends to complicate the recourse to peat as a sign of increasing wetness at a particular date. The initiation of blanket bog at one location may not be representative for that event over an area as a whole (Whittington and Ritchie 1988; cf. Solem 1986). Thus, before peat initiation can even be offered as a sign of increasing wetness, it is vital to undertake a widespread areal survey.

Although climatic features are crucial in peat formation, they also interact with topographic and soil porosity conditions in determining the start and rate of peat growth. It is notable that peat inception is mainly a late Holocene phenomenon (Watts 1988) and thus commonly occurs during the period of human land exploitation. Prehistoric dwellings, for example at Dalnaglar in south-east Perthshire (Stewart 1962), at Cùl a'Bhaile on Jura (Stevenson 1984), at An Sithean on Islay (Barber and Brown 1984) and Scord of Brouster, Shetland (Whittle *et al.* 1986), have associated field systems or land that was used for arable agriculture, all of which lie buried under blanket bog. It is conceivable that over-exploitation of the soil in such areas, perhaps also engendering soil compaction which hampered drainage, may well have induced peat initiation and accumulation. Thus peat inception may be due, not to climate change, but to alteration to the local hydrology (Moore 1993).

The existence of more than a single recurrence surface in a stratigraphic sequence could be interpreted as evidence of cyclic changes in precipitation amounts. Again, great care has to be taken before such a conclusion is drawn. In areas where wood has been scarce, peat has commonly been employed as fuel. Hiatuses identified in the accumulation of peat which are followed by regrowth, may be no more than the result of human interference with the natural processes of the peat formation.

The potential of bog structures and characteristics has been reviewed by K. E. Barber (1982) and together with later work (e.g. Barber 1985, 1994; Haslam 1987; Stoneman et al. 1993; K. E. Barber et al. 1994b), especially with regard to the use of recurrence surfaces, has shown that the results obtained may be best when used at a local scale and for determining short-term changes. New methods are being used to explore bogs as providers of proxy evidence for climate change (Blackford 1993). Until they are realized, it would be wise not to rest the existence of major shifts in the pluviality of Scotland's climate, especially such as has been envisaged for the supposed Atlantic period, upon evidence drawn from sources derived from the peat record. Even in the unlikely event that peat growth could be shown to have been initiated over Scotland synchronously, especially at the start of an Atlantic period, that still might not be evidence for greater rainfall but merely of lowered rates of evaporation. Between 12 000 and 6000 BP (12 040 and 4870 cal BC), the tilt of the earth's axis was greater than at present and perihelion took place in the northern summer (Kutzbach and Guetter 1986). This would have led to intense anticyclonic conditions and, although the winters would have been colder, the summers would have been much warmer than today with a comcomitant increase in evaporation rates. A climatic change took place but not necessarily one which involved an excessive increase in rainfall.

The evidence of faunal change

Many fauna operate within distinct climatic limits and so can be regarded as surrogates for climatic data. Certain mammals fall into this category, but probably only the reindeer would be of value for Scotland. Suggestions have been made for their survival into the late Postglacial (Lawson 1984; Whitaker 1986). This would create a potential conflict with other proxy climatic data which show the early Holocene to have been a period of considerable warmth, a condition which would have had major implications for the reindeer's food source. Proof that reindeer have been present at all since the last Ice Age is difficult to acquire as the survival of bone or antler in Scotland's acidic soils is unlikely. Cave deposits provide a better preservation matrix and thus focus interest on the cave system of Creag nan Uamh in Assynt. Reindeer remains exist there and examination has shown that the latest of them dates to 8300±90 BP (7480–7100 cal BC) (Lawson 1981; Murray et al. 1993). A review of all other potential remains from Scotland has failed to provide a later occurrence other than from those which are intrusive (Clutton-Brock and MacGregor 1988). Thus, this faunal evidence corroborates that from other proxy sources which indicate temperatures greater than those of today in the early Holocene.

In contrast to those of mammals, insect remains preserve well, and can be abundant, in all organically rich deposits that are waterlogged (Coope 1975). Thus, there is a certain similarity between fossil insects and pollen as climatic data surrogates but the former possess a distinct advantage in this category. The greater speed of reproductive maturation, certainly compared with trees, and the capability for individual movement, which is not dependent upon an indirect dispersal mechanism, mean that the problem of the lag effect observed in the case of pollen is less significant. There are also insects which are not dependent upon particular plants for their food supply so that they can respond to shifts in climatic limits very quickly. Where pollen proves to be a superior source of evidence is in its ready availability in small quantities of sediment. While the Chironomidae (midges) are similar in this respect, the Coleoptera (beetles) can only be obtained in an adequate number from several kilograms of material, which demands the availability of relatively rare and suitable sediment exposures.

As a result, climatic evidence drawn from fossil insect study currently available for Scotland is very limited. An investigation carried out in the South-West has, however, provided further confirmation that there had been a rapid climatic amelioration by about 10 000 BP. The presence of the Carabid *Odacantha melanura* at Brighouse Bay in the Stewartry of Kirkcudbright (Bishop and Coope 1977) allows the inference that temperatures as great as those of today had been attained by 9640 BP (8950 cal BC). This method clearly possesses great potential for insights into Scotland's palaeoclimates, but unfortunately palaeoentomologists are even rarer than the exposed deposits upon which they depend.

SHORT-TERM CHANGES

The proposal that Scotland's climate may well have witnessed long-term macro-stability during the Holocene does not preclude intermittent or even periodic changes of a lesser nature, on national or regional scales. Many such events are likely to remain undetected as the proxy evidence has too low a resolution status. There are, however, indications from surrogate climatic data that episodic fluctuations have occurred. For instance, records of the ratios of stable isotopes of oxygen ($\delta^{18}O$) from Lundin Tower, Fife, reveal probable significant oscillations in temperature in the early Holocene (Whittington *et al.* 1996).

Vegetational evidence

It was suggested above that the constituents of Scotland's woodland cover have shown no distributional change since early in the Holocene. Attention has been drawn recently (Gear and Huntley 1991) to the behaviour of pine forests in the north of Scotland. During the period around 4000 BP (2490 cal BC), and within a period of 400 years, the tree line expanded over a distance of 70–80 km and then retreated. It is proposed that marked variation in precipitation totals is at the root of this phenomenon. Perhaps confirmation of this comes from the application of isotope analysis. Dubois and Ferguson (1985), using isotope ratios of deuterium in pine stumps from the Cairngorms, have concluded that a wet episode did occur between 4250 and 3870 BP (2880 and 2330 cal BC). They have also isolated other periods with higher rainfall, at 6250–5800 BP (5220–4690 cal BC) and 1330 BP (cal AD 680). The establishing of these wet episodes does not necessarily mean periods of greater activity on the Polar Front. Possible alternative explanations for the changes in rainfall, and also in pine distribution, on either side of 4000 BP have been advanced (Blackford *et al.* 1992). These involved the eruption of Icelandic volcanoes and the injection of tephra into the atmosphere. The effect of this would not only have been to increase the availability of hygroscopic nuclei but also to have ensured enhanced acidity in any ensuing rainfall, either or both of which may have had a damaging effect on pine trees growing close to their range limits. If this connection is justified, any extrapolation of this event to support climate change over Scotland as a whole is dangerous. The crucial need is to establish the spatial extent of the tephra effect, and even then it may only produce evidence from changes in arboreal cover in areas where the trees were near a critical threshold (Edwards *et al.* 1996).

The evidence of peat development

Large areas of Scotland, for example the uplands to the north of Strathmore (Harris 1984) and the region confined by the Rivers Brora, Naver, Halladale and Helmsdale in the extreme north, still bear the impression of past cultivation practices where altitude or soil conditions now make such activity unrealistic. The peat cover encountered in such areas might still be regarded as involving climate change despite the complications noted in peat generation, especially those related to agricultural exploitation. That having been said, the need to invoke long-term climate change may still be unnecessary. The tephra from the Icelandic eruptions may not only have increased rainfall, making agriculture in the short-term very unrewarding, but its deposition might have also increased the acidity of the soil (cf. Grattan and Gilbertson 1994) to a point where peat initiation became inevitable and irreversible. The climatic phenomena in such an incident would have had long-term consequences but should not be interpreted as arising from an equally long-term climatic shift.

The evidence of sand movement

Evidence for periods of increased storminess, and therefore of increased activity along the Polar Front, or even a probable temporary shift in the location of the Polar Front, is difficult to identify from proxy sources. One of the most obvious features of such events are wind-thrown trees but it is rare for conditions to exist which would preserve the evidence (Lamb 1966). Most important as a source might be sand; it is highly mobile and located in unconsolidated masses along shorelines, which are also vulnerable to exposure to high winds. Some 20% of the Scottish shoreline displays sand as either dunes or links, as at Forvie (Aberdeenshire), along the Atlantic shores of the Outer Hebrides and Luce Bay (Wigtownshire). Only under extremely favourable circumstances can sand movement be used as a climatic surrogate. Response to stormy conditions emanating from one direction can be negated by winds from a contrary direction, a feature which can occur very rapidly during the passage of deep depressions across Scotland. It is also necessary to have a datum against which the onset of a sand blow can be measured. Fortuitous events may make this possible. Recent erosion, caused by deflation, has allowed a study (Whittington and McManus 1996) in the coastal dune area of Tentsmuir in Fife. This has shown that until the middle of the first millennium AD, the dune belt with its *Calluna* (heather) cover was being exploited as managed pasture. After that date, a massive sand blow engulfed the area. This change occurred in the same period as the study of bog deposits in England, Ireland and Wales (Blackford and Chambers 1991) indicated as having been a period of increased storminess and wetness.

The coastal sands of the northern islands appear to offer clear evidence to support statements as to periods of increased storminess, in that deep deposits of sand frequently overwhelmed human occupation surfaces. Thus, a climatic change is argued for at 5000 BP (3780 cal BC) (Keatinge and Dickson 1979) and that line of reasoning can be extended to other sites; for example, at Northton on Harris (Simpson 1976), Udal on North Uist (Crawford and Switsur 1977), Rosinish on Benbecula (Whittington and Ritchie 1988) and Knap of Howar on Papa Westray (Ritchie 1983). Unfortunately, the relationship between increased storminess and major sand movements is not one of direct cause and effect. Also involved in this process are sand supply to provide the movements, the erosional effects of human activity in creating conditions for sand deflation, and changes in sea level during the mid Holocene marine transgression which altered not only the location and extent of exposed sand, but also its mobility (Ritchie 1979; Whittington and Ritchie 1988). Periods of major sand movement may have been no stormier than those in which the sand was stable. Even a conjunction of dates for major sand blow across a wide spatial range of sites could well be related to coastal woodland clearance which molluscan evidence has shown as occurring during the Neolithic (Evans 1979).

CONCLUSION

The description of and explanations for changes in the Holocene climate of Scotland must be regarded not only as being in an interim stage but also in a state of flux. Much of the proxy evidence available is still at a crude level of interpretation and perhaps also suffers from misinterpretation. The principal source, that of pollen data, has to be viewed much more critically and it has to be recognized that its use, in establishing long-term climate change, has in the past involved a large degree of circular argument. The infusion of climatic modelling, although currently only at the macro-scale, is not only bringing new insights but is also encouraging a radical examination of long-held and rather sacrosanct conventions. The extension of isotopic studies also contains great potential. For Scotland, however, it must be confessed that palaeoclimatology is still in its infancy.

3 Geomorphology and Landscape Change

COLIN K. BALLANTYNE AND ALASTAIR G. DAWSON

INTRODUCTION

Throughout the period of known human occupance, the magnificent variety of the Scottish landscape has posed both opportunities and constraints for its inhabitants. During this time, moreover, the physical characteristics of the landscape have undergone significant changes, some of which have strongly influenced the pattern of human activity whilst others reflect the consequences of such activity. This chapter explores the geomorphological background to the early human occupance of Scotland. Its focus is threefold: the first part of the chapter outlines the geomorphological characteristics of the Scottish landscape; the second is devoted to the nature of landform changes during the period since the disappearance of the last glaciers; and the third considers ways in which human activity has affected the form of the Scottish landscape during the first nine millennia of the present interglacial period.

PHYSICAL CHARACTERISTICS OF THE SCOTTISH LANDSCAPE

The physical landscape of Scotland is the product of a long, complex and sometimes violent geological history (Harris 1991). An understanding of its present form involves three main elements: first, the characteristics of the underlying rocks, and in particular their variable resistance to erosion; second, the denudation of the Scottish land mass during the last 70 million years; and finally the effects of glaciation by ice sheets during the Pleistocene Epoch.

Geology and relief

Traditionally, Scotland has been seen as comprising three major relief units (i.e. the Highlands, the Midland Valley and the Southern Uplands), but geologically a fivefold division of the landscape is more appropriate. The oldest of these units is the Hebridean Craton in the extreme north-west (Figure 3.1). A fragment of a

Scotland: Environment and Archaeology, 8000 BC – AD 1000. Edited by Kevin J. Edwards and Ian B. M. Ralston.

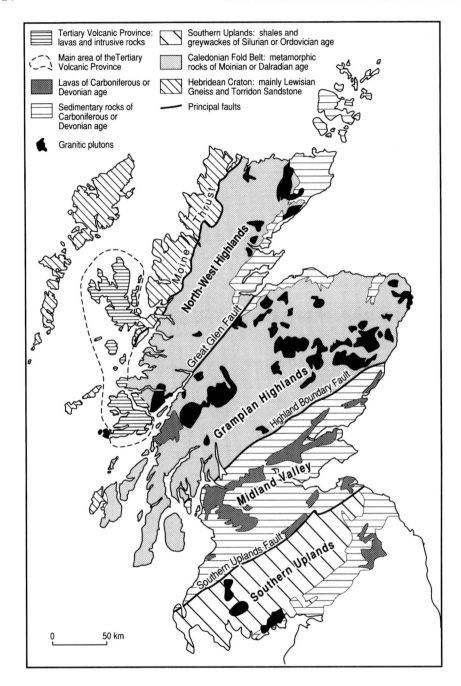

Figure 3.1 The structural provinces of Scotland, showing the distribution of major rock types. Geological boundaries based on Craig (1991)

Plate 3.1   Quartzite mountains rising above a platform of glacially scoured gneiss bedrock, Arkle, Sutherland. Copyright: C. K. Ballantyne

primeval continent, this comprises undulating rocky lowlands underlain by resistant crystalline rocks (Lewisian Gneiss) which are locally surmounted by bold sandstone and quartzite mountains 700–1100 m in altitude (Plate 3.1). A thrust fault zone known as the Moine Thrust separates the Hebridean Craton from the second of the five structural units, the Caledonian Fold Belt. This makes up most of the Scottish Highlands and represents the eroded remnants of an ancient mountain range of alpine stature. It is divided into two major subunits, the North-West Highlands to the north of the Great Glen Fault, and the Grampian Highlands to the south and east (Figure 3.1). Both are dominated by folded and faulted metamorphic rocks, primarily schists. Interspersed amongst these are broad outcrops of granitic rocks that sometimes form high plateaux such as the Cairngorms, and less often underlie broad basins such as Rannoch Moor. Most of the Caledonian Fold Belt is dominated by mountains 800–1200 m in altitude. In the West these form ridges and intervening valleys that trend W–E or SW–NE, but the eastern Grampians are dominated by broad plateaux that constitute a formidable barrier to communications. In Buchan, the Moray Firth area and Caithness, however, the metamorphic rocks of the fold belt form undulating lowlands and are partially overlain by sandstones of Devonian age.

   To the south of the Caledonian Fold Belt, and separated from it by the Highland Boundary Fault, lies the Midland Valley. This represents a broad trough of land that foundered in Carboniferous times as a result of crustal stretching. It is not a valley in the conventional sense, but an area of undulating lowlands underlain by sedimentary rocks. These lowlands are interrupted by upland plateaux composed of resistant lavas (e.g. the Ochil Hills and Campsie Fells) or capped by equally resistant sill rocks. Farther south, between the Southern Uplands Fault and the English border, lies the fourth major structural unit, the Southern Uplands. This area comprises mainly rolling hills under 800 m in altitude, interrupted by the broad valleys of major rivers such as the Clyde, Nith and Tweed. Most of the Southern Uplands area is underlain by shales and greywackes of Ordovician or Silurian age, though several areas of granite occur in south-west Scotland and an area of Carboniferous sedimentary rocks in the south-east underlies the fertile lowlands of the lower Tweed Basin. The final (and geologically most recent) structural unit is the Tertiary Volcanic Province, the main component of which comprises the Hebridean islands that lie between Skye and Mull, together with parts of Arran and the Ardnamurchan Peninsula (Figure 3.1). Most of these areas are dominated by stepped plateaux built of resistant basaltic lavas of Eocene age, together with smaller areas of intrusive rocks that form mountains over 700 m in altitude, such as the Cuillin Hills on Skye and the granite mountains of northern Arran.

Tertiary landscape evolution

Although the main morphotectonic units of the Highlands, Northern Isles and Southern Uplands were probably established by the end of the Palaeozoic era (Hall 1991), much of the present configuration of the Scottish landscape reflects differential erosion during the Tertiary period, which began about 70 million years ago. In early Tertiary times, renewed sea-floor spreading in the North Atlantic was accompanied not only by eruption of lavas in the Tertiary Volcanic Province, but also by widespread uplift and erosion across much of the Scottish mainland and Northern Isles. The North-West Highlands and Grampians appear to have been uplifted *en masse*, with dislocation and downwarping of peripheral areas. Later Tertiary uplift was more modest and episodic. Such periods of uplift were accompanied by vigorous denudation, often concentrated along lines of geological weakness. Intervening phases of tectonic stability favoured deep bedrock weathering that enhanced pre-existing relief, together with widening of valleys and basins and the development or extension of erosion surfaces (Hall 1991).

The glacial legacy

Whilst the broad outlines of the present Scottish landscape had been established by the end of the Tertiary, its detailed configuration owes much to the events of the Quaternary period. This is subdivided into two geological epochs: the Pleistocene ($\sim$2.6 million to 10 000 years BP) and the Holocene (10 000 years BP to the present). The Pleistocene, popularly known as 'the Ice Age', comprised several prolonged periods of extreme cold (glacial stages) separated by briefer temperate interglacial stages. The Holocene epoch is coincident with the present interglacial stage, also known in Britain as the Flandrian.

During successive glacial stages, glaciers developed in the mountains of the Highlands and Southern Uplands and spread over the surrounding low ground, ultimately coalescing to form ice sheets that buried much or all of the present land surface. Our knowledge of the timing and dimensions of most of these glacial events is very incomplete, as each successive glaciation tended to remove or obliterate the evidence left by its predecessors. In consequence, only the final events of the most recent glacial stage have been reconstructed in any detail. These relate to the period known in Britain as the Late Devensian (*c.* 26 000–10 000 BP), which is subdivided into three chronozones: the Dimlington Stadial (cold) of *c.* 26 000–13 000 BP, the Windermere Interstadial (temperate) of *c.* 13 000–11 000 BP, and the Loch Lomond Stadial (cold) of *c.* 11 000–10 000 BP.

During the Dimlington Stadial, the last ice sheet extended southwards into England and Wales, occupying two-thirds of the present land area of Britain. Almost all of Scotland was covered by this ice sheet, though some mountain summits may have remained above its surface (Ballantyne 1990; Boulton *et al.* 1991). Some commentators have also suggested that Buchan and Caithness may have escaped glaciation during the Late Devensian (Sutherland 1984; Bowen *et al.* 1986), but others have argued that the ice-sheet limit lay offshore from these areas (Hall and Bent 1990; Figure 3.2). Deglaciation of Scotland appears to have been well advanced by *c.* 14 500 BP, and by 13 000 BP the remnants of the last ice sheet were confined to the mountains of the western Highlands (Sutherland 1991). Although most of the retreat of the last ice sheet apparently took place under very cold, arid conditions, the period 13 000–12 700 BP witnessed rapid warming of the British climate (Atkinson *et al.* 1987), and by about 12 500 BP July temperatures in southern Scotland were similar to those of the present (Bishop and Coope 1977). From 11 400 BP on, however, renewed cooling marked the transition to the Loch Lomond Stadial, the last period of extreme cold in Scotland. During this relatively short-lived cold phase, a large icefield developed across the western Highlands, together with several smaller icecaps and icefields and numerous small valley and corrie (cirque) glaciers (Figure 3.2). These appear to have reached their maximum extent between 10 500 and 10 000 BP (Rose *et al.* 1988; Peacock *et al.* 1989), before experiencing a period of intermittent retreat that was terminated by rapid warming after *c.* 10 100 BP (Atkinson *et al.* 1987; Benn *et al.* 1992).

Although information concerning Pleistocene events prior to *c.* 26 000 BP is limited, the effects of successive glaciations are nevertheless manifest throughout Scotland. Glacial erosion has sculpted the mountains of the North-West Highlands, western Grampians and the Hebrides into a landscape of cliffs, corries, arêtes and U-shaped troughs (Plate 3.2). Erosion of the eastern Grampians has been less pronounced, with deep troughs intersecting broad plateaux that represent erosion surfaces inherited from the pre-Quaternary landscape (Linton 1959). Across lowland areas underlain by resistant rocks, such as the Lewisian Gneiss or Tertiary lavas, glacial erosion resulted in the formation of 'knock-and-lochan' topography, consisting of low rocky hills interspersed with numerous small lochs. In the Midland Valley, the prime effect of glacial erosion was relief enhancement: areas of sedimentary rock were readily eroded, leaving those underlain by resistant igneous rocks standing proud.

Complementing such erosion has been the deposition of unconsolidated sediments, collectively referred to as drift deposits. These have two main components: tills, which are sediments deposited directly by glaciers; and glaciofluvial or outwash deposits deposited by meltwater rivers at the margins of retreating ice masses (Plate 3.3). Scottish tills are highly variable in terms of their composition, but most

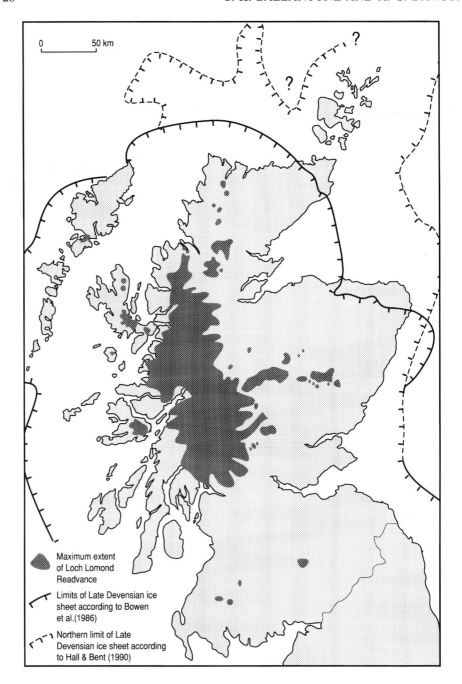

Figure 3.2 Limits of Late Devensian glaciation in Scotland

Plate 3.2 Glaciated landscape of the Loch Quoich area, northwestern Highlands. Copyright: C. K. Ballantyne

Plate 3.3 Fertile deltaic and outwash deposits amid the mountains of the northwestern Highlands, Loch Hourn. Copyright: C. K. Ballantyne

comprise a mixture of stones and boulders of mixed geological provenance embedded in a matrix of sand, silt and clay. Some, particularly in sandstone areas, are fairly loose and free-draining, whilst others are clay-rich, well-consolidated and tend to retard drainage. Glaciofluvial deposits are dominantly composed of alternating beds of sand and gravel, and hence are generally free-draining. Together, such drifts form the dominant parent materials for soils in lowland areas, where their potential agricultural productivity is dictated largely by lithological content (Chapter 4). Tills and glaciofluvial deposits are most widespread across the lowlands of eastern, central and southern Scotland, particularly in areas underlain by sedimentary rocks (Figure 3.3). Many such areas are mantled by a thick, undulating till sheet that completely covers the underlying bedrock. In some places, such as the Glasgow area and lower Tweed Valley, the till has been moulded into low hills (drumlins) with intervening depressions. Glaciofluvial deposits characteristically occupy lowland valley floors, originating as floodplains of sand and gravel that have frequently been modified by the melting of buried ice to form a landscape of small hills (kames) and intervening hollows (kettle holes). Elsewhere, such as in the Spey Valley, thick glaciofluvial deposits have been terraced by successive episodes of river incision. Throughout the western Highlands and Hebrides, however, glacial and glaciofluvial deposits are largely confined to valley floors. Within the limits of the Loch Lomond Reavance, till deposits often form 'hummocky moraines' (Plate 3.4) which were deposited at the margins of retreating valley glaciers (Benn 1992; Bennett and Boulton 1993).

Glaciation and early human occupance

The earliest human artefacts hitherto found in Scotland relate to the early Holocene (Morrison and Bonsall 1989; Wickham-Jones 1990; Chapter 7). It is possible, however, that the archaeological record has been truncated by the advance of the last ice sheet across the country during the Late Devensian, and that the earliest human occupance may have occurred much earlier than the archaeological evidence suggests. In those parts of England that escaped glaciation during the Late Devensian, there is abundant evidence for Palaeolithic settlement, some of it dating back half a million years to the Early Pleistocene (Wymer 1988). As Morrison (1983) has pointed out, some Pleistocene interglacials appear to have been climatically more hospitable than the Holocene, and even during cool interstadials it is plausible that groups adapted to tundra environments could have settled in Scotland. However, organic remains predating the last ice sheet have been found at only a limited number of sites (Lowe 1984), and none of these has yielded evidence of human settlement. As glacier ice appears to have occupied all or almost all of the Scottish lowlands during the Late Devensian, the chances of recovery of pre-Holocene artefacts appear very slender.

It is worth noting the ways in which the effects of glaciation on the Scottish landscape have influenced settlement patterns, agriculture and communications throughout history and prehistory. Valleys incised through the main north–south watershed of the western Highlands provide routeways (such as Glen Shiel and Glen Coe) across an otherwise virtually impenetrable mountain barrier (Linton 1951). Equivalent routeways in the eastern Highlands are rare, as here successive ice

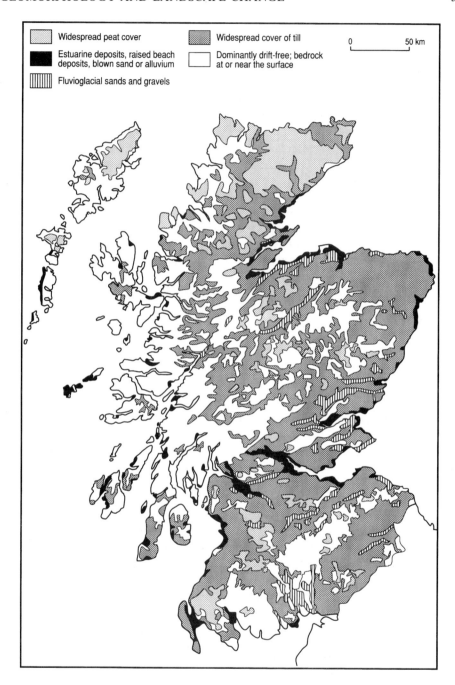

Figure 3.3 Generalized distribution of drift deposits in Scotland. Partly adapted from Geological Survey maps

Plate 3.4 Chains of hummocky recessional moraines deposited at the margins of retreating glacier ice at the end of the Loch Lomond Stadial, Luib, Isle of Skye. Copyright: D. I. Benn

sheets moved sluggishly, being partly frozen to the underlying ground and hence incapable of erosion except along the floors of pre-existing valleys (Sugden 1968; Hall and Sugden 1987). In the North-West Highlands and western Grampians, glacier ice excavated deep rock basins along many valley floors. In inland areas, these are now occupied by ribbon lakes such as Loch Tay, Loch Lomond and Loch Ness; off the west coast such rock basins are represented by a network of fjords and sounds that offer sheltered passage and access to the interior for groups possessing seagoing craft, however simple. Differential glacial erosion of the Midland Valley left the resistant vents of former volcanoes (volcanic plugs) standing proud of the surrounding lowlands, and some of these, such as Traprain Law, have provided defensible places for occupation from at least the first millennium BC. On a smaller scale, partially drowned hummocky moraines in Highland lochs provided the foundations of some timber-built crannogs or lake settlements, a form of dwelling that may have its origins as far back as the end of the Bronze Age (Morrison 1983, 1985).

Of critical importance to our understanding of the patterns of farming settlement is an appreciation of the agricultural potential of different types of glaciated terrain. Although the glacial deposits of lowland Scotland must certainly have offered early farmers the advantages of deep soils rich in nutrients, in their pristine state these drifts must have been strewn with boulders and often poorly drained. Except in particularly favoured locations, such as sandy river terraces, extensive clearance of boulders must have been necessary before even limited cultivation could be carried out. In the Highlands, the potential for arable agriculture has always been limited to the floors and lower slopes of some of the larger glens. In areas re-occupied by glacier ice during the Loch Lomond Stadial, the hummocky recessional moraines deposited during the retreat of the last glaciers constitute a formidable obstacle to cultivation. Similarly, the ice-scoured driftless lowlands underlain by resistant rocks such as Lewisian Gneiss and the basalts of the Tertiary Volcanic Province offer very limited potential for agriculture. In such areas cultivation has of necessity been concentrated along the floodplains of rivers and along the coastal fringe, where raised beaches provide pockets of cultivable land (cf. Bohncke 1988; Wickham-Jones 1990).

Glaciation and sea-level change

Ice sheet growth and decay also had an important indirect effect on the evolution of Scotland's landscape through its effects on changes in relative sea level, both before and during the period of known human occupance. The build-up of the last ice sheet depressed the level of the land surface, such depression being greatest near the centre of the ice sheet where the ice was thickest. During and after the shrinkage of the ice, the land underwent differential glacio-isostatic rebound, with those areas near the centre of the last ice sheet rising more rapidly than those near its former margins. Isostatic uplift was rapid at first, then progressively slowed throughout the Lateglacial and Holocene. Such vertical movements of the Scottish land mass were accompanied by glacio-eustatic changes in global sea level. These took the form of a world-wide drop in sea level as the last great ice sheets expanded, increasing the volume of water stored as glacier ice, and a subsequent rise in sea level as the ice sheets melted, feeding the oceans with glacial meltwater. Because the ice sheets in North America and Eurasia did not melt completely until *c.* 7500–6500 BP (6270–5440 cal BC), it was not until then that global ocean volume reached its present capacity. The process of the severance of Great Britain from the European continent occurred during early Holocene sea-level rise, around 8500 BP (7530 cal BC), and was probably complete by *c.* 7000 BP (5840 cal BC) (Funnel 1995).

The interplay of glacio-eustatic sea-level rise and differential isostatic uplift during the last 15 000 years profoundly influenced the changing configuration of the Scottish coastline. Moreover, because isostatic uplift has been both greater and more rapid in areas close to the centre of the last ice sheet (Figure 3.4), the nature and rate of relative sea-level change has varied spatially with distance from the centre of isostatic uplift, which lay in the vicinity of Rannoch Moor in the western Grampians. Where the rising seas and rising land were temporarily in equilibrium, shorelines were formed. In areas where subsequent uplift outstripped eustatic sea-level rise, these shorelines now take the form of raised beaches, raised deltas, raised estuarine deposits (known in Scotland as carse) and raised rock platforms eroded across bedrock (Plates 3.5 and 3.6). In areas far from the centre of uplift, however, contemporaneous shorelines are sometimes buried under deposits laid down during a later marine transgression, or indeed lie below present sea level.

The earliest Lateglacial raised beaches in Scotland are those that formed in the east of the country as the land emerged from under the last ice sheet (Sissons 1983). Of greater archaeological importance, however, is a pronounced coastal rock platform and backing cliff known as the Main Lateglacial Shoreline, thought to

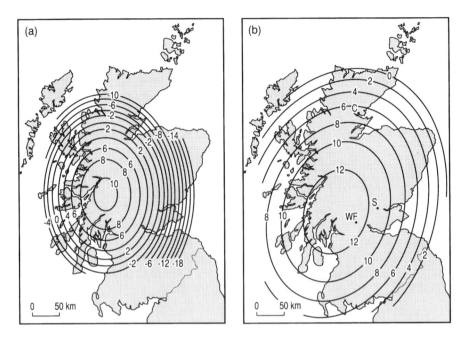

Figure 3.4 (a) Quadratic trend surface map for the Main Lateglacial Shoreline in Scotland. Contours in metres. After Firth (1992). (b) Quadratic trend surface map for the Main Postglacial shoreline in Scotland. Contours in metres. C: Creich; S: lower Strathearn; WF: western Forth Valley (see Figure 3.5). After Firth (1992)

have been formed by frost weathering and wave action during the Loch Lomond Stadial (Sissons 1974; Dawson 1980). The platform is typically 50–150 m wide and is present along considerable stretches of the coastline of western Scotland (Dawson 1984). It has a maximum altitude of over 10 m in the Oban area, and gradually declines in altitude away from the centre of isostatic uplift, passing below present sea level in Islay, western Mull, south-east Skye, Ayrshire and the Inner Moray Firth (Figure 3.4). This shoreline is of particular archaeological interest as caves and fissures in the backing cliff contain Mesolithic remains (Movius 1940a; Lacaille 1954; Bonsall and Sutherland 1992; Chapter 7). Sea-level changes throughout the Holocene, however, continued to affect the configuration of the Scottish coast, the settlement patterns of coastal dwellers and the continuity of the archaeological record in the coastal zone; such topics are considered below.

GEOMORPHOLOGICAL CHANGES DURING THE HOLOCENE

Although the legacy of glaciation still dominates the Scottish landscape, in the 10 000 years since the disappearance of the last glaciers the land surface has been slowly and subtly adjusting to non-glacial conditions. Such Postglacial changes are particularly evident in areas of high relief, and are best documented for the Scottish Highlands (Ballantyne 1991a), where the retreat of the last glaciers exposed a

Plate 3.5   Relict cliffline in western Jura uplifted as a result of glacio-isostatic processes. The clifftop is mantled by several metres of Lateglacial beach sediments. In the foreground is a cave eroded in quartzite and lamprophyre that was available for human occupation for most of the Holocene. Copyright: A. G. Dawson

landscape characterized by numerous glacially-steepened rockwalls. Some of these have adapted gradually to Postglacial conditions by a combination of slope decline, intermittent rockfall and the gradual accumulation of basal talus slopes (Ballantyne and Eckford 1984). Others, however, have experienced major landslips. Nearly 600 sites of rock slope failure have been recorded in the Highlands. These take the form of major rockfalls, topples, rock sags, translational slides, rotational slides and complex failures involving two or more of the aforementioned, and range in size from small failures involving a few tens of cubic metres of rock to great landslips involving the failure of entire mountainsides (Ballantyne 1986). Some are ancient features that pre-date the Loch Lomond Stadial, but the majority seem to have occurred in the early Holocene between *c.* 10 000 and *c.* 5000 BP (9230 and 3780 cal BC). A possible triggering mechanism was seismic activity caused by crustal displacement along ancient fault lines (Ringrose 1989). Davenport *et al.* (1989) have suggested that a combination of glacio-isostatic uplift and tectonic uplift resulted in appreciable seismic activity in the Highlands throughout the Holocene, possibly producing major earthquakes as large as magnitude 6.5–7.0 at the end of the Loch Lomond Stadial, with magnitude 5.0–6.0 events as recently as 3000–2000 BP (1240 cal BC–cal AD 10). Archaeologists seeking explanations for the destruction or abandonment of some early settlements in Scotland would be wise to consider the possible effects of such events.

Plate 3.6 Fossil clifflines with Lateglacial unvegetated raised beach ridges (left) in western Jura. Copyright: A. G. Dawson

One of the most widespread processes that has modified steep soil-mantled slopes during the Holocene has been debris flow, the rapid downslope flow of debris mixed with water. Two types occur in Scotland: hillslope flows that move down open mountainsides, depositing parallel levées of debris, and valley-confined flows that are largely restricted to the floors of gullies or valleys. Deposition by valley-confined flows has resulted in the formation of debris cones at the mouths of gullies, particularly along the flanks of glacial troughs such as Glen Coe and Glen Etive. Most debris cones in the Highlands are completely vegetated, with little or no evidence for recent deposition. Many of those outside the limits of the Loch Lomond Readvance seem likely to have formed under contemporaneous periglacial conditions, but within these glacial limits others appear to have accumulated mainly in the early Holocene as a result of reworking of glacigenic deposits; as such deposits became exhausted, debris flow activity ceased (Brazier *et al.* 1988). Little evidence has emerged for debris flow activity in the middle Holocene, but there are strong indications that both hillslope and valley-confined flows have been much more frequent within the past three centuries than at any time since the early Holocene (Innes 1983; Brazier and Ballantyne 1989).

The floors of many valleys in Scotland were modified during the Holocene by both fluvial deposition and fluvial incision. In the Highlands, alluvial fans within the limits of the Loch Lomond Readvance represent deposition of coarse sediment eroded from mountain catchments after 10 000 BP (9230 cal BC), but most are vegetated, relict landforms, deeply incised by their parent streams. It is tempting to attribute such fans to rapid sedimentation in the early Holocene, when mountain streams had access to abundant sediment and before vegetation cover became fully established on adjacent slopes, but nowhere in the Highlands have fan deposits been securely dated. At a site in the Southern Uplands, however, Tipping and Halliday (1994) have shown that renewed fan accumulation occurred as recently as the eleventh century AD, though the cause of the implied increase in sediment supply could not be determined. Similarly, only limited information is available on the ages of the abundant river terraces that occupy valleys in both upland and lowland Scotland (Macklin 1993; Tipping 1994c), though Robertson-Rintoul (1986) has shown that terrace fragments in Glen Feshie (western Cairngorms) represent five major phases of terracing. The highest terraces she attributed to two periods of terrace development during the Lateglacial, the much lower Holocene features to terrace development at *c.* 3600 BP (1940 cal BC), *c.* 1000 BP (cal AD 1020) and *c.* 80 BP (recent). The causes of terrace formation at these times remain unexplained.

In sum, many Holocene depositional landforms such as talus slopes, debris cones and alluvial fans are relict, vegetated landforms that have effectively ceased to accumulate and are often subject to current erosion or reworking. It seems likely that these represent essentially paraglacial features that accumulated soon after deglaciation as a consequence of rockfall from unstable cliffs or reworking of sediments by debris flows or rivers. Similarly, it appears that most rock-slope failures took place early in the Holocene, possibly in response to the greater magnitude of seismic events during the earlier stages of glacio-isostatic recovery. These interpretations, however, must remain conjectural until further data become available on the age of relict Holocene landforms and deposits (Ballantyne 1991a).

HOLOCENE SEA-LEVEL CHANGES AND THEIR CONSEQUENCES

The pattern of Holocene sea-level change

Because the rate of isostatic recovery diminishes with distance from the centre of uplift, the pattern of Holocene sea-level changes in Scotland is complex. Coastal sites located great distances from the uplift centre tend to have been dominated by progressive submergence throughout the Holocene, and in some peripheral areas, such as the Northern and Western Isles, crustal subsidence may have augmented the rate and extent of marine encroachment. Conversely, coastal sites located closer to the centre of uplift display a more complex sequence of relative sea-level movements in which four distinct phases can be identified (Figure 3.5).

1. Before *c.* 8500–8000 BP (7530–6840 cal BC), the rate of isostatic uplift exceeded that of eustatic sea-level rise, producing a fall in relative sea level.
2. There followed a period of relative marine transgression known in Scotland as the Main Postglacial or Flandrian Transgression, when the rate of eustatic sea-level rise exceeded that of isostatic uplift. This part of the early Holocene coincided with the disintegration of the last mid-latitude ice sheets, and may have been associated with particularly rapid rises in relative sea level.

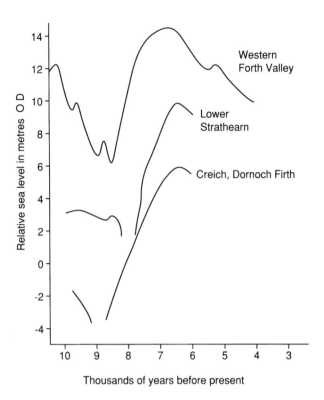

Figure 3.5 Time–altitude graph depicting relative sea-level curves for the western Forth Valley, lower Strathearn (Tayside) and Creich (Dornoch Firth). The differences in relative sea level reflect differences in distance from the centre of glacio-isostatic uplift. Based on Sissons and Brooks (1971), Cullingford *et al.* (1980), Haggart (1987) and Smith *et al.* (1992)

3. The culmination of the Main Postglacial Transgression occurred as global ocean volume approached present-day levels, so that little further eustatic sea-level change took place. Because of differential rates of isostatic uplift, maximum relative sea level occurred in different places at different times within the period 7200–6000 BP (6000–4870 cal BC), though there is some evidence that the date of culmination becomes more recent with increasing distance from the centre of isostatic uplift (Sutherland 1984). The shoreline associated with the culmination of the transgression is known as the Main Postglacial Shoreline and is now raised above present sea level as a result of subsequent isostatic uplift. It reaches a maximum altitude of 13–14 m on the shores of Loch Etive, the upper Forth Valley and southern Loch Lomond, and declines in altitude away from the centre of uplift to less than 2 m above present sea level in north-east Scotland (Figure 3.4).

4. The culmination of the Main Postglacial Trangression was succeeded by gradual relative marine regression as isostatic uplift continued, interrupted by minor periods of relative stasis or transgression caused by small (< 2 m) fluctuations in eustatic sea level.

Implications of Holocene sea-level changes

The inroads made by the sea at the culmination of the Main Postglacial Trans-gression have important implications for the interpretation of the pattern and chronology of Mesolithic settlement, particularly in western Scotland. Here the transgression rose above the level of the rock platform that represents the Main Lateglacial Shoreline, thereby probably destroying shell middens and other early Postglacial archaeological evidence. It is significant in this context that many of the earliest dated Mesolithic remains (> 7000 BP; > 5840 cal BC) relate to coastal sites just above the Holocene marine maximum, e.g. on Rhum (Wickham-Jones 1990) and in the Oban area (Bonsall and Sutherland 1992). Conversely, although the Main Postglacial Shoreline is a time-transgressive feature, it does provide a general 'chronomorphological' datum for constraining the ages of *in situ* archaeological evidence recovered on or below this level, in the sense that such evidence must post-date the formation of the shoreline. In some parts of western Scotland, for example, Mesolithic sites (including shell middens) are located on the former coastline immediately adjacent to the Main Postglacial Shoreline. The proximity of such sites to the Holocene marine maximum suggests occupation at the time of the culmination of the Main Postglacial Transgression (e.g. Jardine 1977; Bonsall 1992). Archaeological interpretations can therefore be influenced by assuming such proximity to the contemporary sea; and thus the suggestion by Mercer (1969) that at a Mesolithic site in northern Jura the Holocene sea reached 15.5 m OD is inconsistent with shoreline data that indicate a Flandrian marine limit of about 12 m (Figure 3.4(b)), and requires reassessment.

The Main Postglacial Transgression resulted in striking changes in the coastal configuration of low-lying areas of Scotland. In the west, the sea encroached onto low ground around the Clyde Estuary and invaded Loch Lomond. In eastern Scotland the sea advanced up the Tay Estuary, flooding lower Strathearn, and inundated the Forth Valley almost as far west as Aberfoyle (Figure 3.6). Forests and peatlands were flooded as the Forth Estuary extended westwards, and whales ventured inland of the present site of Stirling. At the maximum of the transgression in the Forth Valley, between *c.* 6800 and *c.* 6500 BP (5630–5440 cal BC), the rising seas almost severed the Highlands from southern Scotland, and a land bridge only 12 km wide linked the north with the south. As the sea receded from the Forth and Tay Valleys, it left behind flat plains underlain by fertile, silt-rich carse clays. Although the heavy, ill-drained soils of these deposits were not cultivated for many millennia after the withdrawal of the sea (Morrison 1983), dugout canoes dating back to Mesolithic times have been discovered near the base of the carse clays near Perth and Falkirk, as well as below similar deposits in the Glasgow area, and numerous hunting implements have been found within the carse of the Forth Valley (Geikie 1894; Clark 1952).

On more exposed coasts, particularly those of the western Highlands, high raised beaches formed at the time of ice-sheet retreat and lower beaches formed during or after the Main Postglacial Transgression have provided sheltered settlement sites and sandy, fertile soils. Amid the ice-scoured coastal lowlands of the North-West Highlands, raised beaches and associated alluvial terraces at the heads of sheltered bays and fjords represent locally the most extensive oases of agrarian opportunity in

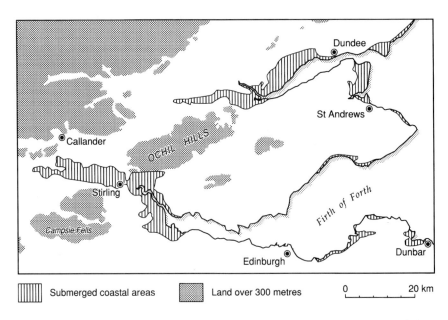

Figure 3.6 Extent of inundation of east-central Scotland at the culmination of the Main Postglacial Transgression

what has become a seemingly sterile desert of rock and bog. Such beaches are absent from Shetland, Orkney and the Outer Hebrides, where submergence appears to have predominated throughout the Holocene. According to Sissons (1967, 210), the Orkney Islands and Outer Hebrides may each have once constituted a single island that became progressively fragmented by rising seas. As evidence of continued relative sea-level rise during the mid and late Holocene, Sissons cited partial submergence in the Uists of Neolithic chambered cairns and of duns dating to the early Christian era (Callander 1929). Progressive submergence of the Outer Hebrides and Northern Isles is also evident from numerous descriptions of Flandrian peat beds below present sea level (e.g. W. Ritchie 1985). Submerged peat near Whalsay (Shetland), for example, indicates that relative sea level has risen there by at least 8 m since 5500 BP (4350 cal BC) (Hoppe 1965).

Although the general pattern of Holocene sea-level change is well understood for Scotland, much less is known about the influence of exceptional storm flooding and its effects on early settlements. Along the east coast, however, there is evidence in the form of a buried sand horizon for a widespread coastal flood in the early Holocene. This was apparently caused by a tsunami generated by a massive submarine landslide, the second Storegga Slide, which took place on the continental slope west of Norway around 7000 BP (5840 cal BC) (Dawson *et al.* 1988). The tsunami is believed to have overwhelmed a Mesolithic occupation site at Inverness and may have flooded other Mesolithic sites at Broughty Ferry on the north side of the Tay Estuary and at Morton in north-east Fife (Dawson *et al.* 1990).

Reworking of raised beach sediments and other sand-rich deposits by wind has in some coastal locations produced extensive spreads of sand dunes, now for the most part anchored by coarse grass or forestry. Examples include much of the area now occupied by Tentsmuir Forest in Fife, Torrs Warren at the head of Luce Bay in south-west Scotland and, most extensive of all, the Culbin Sands on the southern shore of the Moray Firth. Such deposits are often rather infertile. Of much greater agricultural potential are the sandy plains or machair that occur along the west coast of the Outer Hebrides, particularly on the Uists and Benbecula, and on Tiree. According to Ritchie (1979), the rising Holocene seas in these areas moved great quantities of sediment landward across the gentle shelves offshore, providing the source of extensive windblown sand deposits. These belts of sand are rich in comminuted shell fragments and tend to occur between a coastal dune belt and the peat-covered areas inland. Organic layers containing evidence of early occupation have been found intercalated with machair sands, including those at Northton on Harris (Simpson 1976), Rosinish on Benbecula (Shepherd and Tuckwell 1977; Whittington and Ritchie 1988), and Cill Donain and Cladh Hallan on South Uist (Gilbertson *et al.* 1995).

HUMAN ACTIVITY AND GEOMORPHOLOGICAL RESPONSE

During the early and middle Holocene, nearly all of Scotland except high ground was covered by a succession of woodland types (Chapter 5). The principal impact of early human activity on the physical landscape stems directly from clearance of these forests and the subsequent use of the land for agriculture. Woodland clearance affects the ground in several ways. By reducing evapotranspiration and other hydrological changes, it increases stream discharge by 10–40% (Moore 1985), and triggers a response to rainstorms in which flash-floods are more prevalent, causing river incision (Ferguson 1981). Removal of vegetation cover also causes a drastic increase in rates of soil erosion by rainsplash, slopewash and wind, leading to increased deposition of colluvium (slope deposits) or alluvium (fluvial deposits) on lower valley sides and on valley floors respectively. In addition, clearance of woodland tends to result in increased leaching of nutrients and alteration of soil structure, rendering hillslopes more susceptible to erosion. Agricultural practices may also cause general degradation of cleared land. Tillage exposes the soil to erosion by wind and water, and grazing may result in a loss of nutrients, breakdown of soil structure, breakage of vegetation cover and hence erosion by water or wind.

Much of the evidence for episodes of increased soil erosion in Scotland comes from sediment cores retrieved from the floors of lakes. Radiocarbon dating of such cores offers the possibility of identifying periods of enhanced inwash of mineral soil, and accelerated erosion of soils within lake catchments may also be evident in terms of an increased concentration of certain metal ions (Pennington *et al.* 1972; Edwards and Rowntree 1980). Variations in the magnetic susceptibility of lake sediments also provide a measure of fluctuations in the overall rate of catchment erosion (Thompson *et al.* 1975), and the coarseness of the sediment influx gives a general idea of changes in the capacity of the transporting agent. Some caution is due in the interpretation of lake sediment sequences, as uneven deposition of sediment on the lake floor ('sediment focusing') may undermine the representativeness of a single sediment core (Whittington *et al.* 1990; Edwards and Whittington 1993). An alternative way of identifying periods of enhanced soil erosion is through radiocarbon dating of organic material, such as soil, peat, wood and charcoal, that has become buried under colluvium or alluvium (e.g. Brazier *et al.* 1988; Hirons and Edwards 1990). Both approaches identify only the approximate timing of phases of enhanced landscape instability, not their causes. Because climate change, natural vegetation change and events such as exceptional storms or forest fires may trigger enhanced soil erosion, additional evidence is required to demonstrate that periods of accelerated erosion reflect human activity. This may be suggested by proximity of coeval archaeological remains, but the most versatile approach has been to demonstrate that enhanced erosion was accompanied by a marked decline in arboreal pollen and the concomitant appearance of agricultural weeds, cereals, and other indicators of clearance or agriculture in the pollen record.

Although there is some localized indication of clearance in Mesolithic times in Scotland (e.g. Bohncke 1988; Bennett *et al.* 1990; Hirons and Edwards 1990), there appears to be no clear evidence for associated soil erosion. Several sites, however, indicate that the advent of the earliest farmers around 5000 BP (3780 cal BC) was accompanied by enhanced erosion. At Braeroddach Loch, west of Aberdeen, the onset of pastoralism after *c.* 5390 BP (4240 cal BC) was accompanied by a threefold increase in sediment deposition (Figure 3.7), and subsequent increases in the magnetic susceptibility and cation concentrations of the lake sediments have been interpreted in terms of erosion of soils in the catchment (Edwards and Rowntree 1980; Plate 4.1). Evidence of a similar nature has also emerged from north-west Scotland (Pennington *et al.* 1972). At Loch Tarff, for example, increases in concentrations of cations and iron in lacustrine sediments coincide with indications of an episode of deforestation after *c.* 5000 BP. Stratigraphic evidence of accelerated soil erosion during Neolithic times is also evident in cores recovered from Loch Cuithir on Skye and Loch of Park in Aberdeenshire (Vasari and Vasari 1968). At Kinloch on Rhum, the first sustained hillwash event recorded in a peat core recovered by Hirons and Edwards (1990) occurred at *c.* 4660 BP (3470 cal BC), and a sustained increase in slopewash is evident after *c.* 3950 BP (2460 cal BC), by which time the associated pollen spectra indicate open conditions and arable cultivation. At a nearby excavation site, radiocarbon ages of 4260±70 BP (2920–2700 cal BC) and 3945±60 BP (2560–2340 cal BC) have been obtained from organic material beneath colluvium, possibly providing the earliest dates for accelerated erosion at this site (Wickham-Jones 1990). At Northton on Harris there is circumstantial evidence that clearance may have enhanced deflation of machair sand prior to occupation by Neolithic settlers at *c.* 4420 BP (3040 cal BC) (Simpson 1976).

Whilst the evidence outlined above suggests that enhanced soil erosion resulting from woodland clearance and agriculture has occurred at some locations since the beginning of the fifth millennium BP, many parts of Scotland did not experience extensive clearance until much later (Turner 1981). At some sites that exhibit evidence for accelerated soil erosion in Neolithic times, sediment discharge continued to increase, presumably in response to continued clearance, expansion of arable land or more intensive cultivation practices. At Braeroddach Loch, for example, the maximum rates of sediment deposition recorded by Edwards and Rowntree (1980) were achieved between *c.* 3405 BP and *c.* 2100 BP (1690 and 100

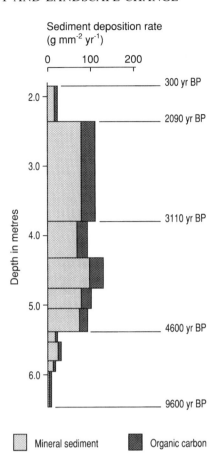

Figure 3.7 Sediment deposition rate at Braeroddach Loch. After Edwards and Rowntree (1980)

cal BC), a period that spans the later Bronze Age and the Iron Age. At Black Loch in north-east Fife, there is some evidence for catchment disturbance and enhanced erosion as early as the Neolithic–Bronze Age transition (*c.* 3890 BP; 2380 cal BC), but the strongest evidence for accelerated erosion relates to the second millennium BP (Whittington *et al.* 1990; Edwards and Whittington 1993).

It is not possible to be certain as to the ways in which woodland clearance and enhanced soil erosion affected the landforms of Scotland during the first nine millennia of the Holocene, though reference to studies carried out in England (e.g. Harvey *et al.* 1981; Harvey and Renwick 1987; Hooke *et al.* 1990; Macklin *et al.* 1992) as well as those relating to disturbance of the Scottish landscape over the past few centuries (e.g. Innes 1983; Brazier *et al.* 1988) suggest that increased runoff resulting from clearance may have resulted in enhanced slope failure, debris flow activity and river incision. Conversely, increased soil erosion is likely to have produced slope-foot colluvial accumulations and to have favoured valley-floor aggradation. Such effects are evident throughout Scotland in the form of relict landslip scars, abandoned alluvial fans, vegetation-covered debris cones and flights of Postglacial river terraces. At present, however, we do not know when such relict landforms developed, or indeed whether they represent the consequences of human activity or the products of landscape instability triggered by Holocene climatic changes (Ballantyne 1991b). Unravelling the complex links between early human activity, environmental change and the development of valley-side and valley-floor landforms presents an exciting challenge for geomorphologists and archaeologists alike.

4 Soils and Their Evolution

DONALD A. DAVIDSON AND STEPHEN P. CARTER

INTRODUCTION

The aims of this chapter are first to provide an introduction to the nature and formation of soils in Scotland, and secondly to review the evidence for soil development since early Holocene times. Though direct evidence is still rather sparse, it is clear that Scottish soils have been subject to considerable change over the last 10 000 years. For at least the last 5000 years humans have had an increasingly important influence on soils; indeed, the arable lowlands of Scotland now contain essentially human-made soils.

Soils exist at the interface of the lithosphere and the atmosphere. They provide the medium for plant growth in terms of physical support and nutrition. As soon as plants colonize a bare surface, the underlying material is subject to change as a result of soil-forming processes which include the incorporation and decay of organic matter in soil, the effects of soil organisms, the movement of soil water with associated solutes and fine sediments and the weathering of rock fragments and minerals by physical, chemical and biological processes. It is the combination of such processes which results in the distinctive form or morphology of soils as expressed in horizon sequences. In essence, soils can be visualized as systems which owe their attributes to present as well as past processes. Soils are thus dynamic entities and their investigation can make a significant contribution to any analysis of environmental and early human history.

THE SOIL RESOURCE

Soil formation

The nature of Scottish soils can be introduced by outlining the dominant soil-forming processes. It is interesting to observe that although there is considerable soil variability in Scotland, there are only a few dominant soil-forming processes: weathering, leaching, podzolization and gleying.

Weathering refers to the *in situ* comminution of minerals, either as individual grains in soil or as rock fragments, by some combination of physical, chemical and

Scotland: Environment and Archaeology, 8000 BC – AD 1000. Edited by Kevin J. Edwards and Ian B. M. Ralston.
© 1997 The editors and contributors. Published in 1997 by John Wiley & Sons Ltd.

biological processes. Chemical weathering is important because it leads to the release of nutrients, e.g. calcium, potassium and phosphorus.

Leaching refers to the downward movement of soil components in solution; thus in well-drained areas, liberated cations in the upper part of soils are flushed downwards. The process of leaching leads to base-deficient soils, especially in areas of high rainfall with well-drained and acid parent materials. In practice this means that many Scottish soils are of low inherent fertility.

Podzolization is a particular form of leaching combined with chemical weathering. The details of podzolization processes are of long-standing controversy in pedology, but the results are very clear: the formation of a spodic B horizon, often also enriched with organic matter in its upper part. The strong brown colour of the B horizon results from the precipitation of iron sesquioxides. The depletion in the upper part of the soil of bases, iron and organic matter may be expressed in the presence of an albic or E horizon. Another characteristic horizon associated with podzolization is an iron pan (Bf horizon) which causes considerable restriction to the downward movement of water. As a result, an iron pan podzol may well demonstrate impeded drainage above the pan whilst it is freely drained below the pan. Podzols are thus soils which pose limitations in terms of low inherent fertility, with soil depth restrictions being a further possible problem.

Gleying is characterized by soil horizons with attributes formed under anaerobic conditions which result in grey or bluish colours. It is caused by the simple situation of soils being unable to shed water quickly, either as a result of topographic position or inherent low permeability.

Although these four processes of weathering, leaching, podzolization and gleying can account for the broad nature of Scottish soils, the role of anthropogenic activity in influencing soils cannot be overemphasized. Most soils in Scotland suffer from one or more limitations of inherent low fertility, drainage, depth or stoniness. Thus farmers since the Neolithic have had to devise strategies for sustaining crop and livestock production which have radically altered large areas of Scotland's soils.

The present-day soils of Scotland

Scotland is fortunate in having comprehensive soil survey cover with 1:63 360 or 1:50 000 maps available; in addition 1:250 000 soil maps have been published for the whole country. Systematic mapping began after the Second World War and the work of the Soil Survey for Scotland was essentially terminated by the mid 1980s following the publication of the 1:250 000 maps. In the Scottish system, use is made of *associations* as a means of grouping *soil series*. Soil associations are distinguished on the basis of parent material types (classified according to lithology, or lithology and stratigraphic age). Soil series display similar sequences of horizons and are formed on particular parent materials. Particular emphasis is given to subdividing soil series within associations on the basis of drainage conditions. Topographic variation in soil drainage conditions, also called the hydrologic sequence, is seen as vital in controlling the spatial occurrence of individual soil series. The Scottish system of soil classification is given in Table 4.1 and the list is useful in indicating the soil types that are present. As can be seen, the nature of leaching and gleying

Table 4.1 Classification of Scottish soils. Source: Macaulay Institute for Soil Research (1984)

Division	Major soil group	Major soil subgroup
1. Immature soils	1.1 Lithosols	
	1.2 Regosols	1.21 Calcareous regosols
		1.22 Non-calcareous regosols
	1.3 Alluvial soils	1.31 Saline alluvial soils
		1.32 Mineral alluvial soils
		1.33 Peaty alluvial soils
	1.4 Rankers	1.41 Brown rankers
		1.42 Podzolic rankers
		1.43 Gley rankers
		1.44 Peaty rankers
2. Non-leached soils	2.1 Rendzinas	2.11 Brown rendzinas
	2.2 Calcareous soils	2.21 Brown calcareous soils
3. Leached soils	3.1 Magnesian soils	3.11 Brown magnesian soils
	3.2 Brown soils	3.21 Brown forest soils
	3.3 Podzols	3.31 Humus podzols
		3.32 Humus-iron podzols
		3.33 Iron podzols
		3.34 Peaty podzols
		3.35 Subalpine podzols
		3.36 Alpine podzols
4. Gleys	4.1 Surface-water gleys	4.11 Saline soils
		4.12 Calcareous gleys
		4.13 Magnesian gleys
		4.14 Non-calcareous gleys
		4.15 Humic gleys
		4.16 Peaty gleys
	4.2 Ground-water gleys	4.21 Calcareous gleys
		4.22 Non-calcareous gleys
		4.23 Humic gleys
		4.24 Peaty gleys
		4.25 Subalpine gleys
		4.26 Alpine gleys
5. Organic soils	5.1 Peats	5.11 Eutrophic flushed peat
		5.12 Mesotrophic flushed peat
		5.13 Dystrophic flushed peat
		5.14 Dystrophic peat

processes plays a major role in influencing Divisions and Major Soil Groups and Subgroups.

The legend for the 1:250 000 soil survey maps of Scotland gives information on all soil associations and a summary is given in Table 4.2. The variable extent of these associations is a reflection of the solid and drift geology of Scotland, which is dominated by the igneous and metamorphic rocks of the Highlands and the Lower Palaeozoic sedimentary rocks of the Southern Uplands. The six most extensive soil associations out of a total of 110 account for 53.5% of the land area. A similarly

Table 4.2 Main soil associations, their extent and parent materials in Scotland. Source: Macaulay Institute for Soil Research (1984)

Soil Associations	% cover	Parent materials
Alluvial soils	1.63	Recent riverine, and lacustrine alluvial deposits, marine alluvial deposits
Organic soils	9.94	Organic deposits
Arkaig	16.22	Drifts derived from schists, gneisses, granulites, and quartzites (Moine Series)
Balrownie	1.83	Drifts from sandstones (Old Red Sandstone age), often water-worked
Corby/Boyndie/Dinnet	3.08	Glaciofluvial and raised beach sands and gravels derived from acid rocks
Countesswells Dalbeattie/Priestlaw	5.75	Drifts derived from granites and granitic rocks
Darleith/Kirktonmoor	3.53	Drifts derived from basaltic rocks
Durnhill	1.60	Drifts derived from quartzites and quartzose grits
Ettrick	9.26	Drifts derived from Lower Palaeozoic greywackes and shales
Foudland	3.25	Drifts derived from slates, phyllites and other weakly metamorphosed argillaceous rocks
Lochinver	4.47	Drifts derived from Lewisian gneisses
Rowanhill/Giffnock/Winton	3.04	Drifts derived from Carboniferous sandstones, shales and limestones
Sourhope	1.71	Drifts derived from Old Red Sandstone intermediate lavas
Strichen	7.98	Drifts derived from arenaceous schists and strongly metamorphosed argillaceous schists (Dalradian Series)
Tarves	2.07	Drifts from intermediate rocks or mixed acid and basic rocks, both metamorphic and igneous
Thurso	1.35	Greyish brown drifts derived from Middle Old Red Sandstone flagstones and sandstones
Torridon	2.25	Drifts derived from Torridonian sandstones and grits
Other soil associations, each < 1%	17.66	
Built-up areas	1.59	

skewed distribution is also found when the occurrence of soil types is examined. The National 1:250 000 soil map legend lists the component soils for each association. No information is given on extents of individual soil types within associations, but the results from taking the first-named soil type are given in Table 4.3. These figures can only be very approximate, but they give some impression of the present-day incidence of different soils. As can be seen, peaty gleys, peat soils and peaty podzols cover 50% of Scotland. Humus iron podzols are often taken to be typical of

Table 4.3 Occurrence in rank order of major soil groups or subgroups in Scotland. Soils which occupy < 1.0% are not shown. Based on figures extracted from the National 1:250 000 Soil Map Legend, Macaulay Institute for Soil Research (1984)

Major soil group or subgroup	% area of Scotland
Peaty gleys	18.8
Peat soils	15.6
Peaty podzols	15.6
Brown forest soils	13.9
Humus iron podzols	11.0
Brown forest soils with gleying	7.3
Non-calcareous gleys	7.0
Subalpine soils	4.0
Alluvial soils	1.6
Rankers	1.6

Scottish soils, yet brown forest soils are more extensive. Soils that occur in very limited areas of Scotland, but are of particular scientific interest include rendzinas, calcareous soils and magnesian soils, which are found on ultrabasic parent materials.

A typical soil landscape in the southern uplands is portrayed in Figure 4.1. The Ettrick Association is extensive in southern Scotland since it is derived from drift deposits dominated by Lower Palaeozoic greywackes and shales. This association covers 9.26% of the land area of Scotland and 30.8% of south-east Scotland (Bown and Shipley 1982). The Ettrick Association consists of a very wide range of soils including freely drained brown forest soils (Linhope Series), imperfectly drained brown forest soils (Kedslie Series), poorly drained non-calcareous gley (Ettrick Series) and a freely drained iron podzol (Minchmoor Series). The landscape pattern of these soil series is shown in Figure 4.1; in addition, a freely drained brown forest soil derived from glaciofluvial gravels (Yarrow Series) is shown to coincide with a terrace, below which alluvial soils occur.

Land resources

It is the combination of soils, topography and climate which determines the potentialities and limitations of the Scottish environment with reference to human occupation and food production. Although Chapters 2–6 give particular emphasis to environmental change through time, the magnitude of environmental gradients on a spatial basis must not be overlooked. Such gradients are expressed in decreases in warmth and increases in precipitation and exposure with elevation. Onto such elevational trends must be superimposed W–E and S–N climatic patterns.

Land capability analysis provides an overall assessment of the Scottish land resource base. Land capability schemes are available for Scotland for agriculture (Bibby *et al.* 1982) and forestry (Bibby *et al.* 1988); results have been published for the whole country at 1:250 000. An outline of the results from the assessment of

50

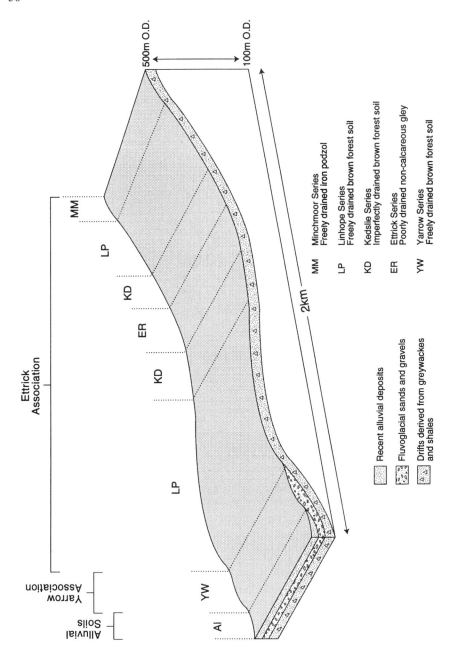

Figure 4.1 Schematic distribution of soils for a typical area in the Southern Uplands

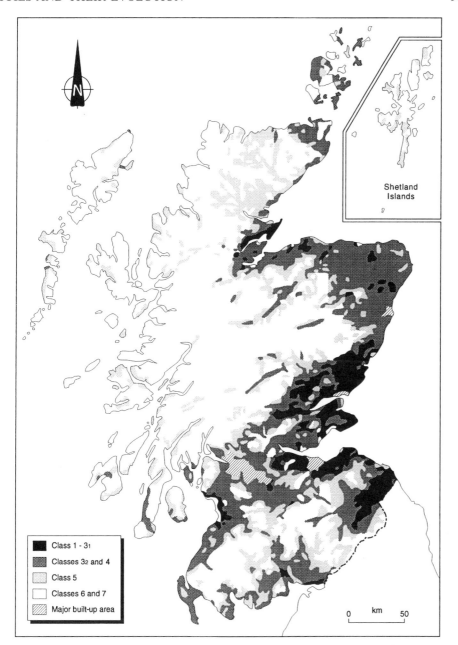

Figure 4.2 Distribution of land capability classes for agriculture. Source: Bibby *et al.* 1982

land capability for agriculture (LCA) provides an integrated evaluation of the land resources of Scotland (Figure 4.2). The LCA scheme combines climate, gradient, soil, wetness, erosion and vegetation attributes. Climatic factors comprise potential soil moisture deficit, accumulated temperature and wind speed. The LCA scheme rates land into seven classes with classes 1–4 being suited to arable cropping and

Table 4.4 Extent of land capability classes for agriculture in Scotland. Source: Macaulay Institute for Soil Research (undated)

Land capability class	Extent of total land area (%)
1	< 1
2	1
3	15
4	11
5	19
6	48
7	3
(built-up areas, quarries, etc.)	2

classes 5–7 to improved grassland and rough grazing. Classes 3 and 4 are further subdivided into two divisions and classes 5 and 6 into three divisions, the last of these being based on assessments of the grazing value of dominant plant communities.

In terms of soils, texture and structure exert limitations primarily through their effects on workability and structural instability. Shallowness and stoniness pose obvious problems. Soil droughtiness is evaluated by taking into account soil–climate–crop interactions. Wetness is another complex soil property with impacts on workability, trafficability and poaching risk. Pattern is another important limiting factor and refers to variation at the within-field scale in such properties as stoniness, depth or slope.

Figure 4.2 shows in generalized form the distribution of LCA classes and selected divisions. Classes 1–3 (Division 1) are taken to correspond to prime agricultural land; in planning there is a presumption against the loss of such land from agriculture. Overall land suitable for arable agriculture (classes 1–4) is restricted to the midland belt between the Highland Boundary and Southern Uplands Faults, north-east Scotland (Buchan) extending along the coast to the Black Isle and Easter Ross, and the basins of the South-East and South-West (the Tweed, Annandale and Nithsdale). As can be seen in Table 4.4, the occurrence of land capability classes is very skew, with 48% of the land area in one class (class 6 – land capable of use only as rough grazing). In contrast, land of prime quality (classes 1–3) is of very limited extent (5.7%). Though these assessments of land capability are made with reference to present-day agriculture, the same relative contrasts in land quality would have existed since the first arrival of people in Scotland. This can be supported by noting that land capability assessment is done on the basis of permanent or semi-permanent land characteristics such as accumulated temperature, soil moisture deficit, soil depth, drainage and stoniness. Whilst the same relative contrasts may have existed, human response to the land was undoubtedly different in the past. These differences in part reflect technology: two of the land characteristics that contribute to the capability classification (gradient and land pattern) can create severe limitations for modern farm machinery but not for earlier tools. Therefore, land now unsuitable for cultivation due to rock outcrops or steep slopes may well have been utilized in the past. Further differences are caused by the change in the economic basis of agriculture: subsistence farmers in the past were in part required, and in part able, to accept lower or more variable crop yields. Therefore land now considered uneconomic for arable agriculture was formerly cultivated. These differences are most clearly seen in land currently graded class 5, which forms a fringe of marginal agricultural land rich in the visible remains of prehistoric and later settlement.

EVALUATION OF SOIL CHANGE

The sources of evidence for soils in the past

The evidence for the nature of soils and their evolution in Scotland during the last 10 000 years is present in three main situations: two of these, surface soils and buried soils, offer direct evidence whilst the third source, accumulated sediments, provides a proxy record of soils.

All present-day soils are the product of development over a period of time and features may survive that reflect either different environmental conditions in the past or an earlier stage in the development of that soil. These fossil features are very unlikely to survive in the biologically active surface horizons of a soil but may occur deeper in the profile. The most commonly described example of a fossil feature in Scotland is the indurated Bx horizon at the base of a soil profile that was formed by freeze–thaw processes under periglacial conditions at the end of the Devensian glaciation (Fitzpatrick 1956). In general, soil B horizons will reflect an earlier period of soil formation whilst A horizons alter rapidly to reflect contemporary processes.

Buried soils provide evidence of soil conditions at the time of burial and soil development up to that date. There are various problems associated with the stability of the properties of buried soils, as the soil is not totally sealed from chemical, biological and physical processes. In general, physical properties are more stable than chemical properties and therefore the analysis of buried soils is now focused on micromorphological characteristics (Kemp 1985; Courty et al. 1989). Published information is available on the nature of buried soils from 20 archaeological sites (Table 4.5 and Figure 4.3). Many more buried soils have been recorded but remain either unstudied or, as yet, unpublished. The distribution of these archaeological buried soils is limited in both time and space: Neolithic ritual monuments, constructed after c. 5000 BP (3780 cal BC), have preserved the earliest soils and thus early Holocene stages in soil development are not represented. There are also clear biases in the spatial distribution of known buried soils from archaeological contexts. On a regional scale, none of the sites listed in Table 4.5 lies in southern mainland Scotland and there is an equally large gap in the central and western Highlands and the Western Isles. On a local scale, human activities tend to occur in areas of freely drained soil which therefore provide the greatest opportunities for the recovery of buried soils in archaeological contexts; thus brown forest soils and podzols are well represented but the equally abundant gleys are rarely preserved.

Table 4.5 Sites for which there is published information on buried soils in Scotland; the location of these sites is shown in Figure 4.3. Radiocarbon dates are expressed as uncorrected years BP

Site	Date of burial	Buried soil	Modern soil	Parent material	Reference
Achnacree	3309±50 BP	Podzol	Peat	Glaciofluvial gravel	Soulsby (1976)
An Sithean	2nd millennium BC	Podzol	Peaty podzol	Glaciofluvial gravel	Barber and Brown (1984)
Tormore	3488±60 BP	Podzol	Peaty podzol	Stony till	J. W. Barber (1982b)
Boghead	c. 5000 BP	Brown forest	Podzol	Glaciofluvial sand	Burl (1984)
Burghead	1st millennium AD	Podzol	Podzol	Sand	Edwards and Ralston (1978)
Castle Hill, Strachan	13th century AD	Podzol	Podzol	Glaciofluvial gravel	Romans and Robertson (1983a)
Cleaven Dyke	c. 5150 BP	Brown forest	Podzol	Glaciofluvial gravel	Barclay et al. (1995)
Cùl a'Bhaile	2950±65 BP	Podzol	Peat	Silt loam till	Stevenson (1984)
Dalladies	5190±105 BP	Brown forest	Podzol	Glaciofluvial gravel	Romans et al. (1973)
Dalnaglar	1st/2nd millennium BC	Podzol	Podzol	Sandy loam till	Stewart (1962)
Iona	12th century AD	Podzol	Podzol	Raised beach gravel	Haggarty (1988)
Kilphedir	2370±40 BP	Podzol	Peaty podzol	Sandy loam till	Fairhurst and Taylor (1971)
Kirkbuddo	3rd century AD	Podzol	Brown forest	Water sorted sandy till	Romans and Robertson (1983a)
Lairg	Mesolithic to post-medieval	Podzol	Peaty podzol	Sandy loam till	McCullagh (in press)
Liddle	2908±45 BP	Gley	Gley	Silt loam till	Hedges (1975)
Monamore	5110±110 BP	Podzol	Podzol	Sandy loam till	MacKie (1964)
North Mains	3735±85 BP	Brown forest	Brown forest	Glaciofluvial gravel	Barclay (1983a)
Scord of Brouster	c. 4000 BP	Brown forest	Peaty podzol	Stony sandy till	Whittle et al. (1986)
Strageath	1st century AD	Brown forest	Brown forest	Raised beach sand and gravel	Romans and Robertson (1983b)
Tofts Ness	Late Neolithic to early Iron Age	Brown forest and skeletal calcareous	Calcareous gley	Windblown calcareous sand	Dockrill and Simpson (in press)

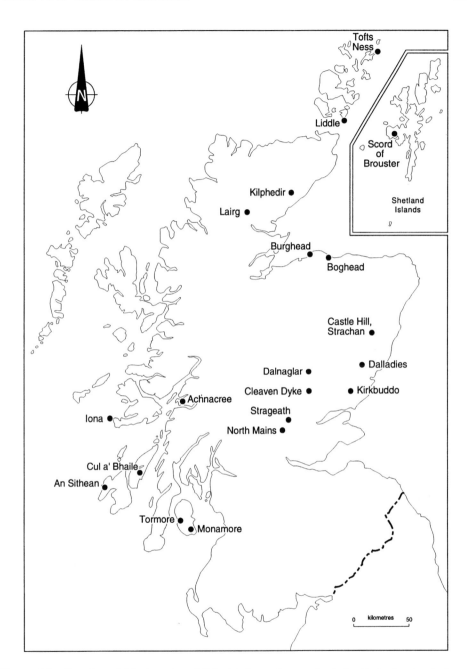

Figure 4.3 Sites for which there is published information on buried soils in Scotland; these sites are listed in Table 4.5

In interpreting archaeological buried soils, the likelihood of human disturbance prior to burial must be considered. Frequently, because of the continuity of settlement in favoured areas, buried soils have been greatly modified by human activity. Whilst this offers the opportunity to study the impact of humans on soil, it limits our ability to determine the progress of soil development in the wider landscape.

In the absence of buried soils, various proxy records of soil characteristics are available. Sediments derived from the erosion of soil (including colluvium, alluvium and lake sediments) offer two types of information. First, the sediment characteristics that derive from the soil (e.g. the texture or magnetic properties), provide information about that soil. Secondly, the quantity and age of the sediments reflect the distribution of soil disturbance processes in a catchment. Information on soil itself also comes from vegetation records, particularly pollen. Using a knowledge of the environmental ranges of plant species it is possible to relate pollen assemblages to soil conditions in the catchment. Similarly, the analysis of diatoms provides useful information through the relationship between soil conditions and water quality in a catchment.

Soil evolution during the Holocene

Since the end of the last glaciation (*c.* 10 000 BP), Scottish soils have evolved in response to the prevailing environmental conditions. Most soils in Scotland have developed in till with physical and chemical properties reflecting its source rocks. Initial differences in soil parent material and topographic position have interacted with climate and vegetation over time, leading to the differentiation of various soil profiles. Various pedogenetic pathways can be identified. In freely draining profiles, leaching and podzolization are the dominant processes. Surface-water gleying may occur in areas of extremely humid climate leading ultimately to blanket peat growth. In poorly draining profiles gleying and, to a lesser extent, leaching processes predominate, with peat accumulation a possible result. These evolutionary stages are closely correlated with changes in soil physical and chemical properties and therefore land capability.

The primary source of information on the nature of soil evolution is the small number of well-dated archaeological buried soils; this is supplemented by more general information from catchment studies. Direct evidence for the nature of soils in the early to mid Holocene is extremely limited as no intact buried soils are known from this period. The structure induced by freeze–thaw processes that survives at the base of some modern soil profiles, whilst undated, is assumed to have developed by 10 000 BP. Coatings of sands and finer material around stones have also been attributed to freeze–thaw mechanisms. The earliest published evidence from a fragmentary buried soil is from Lairg, Sutherland (McCullagh 1996). Here, a podzolic soil B horizon has been identified predating a tree-throw hollow dated to 6200 BP (5140 cal BC). This confirms that podzols had developed by this date, at least on freely draining coarse-textured parent materials.

All other information has to be derived from the interpretation of pollen, diatom and sediment data. One example of this type of study comes from a study of the sediments in Loch Sionascaig, Sutherland (Pennington *et al.* 1972). Here, acidification of initially base-rich soils occurred by 9000 BP (8030 cal BC) in a soil parent material of drift derived from Torridonian sandstone and Lewisian gneiss. An increase in iron and manganese in the lake sediment from roughly 6000 BP (4870 cal BC) was interpreted as the product of progressive waterlogging of soil surfaces. This solutional transport reached a maximum at around 5000 BP (3780 cal BC), by which time blanket peat accumulation had begun in at least three sites in the catchment. The sequence and timing of events at Loch Sionascaig is corroborated by pollen records from elsewhere in the highlands of western and northern Scotland. The pattern of early acidification followed by gleying and peat accumulation does not seem to have occurred in the eastern Highlands, where the drier climate maintained larger areas of freely draining acid soils and promoted the development of podzols.

The overall pattern of soil change through the first half of the Holocene is therefore assumed to be one of profile development under the control of parent material, drainage, climate and vegetation – the classic soil-forming factors. Freely draining coarse-textured parent materials experienced rapid leaching in the humid climate, and developed acidic soils that differentiated into podzolic soils under vegetation producing acid litter (heathland and coniferous forest), and brown forest soils with inputs of more base-rich litter (deciduous forest). Poorly draining fine-textured parent materials were leached more slowly to form non-calcareous gleys. In both coarse- and fine-textured parent materials, soil surface waterlogging led to the accumulation of peat in the more humid uplands. By the time of the first widespread and long-term human impact on soils, which came with the appearance of Neolithic farming communities, the present-day pattern of soils in Scotland had been established.

For the next 5000 years, the further evolution of Scotland's soils is complicated by the increasing influence of human activity. The role of humans in the acceleration of podzolization and peat accumulation has frequently been discussed (Chambers 1988; Moore 1988, 1993) and explanations have focused on the impacts of vegetation change. Loss of tree cover is thought to have led to increased leaching and surface wetness. Heath vegetation, which replaced forests in much of Scotland, produced organic matter that promoted podzolization and peat formation.

The spread of a peat cover is documented by a widespread group of sites from the Scord of Brouster in Shetland, southwards through Kilphedir and Lairg (Sutherland), to sites in Argyll including Achnacree and nearby islands, e.g. Cùl a'Bhaile (Jura), An Sithean (Islay) and Tormore (Arran). All of these sites contain buried podzols in areas now covered by peat or peaty podzols. Most basal peat dates show that the current peat cover was initiated in the first millennium BC or first millennium AD (cf. Carter 1994). All of the sites were formerly under cultivation and therefore it is not clear whether these dates reflect a regional climatic trigger or simply localized land abandonment. The initiation of peat growth over freely draining soils requires the establishment of waterlogged conditions at the soil surface – surface-water gleying. This process only occurs in areas of low moisture deficit (surface-water gleying is not found in the soils of lowland eastern Scotland), so overall climatic control is indicated. However, continued disturbance of the soil through cultivation and other human activities can prevent this evolution, therefore actual dates of blanket peat initiation may relate to land abandonment.

The progress of podzolization in the drier eastern lowlands is not so easily demonstrated, although a number of relevant buried soils have been studied. The

transition from buried brown forest soil to present-day podzol is recorded at
Boghead (Moray), Dalnaglar and Cleaven Dyke (Barclay *et al.* 1995) (Perthshire),
but the reverse was found at Kirkbuddo (Angus); no clear change was recorded at
North Mains, Strathallan, or Strageath (both Perthshire). This confused picture is,
at least in part, a product of the soil terminology. The main difference between a
brown forest soil and a podzol, as defined by the Soil Survey of Scotland, is the
nature of the A horizons. A brown forest soil has a single brown A horizon but a
podzol (iron podzol) has an upper dark-coloured A horizon overlying a pale-
coloured E horizon. Both soil profiles may have a B horizon relatively enriched in
sesquioxides of iron and aluminium. As a result of this classification it is possible to
create brown forest soils by the cultivation of podzols. This process has been
recognized for example in two widespread soil associations (Balrownie and Forfar)
in Strathmore (Laing 1976). In both associations, the uncultivated profile has the
characteristics of a podzol and the cultivated profile, a brown forest soil. However,
Forfar Series profiles have been mapped as podzols and the Balrownie Series as
brown forest soils. Thus, it is possible that soils in the eastern lowlands with brown
forest profiles under deciduous woodland at around 5000 BP (3780 cal BC),
repeatedly developed iron podzol and brown forest profiles as land went in and out
of cultivation.

To summarize, most of the evolution of Scotland's soils (in terms of natural
pedogenesis) occurred in the first 5000 years of the Holocene, and by the time the
oldest buried soils were preserved, the broad patterns of soil types had been
established. Since that time, natural soil evolution is hard to detect, with the
exception of the spread of organic soils in the humid north and west. Much more
apparent is the impact of humans on the soil, both indirect, by interference with
vegetation, and direct, as a result of agricultural practice. These modifications are
increasingly being recognized as the major factor in the development of soils since
5000 BP.

Human impact on soils since 5000 BP

Limited manipulation of the forest cover by Mesolithic communities must have had
an indirect impact on the soil, but the spread of agriculture in the Neolithic period
caused the first substantial human interference with soils in Scotland. Evidence for
the anthropogenic impact on the soils between 5000 and 4000 BP (3780–2490 cal
BC) is largely indirect, coming primarily from pollen and lake sediment records.

So far, only one detailed study of the soils from an early site has been published
in full. This is the Scord of Brouster in Shetland (Whittle *et al.* 1986), where
excavation of houses and associated fields of the Neolithic period was supported by
the micromorphological analysis of buried soils and sediments and a substantial
programme of pollen analysis. The pollen record indicates clearance from 4700 BP
(3420 cal BC) and the earliest house and fields are dated soon after. Soils under this
and a later house demonstrate progressive podzolization before 4000 BP (2490 cal
BC), but the most outstanding soil changes are those directly linked to agriculture.
Cultivation for barley caused substantial erosion in the sloping fields, with the
accumulation of soil lynchets up to 0.5 m deep at their downslope margins. Erosion
led to increasing stoniness in the ploughsoil and this is proposed as a possible cause

of land abandonment. The damaging impact of cultivation was countered by the application of fertilizers, indicated by the presence of hearth ashes in the plough-soils. Thus, at this one site, there is evidence both for the substantial damaging impact of Neolithic cultivation on soils and the early development of agricultural methods to counter this and maintain land capability.

Preliminary results show that similar evidence has been obtained from the multi-period site at Tofts Ness, Orkney (Dockrill and Simpson 1996). Here, from the late Neolithic, turf and associated organic manures were being added to cultivation soils, both to sustain crop yield and to minimize erosion of the windblown sands under cultivation. Further evidence for soil amelioration comes from the Early Bronze Age site of North Mains, Strathallan (Perthshire), where a field of cultivation ridges was preserved beneath a mound constructed in the early second millennium BC (Barclay 1989). Ridging can have various beneficial effects including the raising of soil temperature, increasing rooting depth and improving drainage (see also Chapter 8).

Catchment studies from elsewhere in Scotland indicate that the events documented at the Scord of Brouster are by no means exceptional. Pollen analysis at various sites in lowland eastern Scotland has identified progressive reductions in tree pollen, reflecting the clearance of land for agriculture. Two sites in Fife, Pickletillem and Black Loch (Whittington *et al.* 1991a,b) both show that this process started around 5000 BP (3780 cal BC), with substantial reductions in tree pollen by 4000 BP (2490 cal BC). Early clearance is reported as far north as Shetland where pollen from a number of sites indicates loss of tree cover from 4700 BP (3420 cal BC) (Whittle *et al.* 1986; Bennett *et al.* 1992). Analysis of the rate of sediment accumulation in lake basins has shown that this clearance could have been accompanied by significant soil erosion. At Braeroddach Loch, Aberdeenshire (Edwards and Rowntree 1980), for example, the rate of sediment accumulation *c.* 5390–3405 BP (4240–1690 cal BC) increased from 0.0020 to 0.0072 g cm^{-2} year^{-1} indicating the impact of Neolithic forest clearance on soil erosion (Plate 4.1).

Despite this evidence, the extent and intensity of Neolithic agriculture is unclear and its overall impact on the soils is unknown. It is assumed that activity was focused on the well-drained soils, with climate imposing an altitudinal limit on cultivation which lowered towards the north and west. Shorter-term effects such as the loss of sediment from bare soil surfaces are indicated by the sediment influx into basin sites. Longer-term change in the soil profile caused by vegetation change seems probable, but is hard to detect or quantify at present.

Between 4000 and 1000 BP (2490–930 cal BC), large areas of upland podzolic soils developed into peaty gleyed podzols and, with further accumulation of organic matter, to blanket peat. The characteristic upland Scottish landscape with numerous houses, cairns and banks partially buried beneath peat, developed during this period. This abandoned landscape has traditionally been interpreted in terms of human-induced soil degradation combined with peat growth, triggered by vegetation and climate change in the late Bronze Age, *c.* 3000 BP (1240 cal BC) (Romans and Robertson 1975). A relatively short timespan for podzolization, blanket peat growth and land abandonment in the uplands has promoted the idea that human impact on the soil was a process of irreversible degradation. This assumes that major human impact was a feature only of the late Bronze Age as the soils could

Plate 4.1   Braeroddach Loch, Aberdeenshire. This small loch has provided a sensitive record of land-use change. Soil erosion, possibly from the steep slopes in the left of the picture, began in the Neolithic period and accelerated in the Middle Bronze Age, *c.* 3000 BP, when cereal cultivation began. Copyright: K. J. Edwards

not sustain intensive agriculture for long. A more recent review of the evidence (Askew *et al.* 1985) noted the early date for some podzols and blanket peats and emphasized the variability of soils at this time. At Lairg, a series of buried podzols, dated between 3800 BP and 2200 BP (2200–320 cal BC), demonstrates a history of persistant erosion but maintenance of cultivation through the use of organic fertilizers (McCullagh 1996). At An Sithean, Barber and Brown (1984) recorded a sequence of land use where cultivation of podzols in the second millennium BC ended at an undetermined date to be replaced by peat growth in the first millennium AD. Cultivation was resumed in the medieval or post-medieval period and may have continued into the eighteenth century AD before the land was returned to pastoral use. These two examples make the point that soil degradation, in itself, need not cause land abandonment and if land is abandoned, it can subsequently be brought back into cultivation.

The key factors in these situations are the social and economic forces that motivate people to invest in the maintenance of the soil. Results suggest that humans were having a significant impact on upland soils at least as early as 4000 BP (2490 cal BC) but responded by adopting land management practices that allowed continued exploitation. The role of podzolization as a cause of land abandonment appears to be unimportant and it is to some extent a reversible process. As noted above, it is recognized that large areas of modern brown forest soils in Scotland have been created by the cultivation and improvement of podzols. Increased use of aerial photography in the lowlands of Scotland has provided evidence for extensive organized agricultural landscapes in the later prehistoric period (Maxwell 1983). This suggests that there was prolonged use of the better-drained soils throughout the period and, as in the uplands, successful agriculture was practised through the maintenance of the soil resource.

The historic period

The historic period provides a valuable perspective on the interactions of soil and humans which may be used to improve our understanding of past soil development. For example, ethnographic and documentary research into the use of turf (Fenton 1970) has demonstrated that very large areas of land were regularly stripped of turf for use as fuel, manure or in construction. These practices must have had a considerable impact on the soil, constantly disturbing and truncating large areas of topsoil and concentrating it on fields and in settlements. They will have particularly affected the accumulation of surface organic matter and the development of distinct near-surface soil horizons. This is clearly relevant to the discussion of podzol E horizons and blanket peat development. Turf manuring occurred as early as the twelfth century AD (Davidson and Simpson 1984) and turf was used for fuel and construction throughout prehistory. Davidson and Smout (1996) demonstrate the considerable historical legacy of manuring to the nature and properties of soils.

Changes in settlement pattern and land use in the highlands during the historic period demonstrate the importance of cultural and economic factors on soils – a link that is hard to demonstrate for the prehistoric period. The development of commercial cattle droving in the seventeenth century raised the value of pasture land and introduced cash into the highland economy. This reduced the dependence on local cereal production as grain could be purchased, with the effect that the area of land under cultivation could be reduced without a decline in population. This change would have promoted the development of stable soils and the accumulation of organic matter. The introduction of the potato in the eighteenth century provided a staple food that gave dependable returns in marginal areas. The result was a population increase and an expansion in the area and perhaps types of land under cultivation. In the early nineteenth century the widespread removal of tenant farmers from land for the creation of extensive sheep farms caused a rapid reduction in the area of cultivated land. It should be noted that none of these significant changes in agriculture was driven by changes in the soil resource, and thus any interpretation of human response to soil change during the prehistoric period can be questioned.

CONCLUSIONS

The first part of this chapter stressed the limited range of extensive soil types in Scotland but the spatial complexity of these soils is also outstanding. The dominant soil-forming processes of weathering, leaching, podzolization and gleying result from the particular mix of climatic, topographic, parent material and vegetational conditions. The latter part of this chapter summarized the evidence for soil change in the Holocene and illustrated the potential impact of social, technological and external economic change on land use and soils. A striking feature is the very small number of sites in Scotland for which data are available on buried soils. This is a surprising result since the dating of many archaeological sites has been done through radiocarbon analysis of buried soils. The investigation of such buried soils ought to be an integral element of any radiocarbon dating. From the few well-documented sites, the results emphasize the degree to which anthropogenic activity can change soils. Such transformations are particularly marked in an environment where marginal conditions dominate. Whilst the speed and magnitude of such changes within the historic period may not be applicable in prehistory, it may be concluded that from 5000 to 4000 BP (3780–2490 cal BC) onwards, soil evolution in Scotland has been substantially influenced by human activity.

5 Vegetation Change

KEVIN J. EDWARDS AND GRAEME WHITTINGTON

INTRODUCTION

Prior to 10 000 BP, the Scottish landscape had supported plant communities that were in accord with temperatures of the Lateglacial period. Although many of the herbs and dwarf shrubs of those communities remained part of the succeeding flora, they were driven to high altitudes as a result of increased temperatures and competition. Woodland assumed vegetational dominance over the ensuing 5000 years, representing the greatest biogeographical change in Scotland since pre-glacial times. Vegetation subsequently underwent massive modification and a full consideration of that should rightly encompass all plant forms, ranging from trees to fungi, seaweed to ferns and diatoms to liverworts. Such an undertaking, even if knowledge of the histories of all plant forms were available, would be impossible in the compass of this chapter. Thus the focus here will be upon the changes that affected the woodland, bringing into existence landscapes in which herbaceous species, arable and pastoral fields and peatlands became increasingly prominent. Foremost in the consideration of these vegetational landscapes will be an examination of the role of people.

OBTAINING THE EVIDENCE

Evidence for the past presence and distribution of the higher plants comes overwhelmingly from pollen analysis (palynology). The analysis of subfossil pollen and spores preserved within peat, lake and soil deposits, coupled with satisfactorily stratified and dated contexts, permits the reconstruction of past vegetation communities and their associated environments (Birks and Birks 1980; Moore *et al.* 1991). Pollen analysis was first undertaken in Scotland by the Swedish palynologist Gunnar Erdtman (1923, 1924) and sites from which polleniferous material has been obtained can now be found throughout Scotland, allowing a consideration of problems to which pollen data can be applied (Edwards 1974; Walker 1984a; Tipping 1994a). The value of the pollen content of deposits may be augmented by other environmental indicators such as microscopic charcoal, sedimentology, volcanic tephra, chemical and magnetic analyses (e.g. Edwards and Rowntree 1980; Bennett *et al.* 1992; Blackford *et al.* 1992; Edwards and Whittington 1993; Fossitt

Scotland: Environment and Archaeology, 8000 BC – AD 1000. Edited by Kevin J. Edwards and Ian B. M. Ralston.
© 1997 The editors and contributors. Published in 1997 by John Wiley & Sons Ltd.

1996). The range of pollen types, while large, does not always permit identification of taxa to species level. Furthermore, pollen may be carried distances considerably beyond its immediate origin. This deficiency may be overcome, in part, by the study of plant macrofossils; these may not only range in size from small seeds to substantial tree stumps, but are often found where they grew. Early wood macrofossil investigations include those of Lewis (1905, 1911) and Samuelsson (1910), with more recent examples including Pears (1970), Birks (1975), Bridge *et al.* (1990) and Fossitt (1996). Such material is often restricted to sites which provide a preservation matrix, especially peat bogs and lakes, and this, in turn, limits the types of plants likely to be found to peat, loch-side or aquatic species. Archaeological sites also provide evidence for plant species, and this applies particularly to cultivated plants or those which formed a vital natural food resource (Jessen and Helbaek 1944; Knights *et al.* 1983; Boyd 1988), in addition to wood (Coles *et al.* 1978).

Archaeological site palynology frequently suffers from problems of incomplete depositional accumulation, poor pollen preservation, down-profile pollen transport within minerogenic soils, inadequate dating controls, a lack of adjacent off-site comparative data from peat and lake deposits, and a lack of methodological research (Dimbleby 1985; Edwards 1991; Whittington and Edwards 1994). Useful site and soil-based studies have taken place in Scotland (e.g. Whittington 1983, 1984; Keith-Lucas 1986; Newell 1988; Affleck *et al.* 1988; Crone 1993a; Tipping 1994b; Tipping *et al.* 1994; Mills *et al.* 1994), but the evidence is often temporally and spatially restricted and archaeological site-based data will not be considered closely here.

EARLY AND MID HOLOCENE VEGETATIONAL DEVELOPMENT

The spread of woodland (*c.* 10 000–5800 BP): Mesolithic to the possible beginnings of agriculture

Following the demise of corrie glacier activity at the end of the Lateglacial period, *c.* 10 000 BP (the end of the Loch Lomond Stadial), there was a warming of temperatures to levels probably greater than those of today within a period possibly to be measured in decades rather than centuries (Atkinson *et al.* 1987; Mayewski *et al.* 1996). This climatic change, coupled with the development of soils, facilitated the spread of woodland across the existing open, herb- and shrub-dominated landscape. Grasses (Poaceae), sedges (Cyperaceae), sorrel (*Rumex*), crowberry (*Empetrum nigrum*), dwarf birch (*Betula nana*), dwarf willow (*Salix herbacea*), juniper (*Juniperus communis*), meadowsweet (*Filipendula ulmaria*), and ferns (Pteropsida) lost their dominance.

Research based on ecology, soils and pollen analysis supplemented by plant macrofossil studies, reveals that Scotland became a predominantly wooded territory (McVean and Ratcliffe 1962; Bennett 1989; Tipping 1994a). Figure 5.1 portrays the variations in the assumed *dominant* tree types at the time immediately prior to major discernible human impacts *c.* 5000 BP (3780 cal BC). It is probable, of course, that a woodland mosaic existed in most areas – the map simply presents the possible overall aspect of the vegetation. Research in peripheral areas (cf. Wilkins

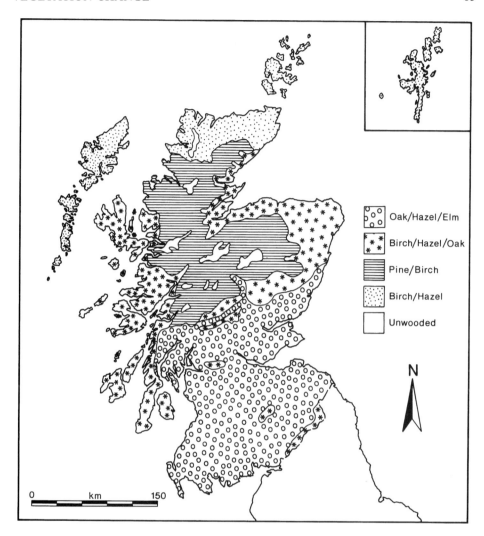

Figure 5.1 Woodland in Scotland *c*. 5000 BP (3780 cal BC). After Tipping (1994a), but with minor updating to the Outer Isles

1984; Bohncke 1988; Bennett *et al.* 1990, 1992; Edwards 1990, 1996a; Brayshay and Edwards 1996; Fossitt 1996) is suggesting that they were wooded for much of the Holocene, although the density of the arboreal cover may be in question (cf. Tipping 1994a).

It is important to stress the time-transgressive nature of the spread of many woodland taxa (Figure 5.2; Birks 1989). For instance, birch (*Betula*) was established over most of Scotland by 10 000 BP; oak (*Quercus*), present in southern Scotland shortly after 8500 BP (7530 cal BC) did not reach Aberdeenshire and Skye until about 6000 BP (4870 cal BC); and the principal area colonized by Scots pine in northern Scotland may have come from a source area close to Loch Maree at around 8500 BP, with pinewoods in south-west Scotland spreading independently

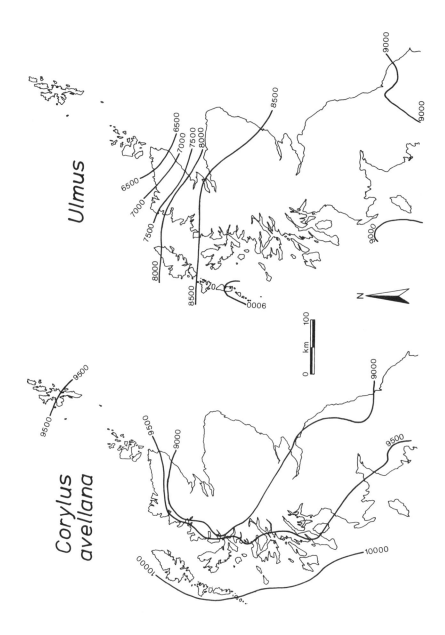

Figure 5.2 Isochrone maps (with dates in years **BP**) showing the extrapolated expansion of *Corylus* (hazel) and *Ulmus* (elm) pollen for Scotland. Based on Birks (1989), with substantial changes

from Ireland. The slow rate of expansion of many woodland taxa in Scotland (approximately 50 and 150 m year^{-1} for oak and pine respectively) compared to rates found in England (350–500 and < 100–700 m year^{-1}), was a function of many factors including climate, soils, topography and competition (Birks 1989). None of these is easy to model, and notions of simple climatic control, for example, are insufficient to explain the spread of trees – variations in seasonality, temperature and precipitation might be involved, as well as combinations of factors that are unknown. Consideration must also be given to chance dispersal, especially at the start of the Holocene, and the fact that reproductive potential decreases close to range margins (Birks 1989).

The broad-scale patterns evident in pollen diagrams represent the general 'surface' (cf. Figures 5.1 and 5.2) with which Mesolithic people interacted. The nature of that interaction could have operated at different scales: first, hunter–gatherer communities could have been totally subservient to the nature of the woodland they encountered; secondly, they may have affected its composition and distribution, but only to a minimal extent; at the most extreme level, major impacts might have induced large clearings or even drastic alterations to the woodland ecology. It is only possible to summarize some of the evidence for this here, but more extended accounts are available (Edwards and Ralston 1984; Edwards 1989a).

A key feature in pollen diagrams is the rise to high levels of hazel (*Corylus avellana*) pollen sometime around 9000 BP (8030 cal BC) (Figures 5.2–5.4). This phenomenon and its maintenance are frequently ascribed to hunter–gatherer impacts and possible resource manipulation in that the coppicing or burning of hazel would promote woody growth, profuse flowering and enhanced hazelnut yields and pollen production (Smith 1970). There are, however, difficulties with this hypothesis. For Scotland, Edwards and Ralston (1984) noted the existence of high hazel values even for areas distant from likely Mesolithic activity; a study of microscopic charcoal at a number of sites in Scotland (Edwards 1990) revealed no correspondence between enhanced fire incidence, as inferred from charcoal, and early maxima for hazel-type pollen. More generally, Huntley (1993) explored a series of hypotheses concerning the migration and abundance of hazel and concluded that climate was likely to be the primary underlying cause. This in no sense denies the usefulness of hazelnuts and hazel wood products to Mesolithic peoples, nor of the later utilization of hazel in a coppicing system (cf. Plate 5.1).

Some uncertainty also surrounds the role of humans in the rise and spread of alder (*Alnus glutinosa*). Smith (1984), following earlier observations by McVean (1956a,b), implicated Mesolithic people in the alder pollen expansion (Figure 5.4). This was held to be subsequent to fire and woodland disturbance noted at a number of sites, and based on a supposition that such activity promoted catchment runoff and valley-bottom waterlogging where alder could thrive. A number of Scottish pollen profiles do display an increase in microscopic charcoal as alder expands (Edwards 1990; Bunting 1994; and for carbonized fragments, Birks 1975; Robinson 1987). This does not prove the involvement of Mesolithic peoples, but suggests that environmental changes common to a number of sites were perhaps happening in areas known to hunter–gatherers.

Many pollen diagrams display temporary and apparently small reductions in woodland of all species. These perturbations are sometimes accompanied by

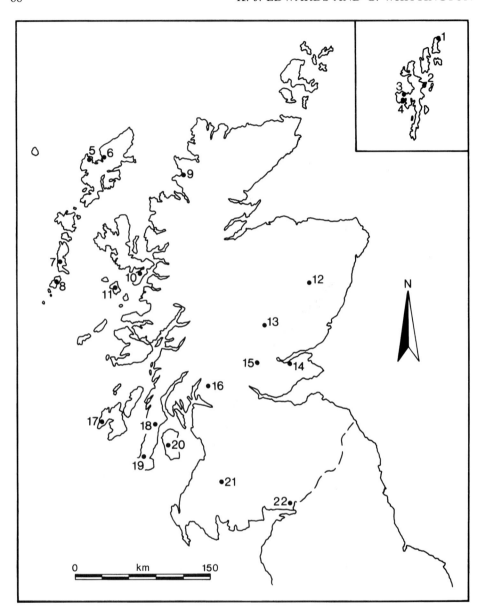

Figure 5.3 Location of sites mentioned in the text.1: Saxa Vord; 2: Dallican Water; 3: Scord of Brouster; 4: Loch of Brunatwatt; 5: Loch Bharabhat; 6: Callanish; 7: Loch an t-Sìl; 8: Lochan na Cartach; 9: Loch Sionascaig; 10: Loch Meodal; 11: Kinloch; 12: Braeroddach Loch; 13: Carn Dubh; 14: Black Loch; 15: North Mains; 16: Loch Lomond; 17: Loch a'Bhogaidh; 18: Loch Cill an Aonghais; 19: Rhoin Farm, Aros Moss; 20: Moorlands, Machrie Moor; 21: Starr, Loch Doon; 22: Burnfoothill Moss

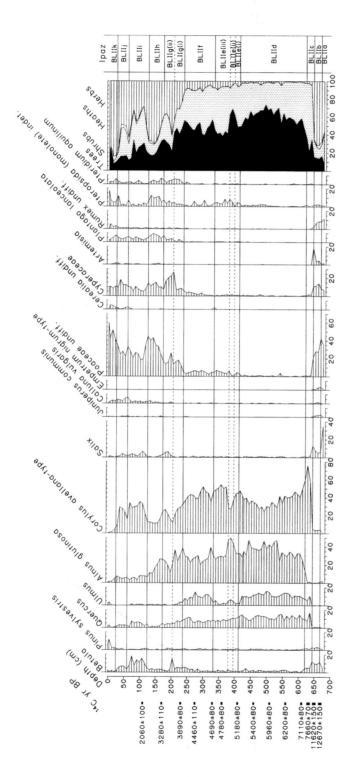

Figure 5.4 Pollen diagram (selected taxa) for Black Loch II, Fife. After Whittington *et al.* (1991b)

Plate 5.1 Coppiced hazel on the Isle of Mull. Coppicing may have been practised in Scotland since Mesolithic times. Copyright: K. J. Edwards

expansions in charcoal values, encouraging the belief that human agency is responsible; indeed, lithics are sometimes known from the pollen sites themselves or their vicinity (e.g. Knox 1954; Edwards *et al.* 1991; Tipping *et al.* 1993) or have been found nearby (e.g. McCullagh 1989; Edwards and Mithen 1995). Despite these indications, it is extremely difficult to separate natural from human causes: woodland always has been subject to disease, death, windthrow and lightning strikes (thus creating openings), while grazing activities by deer, for example, could have encouraged the development and extension of woodland clearings for many hundreds of years (Buckland and Edwards 1984). By the same token, human communities, in using woodland resources for food and shelter, would have disturbed woodland, although the continuity, duration and scale of interference may remain uncertain.

Two studies which demonstrate plausible impacts upon woodland come from island locations in the west of Scotland. Archaeological excavations at Farm Fields, Kinloch, Rhum, have produced the earliest known Mesolithic occupation site in Scotland, with dates on carbonized hazelnut shells extending back to 8590±95 BP (7700–7500 cal BC) (Wickham-Jones 1990). Palaeoecological studies from a site located 300 m from the excavation area reveal sharp and sustained changes in the pollen of alder, hazel, grasses and willow, together with apparently associated peaks in microscopic charcoal (Hirons and Edwards 1990). Although the interpretation of the patterns at Kinloch is very difficult, they do not clearly represent a natural vegetational succession and human involvement seems likely. The second study comes from South Uist, where close sampling of Mesolithic age levels at Loch an t-Sil reveals two phases of woodland removal, mainly involving birch and hazel, at

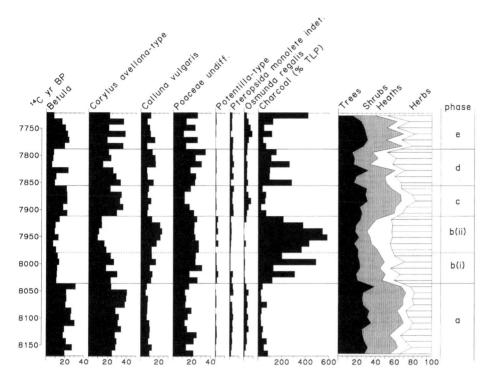

Figure 5.5 Mesolithic age pollen spectra (selected taxa) from Loch an t-Sil, South Uist. After Edwards (1996)

c. 8040 BP (7010 cal BC) and 7870 BP (6620 cal BC), lasting 130 and 70 radiocarbon years respectively (Figure 5.5; Edwards 1996a). These are associated with expansions in Poaceae, *Calluna* and charcoal, and reductions in ferns. The removal of birch and hazel may have an anthropogenic origin and the expansions in grass and heather could indicate their spread into cleared areas. Whether the extension of browse in order to attract grazing animals was the intention or a useful by-product of cropping woodland, remains unknown. The reduction of ferns is similar to features observed in the Shetland pollen sites of Dallican Water (Bennett *et al.* 1992) and Loch of Brunatwatt (Edwards and Moss 1993). At Dallican Water this is taken to indicate possible grazing by red deer which may have been transported to Shetland by hunter–gatherers intent on introducing a valuable resource. The inferences from Loch an t-Sil and from Shetland are of particular interest because there are no proven Mesolithic artefactual remains from either area – their known archaeological record begins with the Neolithic (cf. Chapter 7). Any Mesolithic finds are likely to lie hidden beneath sea, sand or peat.

The sustained charcoal peaks in these and other sites do not have to indicate woodland removal by fire or the driving of game, but may simply result from the burning of felled wood for heating or cooking purposes (cf. Edwards 1990). None the less, the fire-related creation or maintenance of heaths as a grazing resource during the first half of the Holocene has long been mooted for England (e.g.

Dimbleby 1962; Simmons 1969; Caseldine and Hatton 1993). This process has also been conjectured at Callanish, Lewis (Bohncke 1988), and also for evidence from sites in South Uist (including Loch an t-Sìl), but only as a possibility (Edwards *et al.* 1995).

The Meso-Neolithic transition, the establishment of agriculture and woodland regeneration (*c.* 5800–4000 BP)

The transition from hunter–gathering to agriculture represents the major economic boundary in prehistory (Dennell 1983; Edwards 1988). It is no longer justifiable, however, to place the end of the Mesolithic at the first major elm (*Ulmus*) decline of *c.* 5100 BP (3830 cal BC) (Godwin 1975; Smith 1981). While the ubiquitous fall in elm pollen frequencies throughout north-western Europe seems to coincide with events such as a major reduction in woodland, the expansion of weed taxa, the first appearance of cereal pollen and radiocarbon dates for the start of Neolithic monuments, these phenomena do not signify the undoubted beginnings of the 'New Stone Age'. Smith (1981) drew attention to the possibility that pre-elm decline disturbance phases may be due to early Neolithic rather than late Mesolithic activities. This possibility was reinforced by Groenman-van Waateringe (1983) and Edwards and Hirons (1984) who drew attention to a number of sites in Britain and Ireland that had produced cereal-type pollen up to a few centuries prior to the elm decline, or increased representation for pollen of taxa with, *inter alia*, agricultural indicator value. Only six years later, Edwards (1989b) was able to refer to 22 sites in the British Isles that had produced early cereal-type pollen grains, three of which were in Scotland and two of which, Rhoin Farm, Kintyre and Moorlands (Machrie Moor), Arran, had been subjected to simple, but effective techniques aimed at optimizing the detection of Cerealia-type grains (Edwards and McIntosh 1988). The third site, North Mains, Perthshire (Hulme and Shirriffs 1985), may suffer from stratigraphic problems. It must be cautioned, however, that cereal-type pollen includes wild as well as cultivated grasses (Andersen 1979; O'Connell 1987; Dickson 1988; Edwards 1989b) and finds cannot be placed within the Cerealia with total confidence unless there is supporting evidence, with that from macrofossils being the most secure (cf. Wasylikowa 1986); there is, as yet, no such evidence from Scotland. It may be noted that a series of cores from the Outer Hebrides (Edwards and Whittington unpublished; Fossitt 1990, 1996) have produced extremely early cereal-type finds (e.g. from the eighth and seventh millennia BP), and these are presumed to derive from wild grasses.

If the finds from Scotland and elsewhere which appear within a few centuries prior to the elm decline are derived from cereal cultivation, then a number of implications arise. First, there is the existence of pioneer agriculture perhaps up to 800 ^{14}C years earlier than the first *Ulmus* decline; such husbandry could have been practised by either incoming agriculturalists or by indigenous hunter–gatherers in the process of adopting agriculture and hence 'becoming' Neolithic. Secondly, early interference phases without cereal pollen finds (which could be due to poor Cerealia pollen dispersal), could be caused by the first Neolithic peoples (cf. Hirons and Edwards 1986). Thirdly, the elm decline would no longer have significance as the undoubted palynological concomitant of the start of the Neolithic, although it would be indicative of the fact that it was underway. A fourth implication is that post-elm decline disturbances may have been brought about by late Mesolithic communities as has been hypothesized for Braeroddach Loch, Aberdeenshire (Edwards and Ralston 1984) and the Cheviot Hills (Tipping 1994a). It is a possibility that the largely closed mid Holocene woodland, prior to its reduction at the *Ulmus* decline, was in fact, at least at the local scale, a managed system, featuring coppicing, leaf foddering and foraging (Göransson 1986, 1987; Edwards 1993a). The forest would have formed an extremely rich resource for plant and animal food, shelter and materials.

What can be said of the elm decline which is evident in many Scottish pollen diagrams, especially those from mainland areas (e.g. Donner 1957; Nichols 1967; Whittington *et al.* 1991c)? In spite of the fact that there is a spread of dates associated with it, which may owe as much to sediment type and sampling resolution as to serious metachrony, the classic elm decline of *c.* 5100 BP continues to furnish a useful chronological marker. The causes have long been debated (Ten Hove 1968; Huntley and Birks 1983; Sturlodottir and Turner 1985; Whittington *et al.* 1991c), with catastrophic (disease, climate), soil change, agricultural clearance and leaf foddering explanations all being employed singly or collectively. The exact cause(s) remain unknown, but for many areas, archaeological (especially funerary) evidence shows that a Neolithic way of life had been established, perhaps close to elm decline times. This would have involved an arable and/or pastoral economy (cf. Chapter 8); proxy records for both are frequent in the pollen record, although the ecological latitude of many so-called cultural indicators makes the specific determination of land use difficult (Behre 1981; Groenman-van Waateringe 1986, 1993). Furthermore, the extent and duration of open areas from around 5100 BP (3830 cal BC) varies spatially. Thus, at Black Loch II, Fife (Whittington *et al.* 1991c; Plate 5.2), elm pollen, accompanied by *Quercus* and to a variable extent by *Alnus* and *Corylus*, is reduced from *c.* 5200 BP (3990 cal BC) for an estimated ~500 ^{14}C years. At Braeroddach Loch (Edwards and Rowntree 1980; Plate 4.1), the elm decline beginning *c.* 5295±155 BP (4340–3960 cal BC) has an estimated duration of ~665 ^{14}C years, but only *Ulmus* values are consistently depressed. As elsewhere in Europe, there are sites where once reduced, elm does not appear to regenerate, or certainly not to the same extent (e.g. Loch Cill an Aonghais, Argyll [Peglar in Birks 1980], Loch Lomond [Dickson *et al.* 1978], Loch Meodal, Skye [Birks and Williams 1983]) and this is conceivably a function of such factors as soil deterioration and resource pressures.

The marked middle Neolithic regeneration of woodland after the elm decline has received the attention of several reviewers (Göransson 1987; Edwards 1993a; Tipping 1994a) and is evident at a number of Scottish sites in addition to Black Loch and Braeroddach Loch (e.g. Scord of Brouster, Shetland [Keith-Lucas 1986] and Machrie Moor, Arran [Robinson and Dickson 1988]). Why should such a phase occur? It might be assumed that climatic conditions improved, that biotic pressures were released as disease loosened its grip (only relevant for *Ulmus*), or that people and their animals abandoned these areas. This last argument supposes that agriculturalists were willing to leave fertile farmland, or that the soils in such areas became temporarily impoverished. Another interpretation is that advanced by Göransson (1986, 1987) in opposition to the orthodox model which assumes that regeneration is primarily a response to lowered human impact (cf. Berglund 1986).

Plate 5.2 Black Loch, Fife. Pollen profiles from this site are dominated by the pollen of woodland trees, especially oak, elm, hazel and alder. From *c.* 5200 BP, woodland underwent a series of reductions and recoveries. Marked clearance, associated with farming, began *c.* 3600 BP, since when the pollen record has been characterized by taxa derived from a mixed arable/ pastoral regime. Copyright: K. J. Edwards

Göransson's forest utilization model suggests that people did not abandon such areas. The recovery in woodland simply masks a forest farming economy in which coppicing and garden plots could thrive. Indeed, he further suggests that population may have increased at this time as indicated by the appearance or expansion of cereal-type pollen in southern Sweden and the construction of monuments. For Scotland, more detailed palynological work and assessments of archaeological monument chronology are required if any realistic evaluation of this model is to be made.

The recognition that the classic elm decline may simply be the first of two or more such reductions, which may or may not have similar causes, serves to complicate the picture (cf. Aaby 1986; Hirons and Edwards 1986; Smith and Cloutman 1988; Whittington *et al.* 1991c). It not only signifies the caution which should apply to palynological interpretation (especially where sampling resolution is inadequate), but also demonstrates the diversity which may arise from different vegetational, pedological and anthropogenic histories.

Major woodland reduction from late Neolithic times onward (*c.* 4000–1000 BP)

Subsequent to the classic elm decline, or in areas following the restoration of tree cover, there began the major reductions in woodland of the late Neolithic and, more particularly, of the Early to Middle Bronze Ages. Well-dated pollen profiles show that dates for the beginning of substantial incursions into the woodland vary:

1. *c.* 4190 BP (2780 cal BC) – Callanish, Lewis (Bohncke 1988)
2. *c.* 4140 BP (2690 cal BC) – Lochan na Cartach, Barra (Brayshay and Edwards 1996)
3. *c.* 4020 BP (2530 cal BC) – Loch Sionascaig, Invernesshire (Pennington *et al.* 1972)
4. *c.* 3950 BP (2460 cal BC) – Kinloch, Rhum (Hirons and Edwards 1990)
5. *c.* 3800 BP (2200 cal BC) – Burnfoothill Moss, Dumfriesshire (Tipping 1995a)
6. *c.* 3650 BP (2000 cal BC) – Carn Dubh, Perthshire (Tipping 1995b)
7. *c.* 3630 BP (1970 cal BC) – Black Loch, Fife (Whittington *et al.* 1991b)
8. *c.* 3600 BP (1940 cal BC) – Loch Bharabhat, Lewis (T. Lomax pers. comm., 1996; cf. Edwards 1996a)
9. *c.* 3590 BP (1930 cal BC) – Loch a'Bhogaidh, Islay (Edwards and Berridge 1994)
10. *c.* 3080 BP (1340 cal BC) – Machrie Moor, Arran (Robinson and Dickson 1988)
11. *c.* 3065 BP (1350 cal BC) – Braeroddach Loch, Aberdeenshire (Edwards and Rowntree 1980)

This selection of sites is probably insufficient to enable assessments to be made of spatial or even chronological disparity. Quite clearly, major landscape changes were underway and although these might be ascribed, in part, to progressive soil deterioration or climatic change (this is, for example, a time of marked decline in the fortunes of pine [cf. Bennett 1984; Gear and Huntley 1991; Blackford *et al.* 1992] as well as extensions to the spread of blanket peat), there is little doubt from the density of the archaeological finds, especially for the Bronze Age (Chapter 9), that the expansion of settlement and agriculture was of tremendous importance to landscape development.

The stepped, or continuous declines in the fortunes of woodland, often beginning by Neolithic times, frequently accelerating through the Bronze Age and into the Iron Age and beyond, were never to be reversed except in exceptional circumstances (Whittington and Edwards 1993) or with the advent of planted woodland over the last two centuries. The pollen data show the increased representation of herbaceous taxa, including cultivars, weeds of cultivation and pastureland, ruderals, as well as acid grasslands and heathlands. Considerable variation exists in pollen diagrams during this timespan: those from more fertile lowlands demonstrate extensive, though decreasing woodland cover (e.g. Durno 1965; Nichols 1967; Turner 1975; Dickson *et al.* 1978; Edwards 1978; Caseldine 1979; Stewart *et al.* 1984; Whittington *et al.* 1991a; Tipping 1995b); the islands demonstrate a good but possibly less varied tree cover (e.g. Vasari and Vasari 1968; Birks and Williams 1983; Bennett *et al.* 1990, 1992, 1993; Bunting 1994; Edwards and Berridge 1994; Brayshay and Edwards 1996; Fossitt 1996); and the upland or more exposed sites embrace an open aspect for much of the Holocene (e.g. Flenley and Pearson 1967; Hawksworth 1969; Jóhansen 1975; Davidson *et al.* 1976; Walker and Lowe 1977; Birks and Madsen 1979; Keatinge and Dickson 1979; Walker 1984b).

Such variations in the record, linked to differences in geographical location, may be highlighted by reference to three sites; Black Loch on the eastern mainland, Lochan na Cartach in the Western Isles, and the north Shetland site of Saxa Vord.

At Black Loch, Fife (Whittington *et al.* 1991a; Figures 5.4 and 5.6; Plate 5.2), the pollen of elm and other arboreal taxa undergoes a sequence of reductions and recoveries from *c.* 5200 BP (3990 cal BC). Following the middle Bronze Age decline, woodland regeneration of the Bronze and Iron Ages lasted for only ~400–500 years of the last 3500 radiocarbon years. Farming activities appear to have been the primary factor in maintaining an open landscape. Cereal-type pollen (*Hordeum* [barley] group) is present from *c.* 3200 BP (1440 cal BC) with *Avena/Triticum* (oats/wheat group) common after 1430 BP (cal AD 640) (Whittingon *et al.* 1990). Accompanying these types are weed taxa which can be indicative of arable agriculture, such as Caryophyllaceae (pink family), Chenopodiaceae (fat hen family), Brassicaceae (cabbage family) and *Artemisia* (cf. mugwort). There is also good representation for taxa which frequent pasture, rough grazing and ruderal habitats (e.g. *Plantago lanceolata* [ribwort plantain], *Plantago coronopus* [buck's-horn plantain], *Plantago major/media* [great/hoary plantain], Ranunculaceae [buttercup family], *Rumex* and *Pteridium aquilinum* [bracken]). The high values for Poaceae throughout the last 3500 years at Black Loch may reflect a strong reliance on pasture as part of a mixed arable/pastoral regime. Of particular interest is the record for the period which spans the Roman incursions (mainly Zone BLId(i), *c.* 1955–1430 BP [cal AD 40–640]), when the pollen suggests a regeneration of woodland and a reduction or an absence of agriculture. This pattern is replicated in two Aberdeenshire sites, and may well indicate that the military presence during the Flavian (late first century AD), Severan (early third century AD) and post-Severan (up to 390 AD) periods was sufficiently punitive to lead to collapses in native agricultural societies or economies (Whittington and Edwards 1993). Further south, around Hadrian's Wall, it has been shown that woodland was reduced or remained low during Roman times (Dumayne and Barber 1994) and this is related to the military construction and food needs of the invaders.

Contrastingly, at Lochan na Cartach, Barra (Brayshay 1992; Brayshay and Edwards 1996; Figure 5.7 and Plate 5.3), a reduction in birch–hazel woodland took place from *c.* 6330 BP (5270 cal BC), and accelerated from *c.* 4140 BP (2690 cal BC). The decline may have been part of a continuing response to a climatic process which also involved the spread of blanket peat, although the pollen spectra from 4140 BP onwards show clear signs of human impact. The herbaceous element expands throughout this period. Although cereal-type pollen was found at three levels, and taxa common in arable and pasture fields was present in low or reasonable amounts (e.g. Chenopodiaceae, Caryophyllaceae, *Rumex*, Ranunculaceae and *Plantago lanceolata*), the profile is dominated by pollen and spores indicative of peat and acid grassland (e.g. *Calluna vulgaris*, *Sphagnum*, Cyperaceae, *Narthecium ossifragum* [bog asphodel], *Drosera* [sundew], *Equisetum* [horsetail], *Succisa pratensis* [devil's-bit scabious], *Potentilla* [cf. tormentil] and *Pteridium aquilinum*). The rise in the curve for microscopic charcoal, especially from *c.* 1940 BP (cal AD 80), closely follows that of *Calluna*, and may be related to muirburn.

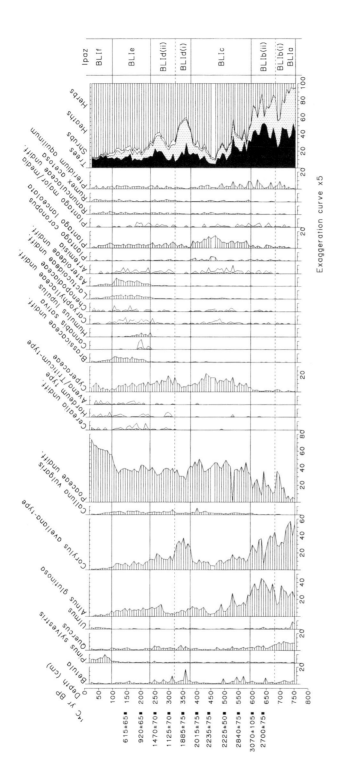

Figure 5.6 Pollen diagram (selected taxa) for Black Loch I, Fife. After Whittington *et al.* (1991b)

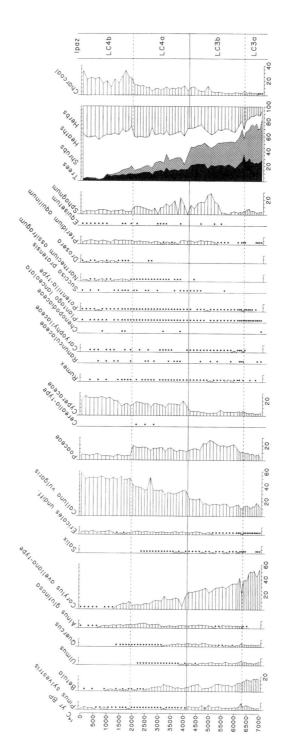

Figure 5.7 Pollen diagram (selected taxa) for the period *c.* 6330 **BP** (5270 cal BC) to the present from Lochan na Cartach, Barra. After Brayshay and Edwards (1996)

Plate 5.3 Lochan na Cartach, Barra. Prior to *c.* 6500 BP, the vicinity of this site was dominated by birch and hazel. Woodland reduction occurred, and since that time the pollen and spores received at the site have been indicative of peat and acid grassland taxa, similar to those within the present-day vegetation. Copyright: K. J. Edwards

The Shetland site of Saxa Vord is located on an exposed hillside at the far northern tip of Unst. The base of this blanket peat profile has been dated to 3760±85 BP (2290–2030 cal BC). The spectra are dominated by grasses and heathers (largely reflecting the vegetation growing on the peat surface), and nowhere in the pollen diagram (Figure 5.8) does tree or shrub pollen exceed 12.6% of the land pollen sum. There is no convincing evidence that cultivation took place or that trees grew in the immediate vicinity of the site at any time. This is entirely consistent with a locality where a gust of 325 km h^{-1} was recorded in 1979 before the anemometer blew away (Berry and Johnston 1980); but such persistent open vegetation is common in a number of upland and island areas subject to exposure, salt-spray or high winds, even where the local topography provides some shelter (e.g. Birks and Madsen 1979; Mills *et al.* 1994). Thus, throughout this period at Saxa Vord, the only potential land use would appear to have been rough grazing.

THE SPREAD OF PEAT

In this chapter the vegetation history of Scotland up to 1000 AD has been surveyed chronologically. In the preceding sections there have been occasional mentions of peat and peatlands. This should not be taken to suggest that peatlands and their attendant heathland vegetation were of minor importance. A problem with these

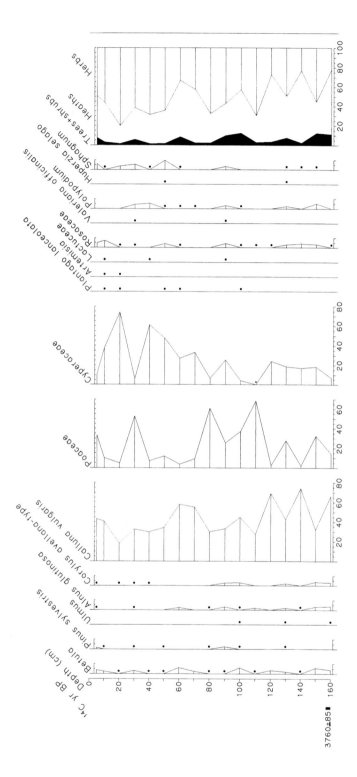

Figure 5.8 Pollen diagram (selected taxa) for Saxa Vord, Unst, Shetland

Plate 5.4 Peat cutting in blanket peat on slopes of Reineval, South Uist. Copyright: K. J. Edwards

wetland features lies in the lack of knowledge as to their date of inception. The possible beginnings of this process can be traced back to soon after deglaciation in the Outer Hebrides (Bennett *et al.* 1990; Edwards *et al.* 1995) and in upland Perthshire (Tipping 1995b) where *Calluna*-dominated heathland seems to have become a component of the landscape. The well-rehearsed arguments for the inception and extension of blanket peat, whether the result of a rather impenetrable mix of climatic, pedogenic, hydrological or anthropogenic processes (Moore 1975, 1993; Kaland 1986; Stevenson and Birks 1995), leave us with one certainty – that blanket peat cloaked the land surface over vast tracts of the country (Plate 5.4). The available dates for peat spread are not easy to evaluate because the general lack of sub-peat surveys make it impossible to assess the topographic origins of the peat (Edwards and Hirons 1982; but see Charman 1992 and Tipping 1995b). Dates of initiation from Scotland vary widely (e.g. 9800–9200 BP [9040–8320 cal BC] at Carn Dubh, Perthshire [Tipping 1995b] and two statistically indistinguishable dates of 2415±25 BP [520–400 cal BC] and 2395±25 BP [420–400 cal BC] obtained from the base of peat sections 20 m apart at Starr, Loch Doon, Ayrshire [Edwards 1996b]; and variations can even be found within very small areas (e.g. > 7270±100 [6180–5980 cal BC] and 4810±60 BP [3650–3520 cal BC] at Callanish for two adjacent profiles on a small peninsula [Bohncke 1988]). As was the case for Ireland (Lynch 1981; Edwards 1985), a continuous process of peat inception and spread would probably have been in operation, and although intuitively it might seem reasonable that times of climatic deterioration would accelerate peat spread, insufficient information is available to substantiate this.

CONCLUSION

Since Erdtman began the systematic exploitation of pollen analysis for the exploration of vegetation history in Scotland, great strides have been made. That the country took on an almost universal tree cover from about 9000 years ago is now fully established. The demands made by prehistoric communities on that vegetation cover, whether it be directly on the plants themselves or on the soils that they occupied, led to an opening up and eventually the virtual demise of the woodland. The floral pattern was changed so that a combination of introduced plants, especially the cereals, and native species such as grasses and many weeds of cultivation became dominant. In the uplands and in parts of the north and west, peatland vegetation became ubiquitous.

The vegetation history of Scotland should not be considered as definitively established. Virtually all new pollen diagrams published are capable of raising new questions, especially as sampling strategies improve. Major problems still remain. The parts played by climatic and pedogenic change, as against or in conjunction with humanly-induced factors, still need further investigation, particularly with regard to the inception of peat. Current knowledge of the status of the woodland cover, especially for the Mesolithic period, remains clouded with doubt, while the identification and characteristics of woodland management practices remain conjectural. The role of fire, the precise timing and nature of the adoption of cereals, and the indicative value for land use of weed-type floras continue to arouse debate. Statistical techniques in the analysis of vegetational change are still under-utilized (cf. Birks and Line 1992; Bennett and Humphrey 1995; Whittington and Edwards 1995).

At a basic level, the substantial corpus of information already accumulated provides an ecological context for human activities in prehistoric and historical times. Quite clearly, this relationship should not be viewed as a passive one; the fossil record is also capable of contributing to a picture of the nature and development of human communities. Beyond this, palynology can be used proactively to predict the existence of human impacts through time (Whittington and Edwards 1994), both when the archaeological record is unknown or absent and where the landscape record is obscure.

6 Faunal Change

THE VERTEBRATE FAUNA

FINBAR MCCORMICK AND PAUL C. BUCKLAND

INTRODUCTION

The rapid warming that opened the present interglacial at *c.* 10 000 BP, changed the Scottish landscape from one of residual icecap, corrie glaciers, snowfields and tundra, with low carrying capacity in terms of vertebrates, to one in which a largely continental steppe rapidly gave way to a succession of birch–pine forest, hazel, and mixed oak forest; only perhaps on a few offshore islands did some form of wooded landscape fail to develop fully (Chapter 5). Through the mid to late Holocene, human impact led progressively to the modification and destruction of the forest cover and its associated fauna. The Holocene history of Scotland's vertebrate fauna is one of extinctions in the native fauna and introductions of aliens, often kept in sufficiently inordinate numbers to lead to further losses and retractions of range in the indigenous biota.

THE NATURE OF THE EVIDENCE

The prehistoric and protohistoric record of the vertebrate fauna of Scotland is, when compared with England, singularly incomplete. The development of raised mires and acid soils over much of the country has removed much of the fossil evidence, and, outside the Midland Valley, the record is largely restricted to coastal locations, usually on and in shell sands, and the limited outcrops of calcareous rocks, such as the caves of the Durness Limestone in the North-West (Figures 6.1 and 6.2). Most assemblages also derive from archaeological contexts, where the selective activities of human and dogs further compound the problems. Antler, tooth ivory and even bone were materials widely utilized for the manufacture of artefacts (cf. Foxon 1991; Hallén 1994; Weber 1994) and are likely to have been dispersed far from the locality where the animal was killed. The indirect impact of human interference is evident in the disturbed nature of many assemblages, particularly from machair and similar coastal sand deposits. Barratt (1995), in a review of Norse and later animal bone groups from Caithness and Orkney, found that rabbit (*Oryctolagus cuniculus*), a medieval introduction to England which only became common and widespread in Scotland with the unrelenting suppression of its

---

Scotland: Environment and Archaeology, 8000 BC – AD 1000. Edited by Kevin J. Edwards and Ian B. M. Ralston.

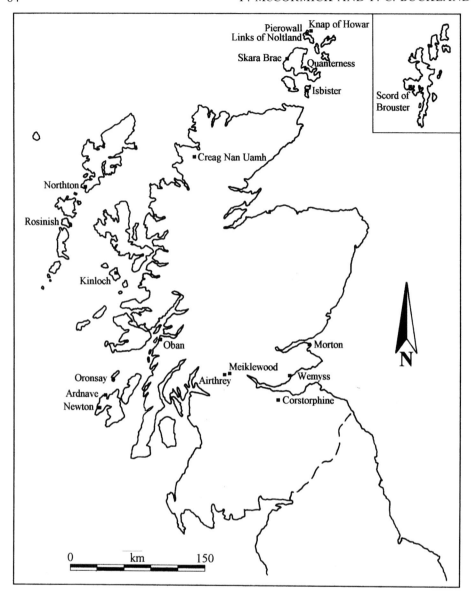

Figure 6.1   Lateglacial to mid Holocene sites referred to in the text

predators, from Scotsmen to wild cats, during the nineteenth century, was frequent in the majority of bone assemblages. Records of small mammals from archaeological sites, therefore, have often to be treated with some circumspection. The net result is that, whilst the history of woodland and its demise is well known from the palynological record, its associated mammals are only known from a few locations, and the archaeological record provides only glimpses of the wild and domestic animals, which not only shaped its destiny but also formed the basis of most of the human activity.

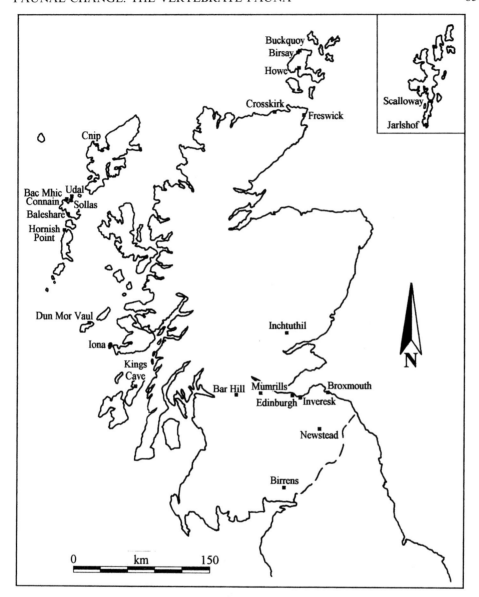

Figure 6.2 Late Holocene sites referred to in the text

A review of the Scottish mammal fauna was provided by Delair (1969), updating
a more exhaustive account by Ritchie (1920). More recent work, largely from
archaeological sites, has added some detail, without substantially modifying the
conclusions. Utilizing variation in skull form, tooth-enamel patterns and pelage,
Berry (1969, 1979; Berry and Rose 1975) has examined the mice and voles of both
the Hebrides and Northern Isles and postulated their origins and dates of
introduction on this basis; only recently has the routine sieving of archaeological
deposits begun to allow the testing of these hypotheses, often with contradictory

results (cf. Serjeantson 1990). Platt (1934) noted the introduced Orkney vole (*Microtus arvalis*) in a number of Neolithic tombs and at Skara Brae, Orkney. Although the problems of obtaining sealed samples from such deposits are considerable (Barker 1983, 143), Corbet (1979) accepted the large numbers of *M. arvalis* from the Quanterness tomb as Neolithic. More recently, Serjeantson (1988) has discussed the human exploitation of seabirds, using both the archaeological and ethnographic record. The fish, amphibians and reptiles are less well served. Along with other groups, their current status was discussed on a national scale in the various papers edited by Hawksworth (1974), but the longer perspective of change through the Holocene was little considered (Wheeler 1974; Prestt *et al.* 1974).

THE LATEGLACIAL AND EARLY HOLOCENE

The Lateglacial vertebrate fauna of Scotland is still inadequately known (Stuart 1982), and most finds were made last century and are poorly documented (Figure 6.1). Arctic or collared lemming (*Dicrostonyx torquatus*) was recovered from the site studied by James Bennie in the nineteenth century at Corstorphine, Edinburgh (Sutcliffe and Kowalski 1976) and it also appears in the mixed assemblages from the so-called Reindeer Cave at Creag Nan Uamh in Assynt. This site, excavated in the 1920s by Callander *et al.* (1927), provides the most extensive list of Devensian and early Holocene mammals from Scotland, but the group is hopelessly jumbled. Accelerator dates upon reindeer (*Rangifer tarandus*) antler from the site all predate the maximum of the last glaciation if the single Lateglacial date based upon a bulk sample of antler fragments is omitted as of doubtful value. A date of 8300±90 BP (7480–7100 cal BC) from a leg bone suggests that reindeer may have formed part of the early Holocene fauna (Murray *et al.* 1993). It is unlikely that suitable habitat for the herds would have lasted long into the present interglacial and Clutton-Brock and MacGregor (1988) have critically reviewed and discounted all later records. Weber (1994) has recently suggested that finds of antler, identified as reindeer, in Pictish and Norse contexts in Orkney, reflect trade in raw materials for comb-making with Norway. Arctic fox (*Alopex lagopus*), lynx (*Felis lynx*) and brown bear (*Ursus arctos*) also occur in the Creag Nan Uamh caves, but, like the reindeer, their stratigraphic relationships are unknown. The suggestion that late Upper Palaeolithic hunters were also present has been largely dismissed in recent work (Murray *et al.* 1993), and the earliest certain occurrence of humans is the Mesolithic artefact assemblage from Kinloch on the Inner Hebridean island of Rhum, where the tools are associated with a phalangeal tooth of a sea wrasse, *Labrus* sp., and a few unidentifiable fragments of bone; the radiocarbon dates from the site lie in the ninth millennium BC (Clarke 1990; Chapter 7).

EARLY TO MID HOLOCENE

Both Stuart (1982) and Yalden (1982) have reviewed the vertebrate fauna, excluding the birds, for the British Isles as a whole, and Stuart (1995, table 5) provides a checklist. Several elements, particularly amongst the more thermophilous reptiles and amphibians, appear not to have reached Scotland, although the inadequacy of the fossil record makes this only a tenuous conclusion. In the same way, ascertaining the dates of extinctions before those in the historical record is difficult. The last wolf (*Canis lupus*) is claimed to have been shot in Durness in 1749, but the dates of the last aurochs (*Bos primigenius*), beaver (*Castor fiber*), boar (*Sus scrofa*), brown bear, elk (*Alces alces*) and lynx are more speculative. Holocene peats and lake deposits have produced evidence of red deer (*Cervus elephus*), roe deer (*Capreolus capreolus*), elk, aurochs, brown bear, boar, beaver and horse (*Equus ferus*), although the latter, like the extinct giant deer (*Megaloceros giganteus*) (Ritchie 1920), is more likely to belong to the Lateglacial. The demise of the elk presents an interesting series of problems. Grigson (in Simmons *et al.* 1981) has argued that the expansion of pine forest would have reduced available habitat and contributed to its early extinction, yet the animal remains characteristic of the large mammal fauna of much of the Boreal forests, feeding on the lush littoral vegetation of lakes and pools, and its loss is more likely to be ascribed largely to hunting activity during the Mesolithic. A rock carving, now lost, at Wemyss in Fife was accepted by Childe (1935, 116) as being of an elk of Bronze Age date, although the animal might equally well be an aurochs. There is one possible later fossil record, of the brow tine and part of the frontal bone from the Roman fort at Newstead in the Borders (Ewart 1911), which suggests that the animal might have survived almost into the historical period, although R. M. Jacobi (pers. comm, 1996) has suggested that the specimen might have been recovered during Roman peat-cutting from much earlier deposits.

The other large cervid, the red deer, is ubiquitous upon Mesolithic and Neolithic sites where bone is preserved, including Orkney. From Lambeck's (1995) reconstructions of sea level through the late Devensian and early Holocene, it is evident that parts of the Orcadian fauna could have crossed the Pentland Firth before it was inundated shortly after 10 000 BP (*contra* Clutton-Brock 1979), but that neither Shetland nor the Outer Hebrides was directly connected to the mainland after the Windermere Interstadial, and the cold phase of the Loch Lomond stadial would have extirpated any thermophilous elements which might have immigrated earlier. Although deer are able to swim easily in the protected waters of the Inner Hebrides, the open waters of the Minch, between Skye and the Outer Isles, and the channels between Orkney, Fair Isle and Shetland present more formidable barriers, even at a slightly lower relative sea level. The question of whether red deer were deliberately introduced or swam over to the islands has been much discussed (e.g. Clutton-Brock 1979, 120; Grigson and Mellars 1987, 246). In a review of the Hebridean evidence, Serjeantson (1990) favours Neolithic introduction for the animals. Bennett *et al.* (1992) have argued that changes in early to mid Holocene vegetation on Shetland probably reflect the impact of grazing by red deer. If this is the case, the apparent inability of deer to colonize Ireland (Woodman and Monaghan 1993, 33), at the closest a mere 20 km from the Scottish mainland, suggests that any terrestrial mammal fauna on Shetland should also be accompanied by Mesolithic hunters. The situation in the Outer Hebrides is similar. Although the earliest red deer fossils belong to the Neolithic, palynological and charcoal evidence have been used to suggest earlier human occupation (Edwards 1990, 1996a). If deer were not introduced, the islands would have offered little in the way of a terrestrially based diet, and it is perhaps not surprising that no certain Mesolithic artefacts have been recovered, since any occupation would inevitably have had to rely upon coastal resources, and sites would now largely lie below present sea level (Chapters 5 and 7).

The few archaeological sites where bone assemblages have been preserved, are essentially mid-Holocene. Material from Mesolithic, 'Obanian' cave sites studied by Turner (1895) included otter (*Lutra lutra*), from MacArthur's Cave, Oban, and Lacaille (1954) also includes domestic dog from the nearby MacKay Cave. Whilst the stratigraphy of these early excavations leaves some doubt as to contexts, dog is also known from other Mesolithic sites in Denmark and England (Grigson, in Simmons *et al.* 1981, 123). The most extensive species list recently published comes from a shell midden in a rock fissure at Carding Mill Bay, Oban (Table 6.1). Although the deposit may have suffered some disturbance by carnivores, the clustering of the radiocarbon dates around the opening of the fifth millennium BP (Connock *et al.* 1993) suggests that the group may be coherent. As well as red and roe deer, and otter, Hamilton-Dyer and McCormick (1993) record field vole (*Microtus agrestis*), common shrew (*Sorex araneus*) and red squirrel (*Sciurus vulgaris*) – a clear indicator of at least some woodland in the region. The bird list, with razorbill (*Alca torda*), guillemot (*Uria aalge*) and various gulls, probably represents human procurement as food, but the swallow (*Hirundo rustica*) presumably nested in the overhang. The fish remains may consist of the food of both otter and humans, and include eel (*Anguilla anguilla*), cod (*Gadus morhua*), pollack and whiting. The dates from the site overlap with the earliest ones on sites with Neolithic pottery, e.g. at Newton on Islay (McCullagh 1989), and it is not possible to distinguish between wild and domestic pig on the single deciduous tooth recovered. The excavations by Mellars (1978, 1987) of the extensive shell middens on the small Inner Hebridean island of Oronsay, however, have also produced a similar assemblage with red and roe deer and boar (Grigson and Mellars 1987). They do cast some doubt upon earlier identifications of pine marten (*Martes martes*) and weasel (*Mustela nivalis*) from the same group of sites (Grigson and Mellars, 1987, 287). The Oronsay bone groups are sufficiently large to consider aspects of the size of the animals being exploited. The small size of at least some of the red deer suggests an origin on one of the adjacent islands, perhaps either Colonsay or Jura, rather than the Scottish mainland, where the more favourable conditions would have supported animals at least as large as present-day woodland populations (Grigson and Mellars 1987, 262).

The marine fauna is well represented in the several shell middens of Oronsay. Otter, largely an animal of the marine littoral around the Scottish coast, is again present, and grey seal (*Halichoerus grypus*) is well represented, the assemblage implying a locally breeding colony; only one bone of the common seal (*Phoca vitulina*) is present. Both small and large whale occur, the former most probably either common porpoise (*Phocaena phocaena*) or common dolphin (*Delphinus delphis*), and the latter possibly the common rorqual (*Balaenoptera physalus*). The dates of the Oronsay sites, largely in the seventh millennium BP (Switsur and Mellars 1987, 144), are again late in the Mesolithic and, whilst, with over 600 identifiable fragments, representing one of the largest assemblages of bone in Britain from the period, the material is still insufficient to provide an overview of the Scottish vertebrate fauna of the mid Holocene. In addition, detailed analyses of

Table 6.1 Animal bone from mesolithic sites at Oronsay, Morton and Carding Mill Bay (Bishop 1914; Coles 1971; Grigson and Mellars 1987; Hamilton-Dyer and McCormick 1993)

Species	Oronsay	Morton	Carding Mill Bay
Mammal			
Aurochs (*Bos primigenius*)	−	+	
Red deer (*Cervus elaphus*)	+	+	+
Roe deer (*Capreolus capreolus*)	+	+	+
Wild pig (*Sus scrofa*)	+	+	+
Otter (*Lutra lutra*)	+	−	−
Marten (*Martes martes*)	?	−	−
Hedgehog (*Erinaceus europaeus*)	−	+	−
Bank vole (*Clethrionomys glareolus*)	−	?	−
Grey seal (*Halichoerus gryphus*)	+	−	−
Common seal (*Phoca vitulina*)	+	−	−
Cetacean	+	−	−
Field vole (*Microtus agrestis*)	−	−	+
Common shrew (*Sorex araneus*)	−	−	+
Red squirrel (*Sciurus vulgaris*)	−	−	+
Birds			
Cormorant (*Phalacrocorax carbo*)	+	+	−
Shag (*P. aristotelis*)	+	+	−
Goose sp.	+	−	−
Shelduck (*Tadorna tadorna*)	?	−	−
Water rail (*Rallus aquaticus*)	+	−	−
Ringed plover (*Charadius hiaticula*)	?	−	−
Tern sp	+	−	−
Gull sp.	+	−	−
Great black-backed gull (*Larus marinus*)	−	+	−
Razorbill (*Alca torda*)	+	+	−
Great Auk (*Pinguinus impennis*)	+	−	−
Guillemot (*Uria aalge*)	+	+	?
Gannet (*Sula bassana*)	+	+	−
Red-breasted merganser (*Mergus serrator*)	+	−	−
Fulmar (*Fulmaris glacialis*)	−	+	−
Puffin (*Fratercula arctica*)	−	+	−
Kittiwake (*Rissa tridactyla*)	−	+	−
Thrush sp. (*Turdus* sp.)	−	+	−
Crow or rook (*Corvus* sp.)	−	+	+
Fish			
Tope (*Galeorhinus galeus*)	+	−	−
Spurdog (*Squalus acanthias*)	+	−	−
Angel shark (*Squatina squatina*)	+	−	−
Thornback ray (*Raja clavata*)	+	−	−
Sturgeon (*Acipenser sturio*)	−	+	−
Conger eel (*Congor congor*)	+	−	−
Salmoid sp.	−	+	−
Cod (*Gadus morhua*)	−	+	+
Haddock (*Melanogrammus aeglefinus*)	−	+	−
Saithe/Pollock	+	−	+
Turbot (*Scophthalmus maximus*)	−	+	−
Eel (*Anguilla anguilla*)	−	−	+
Whiting (*Merlangus merlangus*)	−	−	+

both the fish and bird remains have yet to be published. Mellars (1978) notes a predominance of one species of fish, the saithe or coalfish (*Pollachius virens*), which remains the dominant inshore species in western and northern Scottish waters at the present day. Over 30 species of bird have been identified from the sites (Mellars 1978, 379), including the extinct great auk (*Pinguinus impennis*) (Bramwell 1983a).

In eastern Scotland, the only sizeable mid-Holocene fauna comes from Morton in Fife, close to the mouth of the Tay, and is dated to late in the seventh millennium BP (Coles 1971). The terrestrial fauna again includes red deer, roe deer and boar, but aurochs is also recorded, the distinct difference in size between the two individuals found indicating the marked sexual dimorphism of wild cattle (Grigson 1969). Hedgehog (*Erinaceus europaeus*) and possibly bank vole (*Clethrionomys glareolus*) are represented by single animals. Cod forms the greater part of the identifiable fish fauna at Morton, with a few examples of haddock (*Melanogrammus aeglefinus*) and single bones of turbot (*Scophthalmus maximus*), sturgeon (*Acipenser sturio*), and a salmonid (*Salmo* sp.). The range of edible birds taken by the late Mesolithic hunters of Morton is considerable. Guillemot and gannet (*Sula bassana*) are dominant, although seven other marine species were also recovered. The predominately coastal nature of Mesolithic sites with bone preservation clearly influences assemblages, and only thrush (*Turdus* sp.) and carrion or hooded crow (*Corvus corone*) hint at the more general avifauna of the mid Holocene. The presence of the fulmar (*Fulmarus glacialis*) is particularly interesting. Fisher and Waterston (1941) note that in the nineteenth century it only bred on St Kilda, where it was extensively utilized by the inhabitants. There are later records from archaeological contexts, including Neolithic, Iron Age and Early Christian sites on Orkney (Bramwell 1977, 1983a), but its habit of nesting in the entrance to abandoned rabbit burrows, walls and ruined buildings causes Serjeantson (1990) to be circumspect about these finds. Overexploitation may have extirpated the fulmar from Scotland, and its rapid expansion around the shores of the British Isles and beyond must reflect a recent change in the food chain. Whilst waste from modern fishing boats may have played some part in this, G. Mar Gíslason (pers. comm., 1989) has suggested that the bird may have in part taken the place of the pelagic-feeding great whales, as they themselves were hunted close to extinction.

As well as the whale bones from Mesolithic sites on the West coast, Clark (1947) has drawn attention to finds within the carse clays of the Forth of Firth. These represent the remains of stranded animals during a period of marine transgression dated by Sissons and Brooks (1971) to between the mid-eighth and mid-seventh millennia BP. Several are associated with red deer antler mattocks, a type familar from 'Obanian' sites in the west (Clark 1956), and used to dismember the carcass. Most finds were made during drainage work in the nineteenth century and are therefore rarely identified to species, but a blue whale (*Balaenoptera musculus*) was found at Airthrey in 1819 and a rorqual (*B.* sp.) at Meiklewood in 1877.

MID TO LATE HOLOCENE

The advent of settled agriculturalists and pastoralists during the Neolithic marks a major break with preceding sites in terms of the material available for a study of the history of the vertebrate fauna. Not only do sites become more visible and therefore more frequently identified, but also a range of domestic animals comes to dominate the fossil assemblages. The record, however, remains predominately coastal and the earliest inland evidence may come from the pollen of cereals in deposits predating the widespread marker of the 'elm decline' (Edwards 1989b; Chapter 5).

The earliest evidence for domestic stock may come from Islay, where Harrington and Pierpoint (1980) have claimed sheep, more strictly ovicaprid bones, dated to around the opening of the sixth millennium BP; McCormick (in Connock *et al.* 1993), however, has suggested that the bones are more likely to be of roe deer. The most extensively studied bone group comes from the north Orcadian island of Papa Westray, from the site of Knap of Howar, where the radiocarbon dates lie in the mid-fifth millennium BP (Ritchie 1983, 57), and as A. Ritchie (1985, 41) notes, the artefactual assemblage includes items more appropriate to a food-gathering economy. Whilst Ritchie parallels one piece, a broken part of a shafted red deer antler artefact (1983, no.189), with later maceheads, the form suggests some continuity with Mesolithic antler mattocks. Noddle's (1983) detailed analysis of the domestic animal bone from the site shows roughly equal proportions of sheep (*Ovis aries*) and cattle (*Bos taurus*), with small numbers of pig; domestic dog also appears. In the absence of pannage, Dent (1977) has argued that the keeping of pigs on Orkney would have been difficult, although a small number of stalled animals could easily have been kept on domestic refuse supplemented by fish and crop waste; the large size of the animals, however, with the relatively small number of individuals, raises the possibility of import of deadstock, along with the few red deer. The cattle are also surprisingly large animals, given their island location, and overlap with wild aurochs perhaps implying relatively recent domestication (Noddle 1983, 97). The sheep are similarly regarded as primitive and Noddle (1983, 99) has raised the possibility of a Scandinavian origin for the animals.

Amongst the other wild large vertebrates, seal and whale were probably scavenged locally. The bird and fish remains, however, indicate extensive use of marine and littoral resources (Tables 6.2 and 6.3). Bramwell's (1983a) bird list includes a range of waders, divers and seabirds, with great auk and guillemot dominating. The fish remains (Wheeler 1983) are again dominated by cod and other gadoid fishes, but conger eel (*Conger conger*), flounder (*Plathichthys flesus*), halibut (*Hippoglossus hippoglossus*) and skate (*Raja batis*) are also evident and the presence of turbot might hint at slightly warmer waters than at present. Also from Westray, the late Neolithic site at Pierowall provides the first stratified evidence of pine marten, presumably an import for its pelt (McCormick 1984).

On Orkney Mainland, the late Neolithic site of Skara Brae has a similar bone assemblage, with cattle and sheep appearing in similar equal proportions, although, in contrast with Knap of Howar, the latter are small, more akin to the modern Soay (Clarke and Sharples 1985, 75). The wild cat (*Felis sylvestris*), is also recorded from the site (B. Noddle, unpublished), perhaps an import for its fur from the Scottish mainland; the bones of horse are now known to be intrusive. Gannet and cetacean bones are more frequent and there is some suggestion of the collection of the eggs of eider (*Somateria mollissima*), the bones of which also appear at Knap of Howar (Bramwell 1983a). The similar site at Links of Noltland on the relatively small island of Westray has a proportion of red deer bones, including the enigmatic

Table 6.2 Distribution of bird fragments from a selection of Scottish sites. **Neolithic:** 1 = Knap of Hower (Bramwell 1983a), 2 = Northton (Finley unpub.); **Early Bronze Age:** 3 = Northton; **Iron Age:** 4 = Hornish (Halstead, forthcoming); 5 = Baleshare (Halstead, forthcoming); 6 = Howe, Phase 7 (Bramwell 1994), **Pictish:** 7 = Howe (Phase 8), 8 = Buckquoy, Phase 1–3 (Bramwell 1977), 9 = Udal, Phases XI–XIII (Serjeantson 1988); **Norse:** 10 = Buckquoy, Phases 3–5

	Site									
	1	2	3	4	5	6	7	8	9	10
Species	Knap of Howar	Northton Neolithic	Northton EBA	Hornish	Baleshare	Howe Phase 7	Howe Phase 8	Bucquoy Phases 1–2	Udal XI–XIII	Buckquoy Phases 3–5
Black throated diver (*Gavia arctica*)	3									
Great northern diver (*Gavia immer*)	1						2	9		5
Red throated diver (*Gavia stellata*)							1		1	
Slavonian grebe (*Podiceps auritus*)										
Little grebe (*Tachybaptus ruficollis*)	1									
Fulmar (*Fulmarus glacialis*)	17				24	3	1	6	2	3
Manx shearwater (*Puffinus puffinus*)	1			2	1	3		3	7	11
Sooty shearwater (*Puffinus griseus*)						1				
Shearwater sp.	1									
Gannet (*Sula bassana*)	24	4	2		2	26	16	27	33	48
Cormorant (*Phalacrocorax carbo*)	9		7		1	15	6	1	4	2
Shag (*P. aristotelis*)	14	2	3			15	6		4	7
Grey heron (*Ardea cinerea*)						1				
Common scoter (*Melanitta nigra*)					1	1	1			
Velvet scoter (*Melanitta fusca*)	1					1				
Goldeneye (*Bucephala clangula*)						4	3			
Smew (*Mergus albellus*)						1				
Red brested merganser (*Mergus serrator*)						2	4			
Gossander (*Mergus merganer*)						3	2			
Eider (*Somateria mollissima*)	3					3	3			
Duck sp.	1					1				
Wigeon (*Anas penelope*)						1	2			
Gadwall (*Anas stepera*)					1?	4?	1?	1		
Shelduck (*Tadorna tadorna*)	1					1	1			4

Species	I	II	III	IV	V
Mallard (*Anas platyrynchos*)			5?		1
Mallard/pintail			1	2	
Pintail (*Anas acuta*)				1?	
Pochard (*Aythya ferina*)			1	1	
Long tailed duck (*Clangula hyemalis*)			1	1	
Teal (*Anas crecca*)		1	2	2	
Greylag goose (*Anser anser*)	12		11		
Greylag/domestic goose			1	1	
Domestic goose			1		13
Barnacle goose (*Branta leucopsis*)	10		2		
Brent goose (*Branta bernicla*)			1	1	
White fronted goose (*Anser albifrons*)			1	1?	
Goose sp.			4		4
Mute swan (*Cygnus olor*)			1	1	
Whooper swan (*Cygnus cygnus*)	1	1	9	5	3
Predator sp.			1		
Osprey (*Pandion haliaetus*)					
Red kite (*Milvus milvus*)			1	1	
White tailed eagle (*Haliaeetus albicilla*)			13	7	1
Goshawk (*Accipiter gentilis*)			1	1	
Buzzard (*Buteo buteo*)	1				
Buzzard/goshawk			1	1	
Golden eagle (*Aquila chrysaetos*)				1?	
Kestrel (*Falco tinnunculus*)			9	5	
Merlin (*Falco columbarius*)			4		
Peregrine falcon (*Falco rusticolus*)			2		
Red grouse (*Lagopus lagopus*)			41	42	
Black grouse					1
Grouse sp.			2		6
Water rail (*Rallus aquaticus*)	1	2			
Spotted crake (*Porzana porzana*)	1		3		
Corncrake (*Crex crex*)	1				
Crane (*Grus grus*)		1			
Stork (*Ciconia ciconia*)	1?		8		

continued overleaf

Table 6.2 *(continued)*

	Site									
	1	2	3	4	5	6	7	8	9	10
Species	Knap of Howar	Northton Neolithic	Northton EBA	Hornish	Baleshare	Howe Phase 7	Howe Phase 8	Bucquoy Phases 1–2	Udal XI–XIII	Buckquoy Phases 3–5
Moorhen (*Gallinula chloropus*)										
Golden plover (*Pluvialis apricaria*)						13	4		3?	1
Grey plover (*Pluvialis squatarola*)	1					1	11			
Lapwing (*Vanellus vanellus*)						1	1			
Oystercatcher (*Haematopus ostralegus*)	1						1			
Stint sp.						1				
Dunlin (*Calidris alpina*)	2				1	1	1			
Snipe (*Gallinago gallinago*)						5	6			
Woodcock (*Scolopax rusticola*)							2			
Godwit sp. (*Limosa* sp.)						1				
Redshank (*Tringa totanus*)	1		2		2?	1			1	
Curlew (*Numenius arquata*)	2					7	3		1	8
Knot (*Calidis canutus*)										1
Greenshank (*T. nebularia*)					2?	2				1
Green sandpiper (*Tringa ochropus*)	1					1				
Spotted redshank (*T. erythropus*)									1	
Turnstone (*Arenaria interpres*)	4					4	8	2		
Wader sp.	1					1	2			
Wader small sp.	6									
Phalarope, grey? (*Phalaropus fulicarius*)										1
Great skua (*Stercorarius skua*)	5									
Skua sp.	1									1
Great black-backed gull (*Larus marinus*)	17						2		2	
Great black backed/glaucous gull							1			3
Lesser black backed/herring gull (*Larus fuscus/argentatus*)	6					5		1		
Herring gull (*Larus argentatus*)		2?			1?		3		7	3
Lesser black backed gull (*Larus fuscus*)							1			

Species							
Black-headed gull (*Larus ridibundus*)	2			1	2		
Common gull (*Larus canus*)	3			1	1	6?	
Gull sp.	1			4	3		
Sandwich tern (*Sterna sandvicensis*)							
Razorbill (*Alca torda*)	9				2	3	2
Kittiwake (*Rissa tridactyla*)						1	
Auk sp.				1			
Little auk (*Alce alce*)		1?		1	5	1	
Great auk (*Pinguinus impennis*)	35	1	2	6	15	2	3
Great auk/gannet				25			
Razorbill/guillemot				1	2		
Guillemot (*Uria aalge*)	39	6	4	6	8	6	6
Black guillemot (*Cepphus grylle*)	4			1	6		
Puffin (*Fratercula arctica*)	3	2	5	2	5	7	3
Rock dove (*Columba livia*)				6	6	5	
Stock dove (*Columba oenas*)					1	1	3
Tawny owl (*Strix aluco*)				1			
Short eared owl (*Asio flammeus*)					17		
Skylark (*Alauda arvensis*)	2			1	4		
Pipit sp. (*Anthus sp.*)					1		
Great grey shrike/eastern european thrush							
Waxwing (*Bombycilla garrulus*)				4	1		
Warbler sp.				1	1		
Ringousel (*Turdus torquatus*)				4			
Ringousel/blackbird				1			
Blackbird (*Turdus merula*)		2?		5	7		
Fieldfare (*T. pilaris*)				8	2		
Song thrush (*T. philomelos*)				19	7		
Song thrush/redwing			3	2	2	1	1
Redwing (*T. viliacus*)				1	5		
Mistle thrush (*Turdus viscivorus*)				9	8		
Thrush sp. (*Turdus sp.*)	1		7	8	13		
Thrush/starling				6	20		
Great tit (*Parus major*)				6			

continued overleaf

Table 6.2 (continued)

	Site									
Species	1 Knap of Howar	2 Northton Neolithic	3 Northton EBA	4 Hornish	5 Baleshare	6 Howe Phase 7	7 Howe Phase 8	8 Bucquoy Phases 1–2	9 Udal XI–XIII	10 Buckquoy Phases 3–5
Swallow (*Hirundo rustica*)							6			
Starling (*Sturnus vulgaris*)	1				1	108	127	1		2
Rook/crow (*Corvus* sp.)						1	7			8
Raven (*C. corax*)	1		1?	1		19	10			1
Chaffinch (*Fringilla montifringilla*)						1				
Snow bunting (*Plectrophenax nivalis*)						1				
Reed bunting (*Embriza schoeniclus*)						3				
Corn bunting (*E. calandra*)						1				
Lark/bunting						2				
Finch (*Carduelis* sp.)							1			
Small passerine							1			
Domestic fowl (*Gallus gallus*)						5	3	1	2	11

ble 6.3 Fish from a selection of sites of varying date: **Neolithic** :1 = Northton (Finley unpub.); 2 = Knap of Howar (Wheeler 1983): **Early Bronze Age** 3 = rthton (Beaker Ph7): **Late Bronze Age/Early Iron Age** 4 = Baleshare (Jones unpub.); **Iron Age**: 5 = Cnip (Hamilton-Dyer unpub.), 6 = Howe, Phase L7 olley 1994); **Pictish** period: 7 = Buckquoy, Phases 1–2 (Wheeler 1977), 8 = Howe (Phase 8), 9 = Udal, Phases 11–13, 10 = Iona Guest House (Wheeler 1983); rse: 11 = Buckquoy, Phases 3–5, 12 = Birsay, Phase C2 (Colley 1989). + indicated that the bones were not quantified

	Site											
	1 Northton Neolithic	2 Knap of Howar	3 Northton Beaker	4 Baleshare	5 Cnip	6 Howe Phase L7	7 Buckquoy Phases 1–2	8 Howe Phase 8	9 Udal Phases 11–13	10 Iona	11 Buckquoy Phases 3–5	12 Birsay Phase C2
ark sp.		+		2				3			+	2
rdog (Squalus acanthias)												
pe (Galeorhinus galeus)				5								
gel shark (Squatina squatina)		+		1								
y sp.								1				
ornback ray (Raja clavata)								1				
nger eel (Congor congor)	1	+	1			1	2	6	2		3	3
(Anguilla anguilla)		+				6		7				
rring (Clupea harengus)									+			
monid						6		2				3
out (Salmo trutta)											1	
gler fish (Lophius piscatorius)												
d (Gadus morhua)		1		29	3	4		28	12	33	8	442
ddock (Melanogrammus aeglefinus)											2	2
iting (Merlangius merlangus)						3						
or cod cf (Trisopterus poutassou)												
rway pout cf (Trisopterus esmarkii)						3						
the (Pollachius virens)		+		1		1036	2?	271	1	12	13?	228
rsk (Brosme brosme)												9
g (Molva molva)		+	9	9			1	5	1		9	59
llack (Pollachius pollachius)				4								4
ckling sp.						5		2				
doid sp.		+		10	4	369	+	160			+	1150

continued overleaf

ble **6.3** (*continued*)

						Site						
	1	2	3	4	5	6	7	8	9	10	11	12
	Northton Neolithic	Knap of Howar	Northton Beaker	Baleshare	Cnip	Howe Phase L7	Buckquoy Phases 1–2	Howe Phase 8	Udal Phases 11–13	Iona	Buckquoy Phases 3–5	Birsay Phase C2
ecies												
ke (*Merlussius merluccius*)				52	17	1	1		2	4		8
rfish (*Belone belone*)						7		2				
ckleback (*Gasterosteus aculeatus*)								1				
ey gurnard (*Eutrigla gurnardus*)						31		6		13	2	1
scorpion (*Taurulus bubalis*)								1				1
ll-rout (*Myoxocephalus scorpius*)								1				2
llhead cf						7		1				
d (*Trachurus trachurus*)						3		7				
am, sea cf												
d sea-bream (*Pagellus bogaraveo*)											4	
asse, ballen (*Labrus mixtus*)		+	1	1			1			1	3	
asse, corkwing (*Crenilabrus melops*)						59		3				18
asse sp.						1		335				
bey, black cf (*Gobius niger*)						1		3				
ckerel (*Scomber scombrus*)				1								
tfish sp.				4		7		1			2	
ice/flounder				3		3		10				1
e (*Solea solea*)									1			
ounder (*Platichthys flesus*)		++										
rbot (*Scophthalmus maximus*)		++										
libut (*Hippoglossus hippoglossus*)		++										
ib						1						

find of 15 completely articulated skeletons. The implication is one of active management of a herd (Plate 4.10 in Bramwell 1983a), but why the carcasses should have been left unused is unknown.

Although assemblages from tombs run the risk of bias by selection processes of a ritual nature, the animal frequencies from the chambered tomb at Isbister on South Ronaldsay are similar to those from most contemporary occupation sites on Orkney, with sheep and cattle in equal proportions and red deer and pig in small numbers (Barker 1983). Although some of the bird and fish remains may reflect the activities of predators (Colley 1983), both fox (*Vulpes vulpes*), no longer present on Orkney, and otter were found in the tomb. Wheeler (1979) preferred to regard the bulk of bone as part of the human component in the tomb at Quanterness on Mainland. Cottid fish, either sea scorpion (*Taurolus bubalis*) or father lasher (*Myoxocephalus scorpio*), are most abundant at Isbister, whilst corkwing wrasse (*Crenilabrus melops*), presently a rather southern species, dominates at Quanterness. Amongst the birds, bones of the white-tailed eagle (*Haliaeetus albicilla*), which was exterminated in Scotland during the late nineteenth century AD and has recently been reintroduced, are well sealed in the foundation deposit of the Isbister tomb, and there are the remains of at least six other birds from the chamber (Bramwell 1983b). Another raptor, not currently resident in Orkney, the goshawk, *Accipiter gentilis*, is present in both the Isbister and Quanterness tombs (Bramwell 1979).

Evidence for the vertebrate fauna, contemporary with Neolithic settlement outside the Orkneys, is surprisingly sparse. There are no useful assemblages from Shetland, although the few fragments identifiable from Scord of Brouster show the presence of cattle, sheep and red deer (Noddle 1986) by the mid fifth millennium BP (Whittle *et al.* 1986). On the Outer Isles, the bone from Northton on Harris has yet to be published in detail (Simpson 1976), but Grigson (in Simmons *et al.* 1981) records a dominance of sheep bones, with hare (*Lepus* sp.), and seal, both grey and common, amongst the wild component; cattle, red deer and pig are also present (Finlay 1984). The presence of part of the skull of a badger (*Meles meles*), on this site may relate to imported pelts since the animal is not recorded from the Outer Isles (Serjeantson 1990); if the record of hare is correct (Grigson, in Simmons *et al.* 1981: 192), its origin is likely to have been similar. On the Scottish mainland, on a group of sites in the Firths of Forth and Moray, a group of shell midden sites contains a few bones of domestic cattle and ovicaprid, which appear to belong to the fifth millennium BP, and it is possible to consider these sites in the context of the well-known contemporary Ertebølle sites of the Baltic littoral (Sloan 1989).

The early Bronze Age evidence for the vertebrate fauna is even more restricted. At Northton on Harris, red deer bones are as numerous as cattle and sheep (Finlay 1984), perhaps suggesting a regular pattern of culling. The Neolithic and later faunas from Northton also include grey and common seal, and the fish remains include ling (*Molva molva*). The earliest evidence for domestic horse in Scotland also comes from the Outer Hebrides, from the small assemblage on the Beaker site at Rosinish on Benbecula. The same site includes field mouse (*Apodemus sylvaticus*), field vole (*Microtus agrestis*) and house mouse (*Mus musculus*), although the sandy nature of the site causes some concern that they may have burrowed in from later, Norse occupation deposits on the site (Serjeantson 1990); *M. musculus*, at least, is usually regarded as a late introduction. From Ardnave on Islay, Harman (1983) has described a small bone assemblage dominated by sheep, cattle and pig, associated with red deer, seal, field vole and a fox skull. The latter is now only present on Skye, and formerly Mull, in the Inner Hebrides, and may represent a trophy or pelt brought from the mainland.

THE LATE HOLOCENE

Whilst the pollen data show the early part of the late Holocene, the Bronze Age and early Iron Age, as the time during which the Scottish landscape rapidly approached its modern form, with the widespread reduction in woodland and expansion of heather moorland and blanket bog (Tipping 1994a; Chapter 5), the evidence of the associated fauna is singularly lacking. At the late Bronze Age/early Iron Age sites of Baleshare and Hornish Point in the Uists (Figure 6.2), sheep and cattle dominate, and deer, pig, seal and otter are less significant (Halstead forthcoming). An extinct large crane (*Grus primigenia*) has been described from late Bronze Age or early Iron Age deposits at King's Cave, on the shore of Loch Tarbert, Jura (Harrison and Cowles 1977), but the remainder of the assemblage has yet to be published. Crane is also present at the Iron Age wheelhouse site of Bac Mhic Connain on North Uist (Hallén 1994) and in the broch at Howe on Orkney (Bramwell 1994); in both cases, the bones have been ascribed to the extant crane (*Grus grus*), now a very rare visitor to the British Isles.

The largest Iron Age assemblages are again from the Northern Isles, from Howe on Mainland, Orkney; the bulk of the material belongs to the latter part of the period, contemporary with Roman occupation to the south. This site shows a decline in red deer through the late Iron Age (C. Smith *et al.* 1994) and, whilst it survived on Lewis and Harris, this species was extirpated from Orkney before the medieval period. The late phases of this site, associated with the broch, show sheep tending to dominate over cattle among the domestic animals, and there are also significant numbers of pig. A small horse, similar to the Shetland pony in size, and evident in the contemporary deposits at Jarlshof on Shetland (Platt 1956), occurs. Dog is evident at Howe, although one skull may have been a wolf rather than a domestic animal, presumably imported with its pelt from the Scottish mainland (C. Smith *et al.* 1994, 146). A possible wolf is also recorded from Jarlshof, Shetland (Platt 1956); in contrast, the articulated skeleton of a dog from the site was of an animal about the size of a terrier. Whilst the Jarlshof cats were thought to be wild, probably an imported source of furs, the Howe ones, with the bones also showing signs of skinning, appear to have been domestic. Other furs at the site were provided by both otter and fox. Compared with the Neolithic animals from Knap of Howar, the Iron Age cattle are small, comparable with material from contemporary sites elsewhere in Britain. The type, usually termed the Celtic shorthorn, is also present on the north coast of Caithness in the material from the broch at Crosskirk (Macartney 1984). Finlay (1991) compared the cattle bone from Sollas on North Uist with the lightly built West Highland or 'black' breed. The sheep from this site more resemble the Shetland type than the Soay. Pig appears more frequent on the mainland than in the island assemblages and goat makes its first definite appearance in the north at Crosskirk (Macartney 1984) and at Dun Mor Vaul on

Tiree in the Inner Hebrides (Noddle 1974). Despite the existence today of feral flocks in Galloway and on several offshore islands, goats do not appear in numbers in any of the Scottish bone groups and they appear never to have formed a significant component of the domestic fauna.

The wild fauna from Crosskirk also includes roe deer and possibly wild cat. The red deer remains are of animals perhaps intermediate in size between the large forest animals and the smaller ones of the present day, an adaptation to the poor range provided by moorland (Macartney 1984). Despite its coastal location, there are few fish, although wrasse and pollack are again present. Although the explanation could be taphonomic, in the way fish bone was disposed of, Howe shows a similar pattern in the earlier phase of its broch (Locker 1994). The fish list from the later phases is extensive (Table 48 in Locker 1994), dominated by saithe probably caught inshore, small wrasse taken from the shoreline, and smaller numbers of gadoid fishes, including cod and ling, both of which would have been taken in deeper waters. Both species are also present at Jarlshof on Shetland (Platt 1956). The marine fauna at both sites also includes cetacean bone, usually used for artefact manufacture, and seals, with grey seal identified at Crosskirk. The Jarlshof assemblage includes both seals, and bones of walrus (*Odobenus rosmarus*) (Platt 1956). Whilst its ivory and skins were widely traded out of the Arctic in the early historical period (McGovern 1985), the few records of bones from Scottish prehistoric sites are likely to reflect the chance killing of the occasional stray animal.

Both Crosskirk and Howe have provided extensive bird lists (Macartney 1984; Bramwell 1994). Gannet, great auk and shag (*Phalocrocorax aristotelis*), cormorant (*P. carbo*), manx shearwater (*Puffinus puffinus*) and razorbill indicate the exploitation of coastal resources, but black grouse (*Lyrus tetrix*), characteristic of open moorland, usually with a few trees, is present at the former site, whilst red grouse (*Lagopus lagopus*), also a heath and bog bird, appears on the Orkney site. Chicken (*Gallus gallus*) occurs in the latest deposits on the site, which are probably post-Roman; there is contemporary material from Howe, where domestic goose (*Anser anser*) and duck (*Anas platyrrhynchos*) are also represented. The great auk is evident in virtually all contemporary Iron Age assemblages in the Northern Isles, but is not included in Platt's (1956) list from the later, Norse deposits from Jarlshof on Shetland. Individual island dates of extinction may vary, but, flightless and an easy prey to hunters, it shows an inexorable decline; it was last seen in Orkney in 1813 and finally became extinct in 1844 (Buckley and Harvey-Brown 1891; Cramp 1983).

In southern Scotland, in the region intermittently controlled by Rome in the first few centuries AD, the record is sparse apart from the excavated Roman forts and *vici*, where the bone either survives or has been studied. Barnetson (1982) has reviewed the evidence for animal husbandry from the hillfort at Broxmouth, near Dunbar in East Lothian. The main occupation of the site ranges in date from late in the third millenium BP into the Roman period. The scatter of bones from the early phases of the site included goat (the earliest certain record in Scotland), sheep, cattle, pig, horse and dog, the wild fauna being restricted to roe deer and a bird bone, perhaps of a heron (*Ardea cinerea*). Later assemblages from the site contain a similar mix, with cattle slightly predominating over sheep, with a few pigs and goats.

The most extensive bone assemblage published from a Roman site comes from James Curle's excavations at Newstead, near Melrose in the Borders (Ewart 1911).

The presence of elk has already been mentioned, and the remainder of the assemblage consists largely of the usual mix of domestic animals. The presence of boar both on this site and at Bar Hill (Macdonald and Park 1906, 530) implies the continued existence of at least some form of forest cover in some areas. The animal is also figured on a number of later Pictish and Early Christian stones (Allen and Anderson 1993, I, 73), although by this period, like the deer also depicted, it may have become largely the hunting preserve of the nobility. Boar is also present in the contemporary native assemblage from Edinburgh Castle (F. McCormick, unpublished). In the West, both red and roe deer are recorded from the Roman fort at Birrens in Dumfriesshire (Robertson 1975, 107), but there is no quantified bone report from the site, a problem which also accompanies most early excavations along the line of the Antonine Wall. With the dearth of bone reports to go on, it is not surprising that Piggott (1958, 25) saw fit to characterize the Iron Age inhabitants of the north of Britain as 'footloose Celtic cowboys'. At Inveresk, on the south side of the Forth, mid second century bone groups from the *vicus* attached to the fort show a dominance of older sheep and cattle (Barnetson 1988), a contrast with most contemporary rural assemblages and perhaps a reflection of stock being driven in to feed the garrison, rather than a greater availability of fodder in the Lowlands. A similar pattern is apparent in the Antonine Wall fort at Mumrills (Macdonald and Curle 1929). Although probably a Roman introduction, domestic cats appear not to be recorded from any fort or *vicus* site, but dog, of about the size of a whippet, is evident as a paw impression from Flavian Inchtuthil on the edge of the Highlands (Pitts and St Joseph 1985, 340), and two sizes of dog are apparent in similar impressions from Newstead (Elliot 1991).

The post-Roman period shows little change in the bone assemblages, although the progressive expansion of the dietary constraints of Christianity could be expected to have some impact. The eating of horse flesh was explicitly prohibited, and fasting reduced the available meat to fish, seal, whale and possibly deer (McCormick 1987) on certain days and during Lent. The high incidence of deer on the Columban monastic site of Iona may reflect this dietary requirement, and the community from the sixth century AD onwards exercised control over neighbouring seal rookeries (McCormick 1981). Although the animal bone evidence is inconclusive, finds of bog butter (Earwood 1991) suggest that dairying had become widespread in the West and the Highlands by the early medieval period; earlier evidence is debatable (McCormick 1992).

Domestic fowls appear to become more common during the Pictish period, but Serjeantson (1988, 212) notes that their place in the islands was frequently taken by exploitation of seabirds; gannet at Buckquoy, Orkney (Bramwell 1977) and the Udal on North Uist, and puffin (*Fratercula arctica*) at Scalloway, Shetland (T. O'Sullivan, pers. comm., 1995). There appears to be an increased emphasis on poultry in some later Norse deposits. Rackham (1989) suggests that the small size of red deer antler, used in artefacts at Birsay on Orkney, indicates animals from Caithness, but the widespread trade in raw materials, including antler (Weber 1994) and walrus and whale ivory, indicated by such items as the Lewis chessmen, requires care to be exercised in the interpretation of unusual elements in vertebrate assemblages.

Although, as in much of the artefactual record, there are no clear differences between late Pictish and early Norse terrestrial vertebrate assemblages, it is evident that fishing becomes increasingly important. In the late ninth to early eleventh century at Birsay on Orkney, inshore fishing for small saithe dominates the marine component, although a range of other gadoids shows deep-water activity (Colley 1989, 258). At Freswick Links in Caithness, the massive quantities of fish bone indicate a community moving towards the commercialization of activities (Morris *et al.* 1992), and the locking of Scotland into the larger world of supply to the nascent urban centres of Europe is evident in such coastal Late Norse and medieval sites as Robertshaven (Barrett 1995) on the same coast.

LAND SNAILS

Stephen P. Carter

The modern land snail fauna of Scotland totals 77 species (Kerney 1976). The distribution of these is only imperfectly known, but they may be divided into two broad categories with distributions that reflect habitiat requirements. The majority (61 species) occur over a wide area and are tolerant of acidic, calcium-deficient habitats. A smaller group (16 species) is more or less restricted to coastal sites. This coastal distribution probably reflects their dependence on the presence of calcareous sediments, in this case shell sand, and the almost total absence of such sediments in inland areas.

The history of the development of this fauna is also limited by the extreme rarity of calcareous sediments in Scotland, which are required to preserve the calcium carbonate shells. Added to this restriction is the fact that none of the slugs (19 species) can be identified from their internal shells, and therefore their history is entirely unknown. Although all fossil assemblages of land snail shells have been recorded in calcareous sand accumulations, largely on the west and north coasts of Scotland (Evans 1979), these assemblages provide a picture of faunal change since 5000 BP (3780 cal BC) in one distinct, and in Scottish terms, exceptional habitat.

Despite these restrictions, two causes of change can be detected: those attributable to habitat change and those due to species migration and human introduction. Nothing is known about the land snail fauna of the early Holocene; it presumably underwent frequent changes in response to rapid climate change and the rather slower development of vegetation. One species of the modern Scottish fauna, *Deroceras agreste*, has a restricted montane distribution, suggesting that it may have been more widespread in the early Holocene. Other species, adapted to boreal vegetation, may have become extinct. The species which made up the early prehistoric fauna are probably all still present today, although some (e.g. *Vertigo pusilla* and *V. lilljeborgi*) are rarely recorded. This probably reflects habitat change, particularly the loss of woodland and the draining of wetlands. The restriction of the early prehistoric fauna has been balanced by the spread of species able to thrive in the coastal shell sand accumulations. New species have progressively colonized this habitat: *Helicella itala* followed by *Cochlicella acuta*, both during the Iron Age, and *Candidula intersecta* and *Cernuella virgata* probably in the medieval period (Evans 1979). These species occur earlier in south-west England (Bell 1990: 247) and there appears to have been a gradual spread northwards, perhaps passively assisted by humans. Other species have benefitted directly from human presence: *Oxychilus draparnaudi* and *Helix aspersa* are widespread in southern Britain but become increasingly synanthropic further north. Both are entirely dependent on the microclimate of human settlement for survival in the north of Scotland.

As a result of the changes outlined, the present-day snail fauna of Scotland probably includes more species than at any other time in the Holocene.

INSECTS

Paul C. Buckland and Jon P. Sadler

The insect fauna of Scotland has not been subjected to the same detailed study that the palynological record provides for the flora. Not only is the insect record patchy temporally, it is also, with the exception of recent unpublished work on archaeological sites, virtually restricted to the Coleoptera (beetles). Although the fauna includes no true endemics, it does contain a number of species not recorded elsewhere in the British Isles. Several may reflect survival from the cold of the Loch Lomond stadial in subarctic mountain top locations, whilst others remain in habitats that have been virtually eradicated from lowland Britain by human activity – particularly old established pine (Hunter 1977) and birchwoods. The extent of late Devensian glaciation (Sutherland 1984) and the intensity of the cold, would appear to preclude the survival of any animals from previous interglacials, despite earlier arguments to the contrary concerning offshore islands (Balfour-Browne 1953; Jackson 1956), and the Lateglacial must have begun *c*. 14 500 BP with a *tabula rasa*.

More assemblages of the Lateglacial in Scotland have been examined than for most of the Holocene, but the sample remains inadequate. Bishop and Coope (1977) examined four in Dumfriesshire (Roberthill, Redkirk Point, Bigholm Burn and Sanquhar). The geologist James Bennie had recovered some beetle remains from sites at Corstorphine and Saughton near Edinburgh late last century, and these have been published by Coope (1968). In addition, a similar small assemblage from Burnhead, near Airdrie, Lanarkshire (Coope 1962) is probably also of Lateglacial origin. More recently, Merritt *et al.* (1990) published a small fauna from beneath Loch Lomond till in the Teith Valley, Perthshire, and Walsh (1992) examined material from a kettlehole at Logie in Fife. No assemblages from the Highlands or Islands have yet been studied and none of the sites provides a complete record from initial warming through the Windermere Interstadial into the early Holocene. Coope (in Bishop and Coope 1977), however, provides a July temperature curve, constrained by radiocarbon dates and derived from the fossil beetle evidence, from the thermal maximum of the Windermere *c*. 13 000 BP through to the early Holocene, and several of the Scottish sites are incorporated in the revised diagram of Atkinson *et al.* (1987). It is probable that some of the more eurythermal elements in the insect fauna (tolerant of a wide range of temperatures), survived from initial immigration after deglaciation, but much of the extensive interstadial fauna, described from such sites as Roberthill (Bishop and Coope 1977), would have been cut out by the return to conditions equivalent to high arctic during the Loch Lomond Stadial.

The rapidity of early Holocene climatic amelioration has long been apparent from the fossil insect assemblages (cf. Ashworth 1972; Osborne 1980). At Brighouse Bay (Stewartry of Kirkcudbright), the fauna is thoroughly temperate by 9640±180 BP (9050–8470 cal BC), but no Scottish site has so far provided insect remains which cover the actual transition. The latest evidence from the Greenland ice cap suggests that warming by as much as 7 °C occurred in less than a decade at

11 500±200 calendar years ago (Alley *et al.* 1993). Such a rapid transition would have led to the extinction of any arctic elements not already close to mountain top refugia, and the consequent rapid dissolution of the Loch Lomond icecap would have provided ice rafts and freshwater for the dispersal of biota to offshore islands (cf. Buckland 1988). The Brighouse Bay faunas include several species of Coleoptera which have yet to be found in Scotland, and even allowing for our inadequate knowledge of the present fauna, their presence in the early Holocene merits some discussion. At least ten species occur no further north than Yorkshire and Lancashire, and cumulatively these might be seen to indicate summer temperatures slightly warmer than at present. The broad-leaved pinehole borer (*Xyleborus* [=*Anisandrus* of Bishop and Coope 1977] *dispar*) is largely restricted to south-east England, boring in various species of deciduous tree. In Scandinavia, Lekander *et al.* (1977) note a correlation of its distribution with the 16 °C July isotherm, implying temperatures up to 2 °C warmer than today. The Brighouse assemblage, however, requires some caution in interpretation. The four species of ground beetle, *Bembidion gilvipes, B. fumigatum, Agonum thoreyi* and *Odacantha melanura*, with the sedge smut beetle (*Phalacrus caricis*), are associated with wetland habitats, all of which have both declined catastrophically and been heavily polluted during the past century. The remainder are associated with woodland habitats, which have similarly been destroyed, and the dates of extinction of these species from Scotland is open to some doubt. Dinnin (1993) records the Colydiid *Cerylon histeroides* from mid Holocene deposits on South Uist in the Outer Hebrides, and the Eucnemid *Melasis buprestoides* survived at least until the post-Roman period at Buiston, north of Kilmarnock, Ayrshire. Whilst otherwise restricted to south of the Border, the large arboreal ground beetle, *Calosoma inquisitor*, is recorded from old woodland at Spean Bridge in Invernesshire by Crowson (Bishop and Coope 1977). Its usual prey are the caterpillars and chrysalises of Geometrid and Torticid moths on oak, but it appears long before the immigration of that tree into Scotland at Brighouse Bay, a pattern of response to climatic amelioration noted by Osborne (1974) for several other forest species. Whilst it is tempting to dismiss the apparent climatic implications of the fauna in favour of anthropogenic factors, Crowson (1981, 630–632) has pointed out that the present Scottish fauna includes a number of species which could be regarded as thermal relicts, either surviving in favourable microclimates or now flightless as a result of declining temperatures. These may, however, reflect the impact of recent changes through the 'Little Ice Age' (Grove 1988), involving thresholds and limited corridors for redispersal during the current warming, rather than the nature of initial Holocene climate. In the absence of several more complete sequences from the Scottish mainland, the nature of the early Postglacial climate remains uncertain beyond the rapid warming to conditions at least equivalent to those at present.

The problems of the development of the Holocene fauna are further compounded when the faunas of the Scottish pine forests are considered (Hunter 1977). Whilst the palynological record is fairly detailed (Chapter 5), the macrofossil is less secure and conclusions based upon modern distribution patterns may be falsified by the study of fossil assemblages. A number of xylophagous species (obtaining nourishment from wood), largely restricted to Scotland in the last century, have been able to take advantage of the activities of the Forestry Commission and expand their range southwards. Some, like the longhorn (*Arhopalus rusticus*) and the Cossonine weevil (*Eremotes ater* [= *Rhyncholus chloropus*]), are, along with a range of species now extinct in Britain (Buckland and Dinnin 1993), present in lowland England at least until the late Bronze Age, and it is apparent that it is the demise of lowland pine forests which have led to their decline. Others, like the birch bark beetle (*Scolytus ratzeburgi*), remain restricted to limited areas of the Highlands (Crowson 1971), despite an abundance of apparently suitable habitat in the lowlands. For this particular species, the fossil record extends as far south at the Somerset Levels (Girling 1977); at Thorne in South Yorkshire, its characteristic exit holes appear in virtually all the birches being pulled from the bog during peat extraction and it is difficult to see any reason for its present reduced distribution other than the destruction of continuous old forest habitat. Caution must be exercised when discussing potential extinctions from Britain. The dytiscid, *Agabus wasastjernae*, previously only recorded as a fossil from early Holocene deposits at Church Stretton, Shropshire (Osborne 1972), has since been found living in waterfilled rotholes in trees in Speyside (Owen *et al.* 1992).

The absence of suitably studied sites on the Scottish mainland is only partly compensated for by recent work by Dinnin (1993) and others on the islands. Faunas from beneath machair on Barra and the Uists in the Outer Hebrides show a wetland assemblage not dissimilar from the present day, but forest elements are evident in mid-Holocene faunas from both South Uist and Harris, where the ant (*Formica lemanii*) dominates the assemblages during the initial phase of paludification of the forest floor. At South Lochboisdale on South Uist, the forest element extends to a number of xylophagous species as well as leaf and litter species (Dinnin 1993) and work on a small wooded island in Loch Druidibeg, South Uist, shows several of these species still surviving in a fragment of this habitat. The impact of human activity is evident in all faunas from the mid Holocene onwards. Changes in the trophic status of Loch of Brunnatwatt, Mainland, Shetland, resulting from the activity of humans in the local catchment, are attested by the fluctuation in the abundance of chironomids (non-biting midges). On the basis of beetle evidence from the Outer Hebrides, Hoy in Orkney and Unst in Shetland, the picture is of expansion of heathland and raised mire habitats, alongside a rising frequency of dung feeders and other probable anthropochores (cf. Girling and Greig 1977). The biogeographically important fauna from Skara Brae illustrate this point well, and implies that the anthropochorous fauna so typical of the 'culture-steppe' landscape of the present day coalesced into a recognizable assemblage as early as the the Neolithic period. At Skara Brae, the predominance of the faunas associated with stored hay, and decomposing plant and animal waste, including dung, are in accord with other palaeoecological evidence which indicates a farming economy based upon pastoral activity. The occurrence of the human flea (*Pulex irritans*) provides the earliest European record of a species which has probable New World origins (Buckland and Sadler 1989).

From their initial habitats, many anthropochorous insects have been transported by humans around the globe and several are now cosmopolitan in distribution. This is very evident in pest species which are associated with stores of food and other resources. For example, the grain pests, the grain weevil (*Sitophilus granarius*) and the saw-toothed grain beetle (*Oryzaephilus surinamensis*), first attain widespread distribution along with the Romans (Buckland 1981). Members of both genera, along with the large ground beetle, *Laemostenus* sp., are recorded from the Roman fort at Bearsden, lying on the Antonine Wall in Glasgow, and dating to about AD 142–158 (Dickson *et al.* 1979). Data from later periods indicate further movement of species (Sadler 1991) and offer a consistent picture of squalid household conditions. The latter is particularly evident in the Dipterous evidence from sites such as the Dark Age crannog at Buiston, where records of the housefly (*Musca domestica*) constitute the earliest positive records of the species in Scotland, although it is recorded from Roman and later deposits in England as far north as Carlisle (Kenward *et al.* 1991). In some levels at Buiston the species reached almost plague populations, and under such conditions, incidences of myiasis and fly-borne diseases may have been common in the human population (P. Skidmore, pers. comm., 1994).

Both Coleoptera and Diptera from the Viking age 'pit' at Earl Thorfinn's farm at Tuquoy on the island of Westray, Orkney, indicate dumped material from a byre. The ethnographic record from Scotland highlights the importance of peat as a domestic fuel, and the suite of acidophile insects from Tuquoy suggest that peat may also have been utilized as litter in the byre, a practice known from Medieval Norway (Fredskild and Humle 1991). Several species from the organic deposits at Tuquoy are no longer recorded from Orkney and the small clambid, *Calyptomerus dubius*, is only known as far north as the Scottish lowlands at the present day. This may reflect the loss of habitats due to changing farming practices during the last century, but further modern collecting is necessary before any firm conclusions can be drawn. A fauna from the Biggins, Papa Stour, Shetland, provides the only other dated Norse assemblage in Scotland. The modern fauna of Shetland has been collated by Bacchus (1980) and it is deficient in anthropochores. Not surprisingly, four species recorded from the Norse deposits are not presently known from the island. Of these, the blind colydiid, *Aglenus brunneus*, has been subjected to detailed discussion by Kenward (1975, 1976). The species is found in accumulations of rotting plant material and sour grain residues. It first occurs in archaeological deposits in the Roman period and is more common in Norse deposits where it is a frequent fossil from Dublin (Coope 1981) and York (Hall *et al.* 1983). The single specimen recorded from the Biggins almost certainly represents a Norse introduction. Smith's (1996) study of a black house on South Uist not only provides a useful analogue for resolving the archaeological assemblages from rural Scottish sites, but affords new records of insects not previously recorded from the Outer Hebrides. The picture, however, remains far from complete.

ACKNOWLEDGEMENTS

This chapter is an attempt to summarize the work of a number of specialists working in Scotland and it is recognized that it is impossible to do justice to all the detailed studies that have appeared. Particularly thanks are due to Barrie Andrian, Ian Armit, John Barber, Lynn Barnetson, James Barrett, Stephen Driscoll, Paul Halstead, Peter Hill, Barbara Noddle, Tanya O'Sullivan, Derek Simpson and Dale Sarjeantson for access to unpublished data. The comments of Kevin Edwards, Paul Halstead and Ingrid Mainland are also acknowledged.

7 The Mesolithic

BILL FINLAYSON AND KEVIN J. EDWARDS

INTRODUCTION

The term Mesolithic is generally used to refer to the period from the onset of Holocene warming, *c.* 10 000 BP, to the local introduction of agriculture associated with the Neolithic. In some areas, Mesolithic hunter–gatherers and Neolithic agriculturalists co-existed within the same region, exploiting different parts of the environment. This may have been the case in Scotland, as indicated by an overlap of radiocarbon dates (Morrison 1980; Bonsall and Smith 1990) and by reconstructions of environmental evidence (Edwards and Ralston 1984; Edwards 1988; Chapter 5). Although the onset of the Neolithic has been considered to be primarily an economic change, it has also been assumed that the social systems of hunter–gatherers and farmers were different and largely incompatible, and that there was also a social transformation (Thomas 1991).

Research on the European Mesolithic has shown it to be part of a phenomenon which extends to the Middle East, marked by an increasing use of microliths (small tools made on deliberately modified flakes of flint or similar materials, often shaped into geometric forms) from the late Upper Palaeolithic onwards. It was a period when society was changing rapidly and producing a great many specifically local adaptations. Some of these accommodations in easterly areas led directly to farming, while on the western fringes of Europe there was an intensive exploitation of the environment. In southern Scandinavia, where particularly good evidence survives, intensification and specialization permitted a degree of Mesolithic sedentism, and arguably the introduction of a number of material and possibly social traits that are also conventionally associated with farmers, such as pottery and social hierarchies (Albrethsen and Brinch Petersen 1976; Larsson 1984; Rowley-Conwy 1985). The difference between hunter–gatherers and farmers appears to be especially diminished amongst communities who depended heavily on marine sources.

Modern Mesolithic research has been characterized by the importance given to environmental archaeology. Clark (1980, 38) argued that after the 1930s, 'It soon became apparent that the most promising way of gaining a picture of the achievements of the inhabitants of Europe between the end of the Ice Age and the adoption

Scotland: Environment and Archaeology, 8000 BC – AD 1000. Edited by Kevin J. Edwards and Ian B. M. Ralston.

Plate 7.1   Mesolithic shell middens on Oronsay. Copyright: J. N. G. Ritchie

of a Neolithic way of life was to adopt an ecological approach and deploy the full armoury of Quaternary Research.' In Scotland this approach can best be seen in the work conducted on the Oronsay shell middens (Mellars 1978, 1987; Plate 7.1), on the reinterpretation of Morton, Fife (Deith 1983, 1986), on recent work at Ulva Cave, close to Mull (Russell *et al.* 1995) and in studies relating past vegetational history to early settlement (Edwards 1989a, 1990, 1996a; Edwards and Ralston 1984; Hirons and Edwards 1990; Newell 1990).

A separate strand of research has focused on Mesolithic chipped-stone technology. This work has had considerable significance given that many Scottish Mesolithic sites are principally composed of scatters of flint or equivalent material. These studies have until recently been dominated by the work of Lacaille, most notably his publication of *The Stone Age in Scotland* in 1954. A new impetus has been given to this line of inquiry by the publication of results from Kinloch Farm, Rhum (Wickham-Jones 1990), the first comprehensive publication of a modern Mesolithic excavation in Scotland. Previously, lithic studies tended to concentrate on technology and the availability of raw material. The database has only recently become substantial enough to permit more elaborate, although as yet tentative reconstructions (Finlayson 1990a).

Modern research has also focused on the earliest evidence for human settlement in Scotland and its extent, and regional contexts rather than single sites are emphasized. This particularly applies to work in the southern Hebrides (Mithen *et al.* 1992; Edwards and Mithen 1995; Finlayson *et al.* 1996; Mithen and Lake 1996) and around Oban (Bonsall and Sutherland 1992; Bonsall 1996). Sites mentioned in the text and key areas are shown in Figure 7.1.

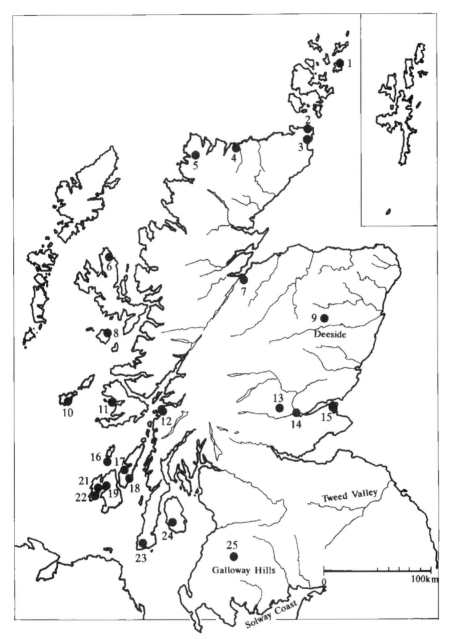

Figure 7.1 Sites and key areas mentioned in the text. 1: Millfield Farm; 2: Bay of Sannick; 3: Freswick; 4: Bettyhill; 5: Inchnadamph; 6: An Corran; 7: Inverness; 8: Kinloch; 9: Braeroddach Loch; 10: Ballevullin; 11: Ulva Cave; 12: Carding Mill Bay and Druimvargie Rockshelter; 13: North Mains; 14: Friarton; 15: Morton; 16: Caisteal nan Gillean I; 17: Glenbatrick; 18: Lussa Bay and Lussa Wood; 19: Bridgend; 21: Gleann Mor; 22: Bolsay Farm; 23: Rhoin Farm; 24: Moorlands; 25: Starr

THE MESOLITHIC IN SCOTLAND

In a recent review, Woodman (1989) painted a bleak picture of the state of research and the lack of evidence. He maintained that there was no workable chronology, no 'feeling for the type(s) of economy', no certainty on the extent of human occupation, types of artefact used, or indication that the Mesolithic was a single economic entity (Woodman 1989, 1). While there are undeniable problems, this picture is perhaps too pessimistic. Woodman (1989, 5), for the sake of argument, has indicated that Scotland would have one of the lowest densities of Mesolithic sites in Europe, perhaps one per 800 km^2, if a strictly defined diagnostic trait such as microliths was chosen as a basis for estimation. An alternative assessment founded on the density of fauna-bearing Mesolithic sites would, however, place Scotland as the fourth richest of 17 west European countries (Andersen *et al.* 1990).

The Mesolithic environment

After deglaciation and rapid afforestation, the natural environment of Scotland consisted of a varied woodland landscape, which, combined with fresh and sea water, provided an abundance of resources for hunter–gatherer groups. The resultant economy was based on harvesting wild resources by hunting, gathering and fishing. Much environmental research is aimed at reconstructing the resource base and at assessing the levels of environmental impact by Mesolithic peoples.

Was impact minor and accidental, or is there evidence for deliberate modification to the natural environment? It has been argued (Smith 1970; Mellars 1976a; Simmons *et al.* 1981) that fire could have been used deliberately, either directly as part of hunting strategies, or to modify the environment, both to improve grazing for deer and to encourage hazel growth for food and the supply of twigs (Morrison 1980). Edwards (1990) examined the evidence provided by microscopic charcoal in Scotland and one conclusion reached was that hazel did not represent a fire-climax vegetation type (Chapter 5). On many palynological sites there are markedly diachronous increases in the quantities of charcoal during the Mesolithic. These may be indicative of local burning, perhaps from anthropogenic fires, either domestic or even for land management (Edwards *et al.* 1995), but natural conflagrations made possible by climatic dryness have also been mooted (cf. Tipping 1996).

Interestingly, the fire record may indicate Mesolithic occupation in areas where no artefactual evidence has yet been recovered, e.g. in the Western Isles (Edwards *et al.* 1995; Edwards 1996a) and Shetland (Bennett *et al.* 1992). In such cases, it is suggested that bone has disappeared in acid soils, while lithics are likely to be hidden beneath peat, coastal sand and sea waters which have risen since the early Holocene (Edwards 1996a; Plate 7.2).

The pollen record is full of other instances, not necessarily fire-related, where both major and minor reductions in woodland occurred during the Mesolithic (Chapter 5). One effect of these would be the provision of open areas attractive to grazing animals. Unfortunately, the inadequate vertebrate fauna record does not provide clear indications of the extent and survival of all species through the Mesolithic. Red deer are well represented (although their presence in Shetland and the Outer Hebrides is only an inference from palynological evidence, which, if

Plate 7.2   The machair at Bàgh Siar, Vatersay. Sea-level rise in mid Holocene times and the movement of calcareous 'machair' sands inland may have concealed evidence for Mesolithic occupation in the Western Isles. Copyright: K. J. Edwards

correct, would imply human transport of deer: Bennett *et al.* 1992; Edwards 1996a). Roe deer, elk, aurochs, brown bear and boar were also present (Ritchie 1920; Simmons *et al.* 1981; Chapter 6).

Coastal locations (Lacaille 1954; Coles 1971; Mellars 1978) have produced evidence for otter and possibly domestic dog, as well as, for example, grey seal, common porpoise or dolphin, field vole, common shrew and red squirrel. Amongst the birds, razorbill and guillemot were probably used for food, and fish remains include cod, saithe, haddock, pollack, whiting and eel.

Colonization and the early Mesolithic

Although it seems reasonable to suppose pre-glacial and Lateglacial human occupation of Scotland, much of the evidence will have been destroyed by such events as glaciation and rises in sea level and there is no definitive evidence in Scotland for any pre-Mesolithic occupation. A review of lower Palaeolithic tools found in Scotland reports that all are most likely recent losses from collections made elsewhere (Saville 1993). Late Palaeolithic occupation had been proposed at Inchnadamph, Assynt, where there is possible evidence for human exploitation of reindeer (Lawson and Bonsall 1986). Radiocarbon dates now indicate that the material in question may rather be part of a sequence of natural deposits (Murray *et al.* 1993), albeit extending back into Lateglacial times.

A flint scraper stratified within a marine core taken between the Shetland Islands and Norway, if not secondarily derived, suggests either human occupation of land

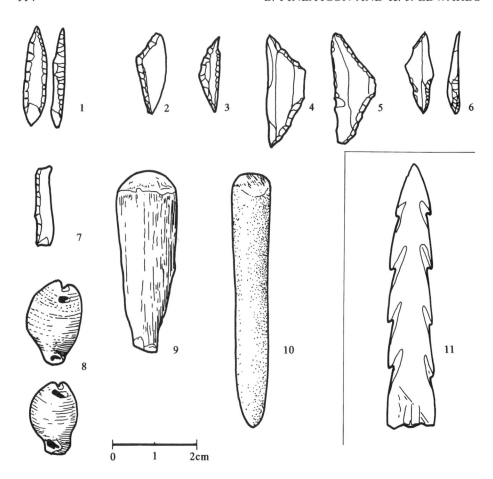

Figure 7.2 Mesolithic artefacts. 1, 2, 3 and 7: narrow blade microliths; 4 and 5: broad blade microliths; 6: tanged point; 8: perforated shell; 9 and 10: bevel ended tools; 11: barbed bone point (number 11 not to scale). After Wickham-Jones (1994)

which had been exposed by low sea levels before the Holocene, at about 18 000 BP, and certainly submerged again by 10 000 BP (Long *et al.* 1986), or a stone tool lost overboard during a fishing expedition. There is evidence for rapid Postglacial colonization in Norway (Bang-Andersen 1989) and occupation north of the Arctic Circle before 9000 BP (8030 cal BC) (Engelstad 1989). This evidence of high-latitude settlement may suggest early colonization of Scotland from the east. This hypothesis is possibly supported by the tanged points of proposed Lateglacial and early Postglacial Ahrensburgian affinity, with similarities to examples initially identified in northern Germany (Taute 1968; Barton 1989). These are recorded as stray finds from Ballevullin (Tiree) (6 in Figure 7.2), Millfield Farm and Brodgar (Orkney), Lussa Wood and Lussa Bay (Jura), and Bridgend (Islay; the only precise findspot) (Livens 1956; Mercer 1980; Edwards and Mithen 1995). Doubt has been cast on the Jura examples because of their broken and heavily rolled condition (Morrison and

Bonsall 1989). None of the finds comes from a radiocarbon-dated context, and pollen evidence from Islay allows, at best, only circumstantial evidence for an immediately Postglacial presence elsewhere on that island (Edwards and Berridge 1994; Edwards and Mithen 1995). The group retains interest as a possible indication of relatively precocious occupation.

Further south in Britain, the early Mesolithic is characterized by non-geometric broad-blade flint artefacts (Jacobi 1973; Mellars 1974, 1976b). Similar artefacts have been recovered from a few Scottish sites, most convincingly from Glenbatrick on Jura (Mercer 1974) (4 and 5 in Figure 7.2). At present it seems reasonable to assume that the appearance of such broad-blade assemblages precedes narrow-blade ones in Scotland also (Myers 1988). Bonsall (1988) has argued from the presence of broad-blade material, albeit in mixed assemblages, at both Morton (Coles 1971, 1983) and Lussa Wood (Mercer 1980), that elements of these sites predate 9000 BP (8030 cal BC). This dating is founded mainly on a consideration of local sea-level change (Dawson 1979; Sutherland 1984), and the location of the sites is considered to result from exploitation of marine resources. It might thus seem reasonable that the broad-blade assemblages date from 11 000 to 9000 BP (8030 cal BC). The distribution of sites with such material, from Jura in the west to Fife in the east, combined with the evidence of tanged points from Orkney and the southern Hebrides, may suggest that most of Scotland was colonized before 9000 BP (8030 cal BC). More secure evidence for this earliest occupation of Scotland remains to be found.

Later Mesolithic (from *c.* 9000 BP)

While an early Mesolithic occupation of Scotland is at present insubstantially attested, there is, by the later Mesolithic, clear evidence for established settlement. The earliest radiometric dates for an occupation site are those from excavations at Kinloch, Rhum (Plate 7.3), extending back to 8590±95 BP (7700–7500 cal BC) (Wickham-Jones 1990). There, as on most Scottish sites, the assemblage is characterized by a geometric narrow-blade microlithic component (1, 2, 3 and 7 in Figure 7.2). Scrapers form the other main artefact type, but it is notable that unlike in England (Mellars 1976b), microliths dominate most assemblages regardless of site location. Narrow-blade assemblages have been found at Inverness (Wordsworth 1985), Morton (Coles 1971), Deeside (Kenworthy 1981), the Solway coast (Cormack and Coles 1964; Cormack 1970), the Galloway Hills (Affleck 1986), the Tweed Valley (Mulholland 1970) and from many west coast islands: Arran (Allan and Edwards 1987; Affleck *et al.* 1988), Jura (Mercer 1970, 1971, 1972; Mercer and Searight 1987), Islay (McCullagh 1989; Mithen 1990) and Colonsay (Mithen 1989; Mithen and Finlayson 1991), as well as Rhum. The most northerly mainland material comes from the Caithness sites of Freswick (Lacaille 1954) and possibly the Bay of Sannick (Pollard and Humphreys 1993), and from Bettyhill, Sutherland (Wickham-Jones and Firth 1990). Material has also been reported from Orkney (Wickham-Jones 1990; and see Saville 1996). Within the limits of admittedly very patchy research, it can now be suggested that later Mesolithic occupation extended throughout Scotland and in some areas was probably fairly dense. At some sites, such as Kinloch (Wickham-Jones 1990) and Bolsay Farm, Islay (Mithen *et al.*

Plate 7.3 Kinloch, Isle of Rhum with the Cuillins of Skye in the background. The setting of the earliest known occupation site in Scotland. The Farm Fields Mesolithic excavation site lay to the centre bottom of the picture, just above Loch Scresort, and contained charred hazelnut shells dated to 8590±95 BP. The Kinloch pollen site (bottom left) indicated a Mesolithic landscape in which hazel and alder were locally prominent, and which experienced regular phases of woodland reduction. Grass and sedge-dominated vegetational communities have existed since 3950 BP, and the appearance of cereal pollen and soil erosion at this time suggests that Neolithic activities were responsible for the landscape changes. Copyright: K. J. Edwards

1992), hundreds of thousands of artefacts have been recovered from the excavated areas alone, although these assemblages have accumulated as the result of recurrent or continuous occupation. The Mesolithic settlement of Scotland is clearly more substantial than occasional foraging trips or 'the wanderings of a single family' (Atkinson 1962, 6).

The later Mesolithic comprises not only the narrow-blade flint scatters, but also the 'Obanian' sites (Turner 1872, 1895; Grieve 1882; Anderson 1895, 1898; Bishop 1914; Lacaille 1954; Pollard 1990; Bonsall 1996). These consist of midden deposits, preserving a wide range of faunal material including shells and the bones of fish, birds and mammals, with fish probably representing the most important dietary element. In addition, a range of organic artefacts not found elsewhere has survived. These include barbed points (harpoons), antler mattocks, and bone and antler bevel-ended tools, often referred to as 'limpet scoops' (8, 9, 10 and 11 in Figure 7.2). Chipped-stone artefacts are also represented, but rarely microliths. This industry has mostly been produced by bipolar knapping, rather than the platform-core technique associated with the narrow-blade assemblages. While it would no longer be argued that the 'Obanian' (a term first used by Movius [1940b, 76]), represents a discrete cultural entity, there is clearly a group of sites with a distinctive set of artefacts. Their relationship with other Mesolithic (and early Neolithic) sites is problematic. The discovery of inter-stratified 'Obanian' and microlithic material in rock shelter deposits at An Corran, Skye (Saville and Miket 1994) promises to be of major importance in furnishing a key sequence pertinent to this problem.

THE NATURE OF THE EVIDENCE

Site types and archaeological perceptions

Much of the available evidence from Scotland is undoubtedly of restricted value for interpretation. The lithic scatters are rarely stratified and apparently have few associated structural remains (although most scatters are known from fieldwalking rather than excavation: cf. Mulholland 1970; Edwards *et al.* 1983). For many sites, it is difficult to judge how many occupations are represented or their duration, and basic parameters such as group size and site function remain elusive.

Lithic scatters are generally open air, while midden deposits occur both in the open (Mellars 1987) and in caves. Open-air midden sites perhaps appear simple to assess, but although these disposal heaps produce an abundance of economic data, these can be hard to interpret. On Oronsay, apparent seasonality on four middens was elegantly demonstrated (Andrews *et al.* 1985; Mellars 1987); it is impossible, however, to quantify the scale of occupation or the frequency of visits. Some cave sites have better stratigraphic records, but contain deposits disturbed since the Mesolithic by subsequent human use.

Study of the period has not been assisted by the division in primary interests between researchers interested in environmental evidence (fullest in midden deposits) and those concerned with information on lithics (best served by flint scatters). The latter sites generally provide much less direct evidence for subsistence, while the lithic industry from sites in the former category is not accorded a high priority for publication (e.g. Oronsay). This distinction is almost certainly an over-simplification. The shell midden sites vary in form and artefact content, while the narrow-blade assemblages vary from large sites such as Kinloch, Rhum and Bolsay Farm, Islay, to small, possibly single occupation sites as at Gleann Mor, Islay (Mithen *et al.* 1992).

Chronological framework

The chronological framework for the Mesolithic has generally been regarded as inadequate because of poor site stratification and the frequent reoccupation of sites. On such palimpsest sites, only radiometric dating of numerous features can date the elements within them; and the temporal resolution is clearly too coarse to distinguish annual or seasonal visits. The difficulties involved are illustrated by the debate surrounding the chronological interpretations of the site at Morton in Fife (Coles 1971, 1983; Myers 1988; Bonsall 1988; Woodman 1988; Clarke and Wickham-Jones 1988).

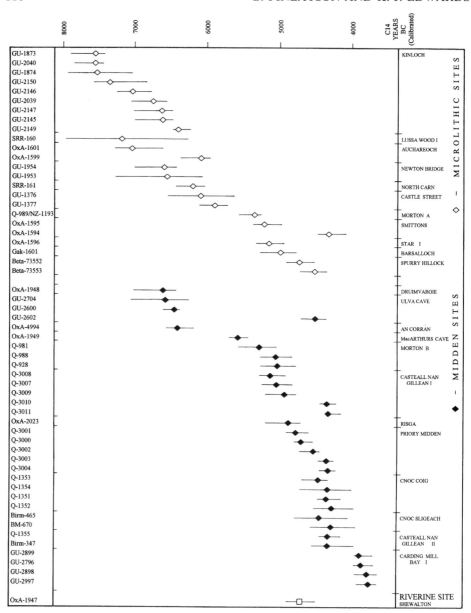

Figure 7.3   Radiocarbon dates for the Mesolithic. Dates are calibrated at 2 σ

Recent dating programmes have resulted in a dramatic move of the 'Obanian' from being an early Neolithic (Jacobi 1982) or a final Mesolithic phase (Woodman 1989). Some workers now envisage it as a long-lasting complex spanning the later narrow-blade Mesolithic (Bonsall and Smith 1989), reinforcing doubts as to its existence as a cultural entity. The main cluster of 'Obanian' dates (Figure 7.3) range from 6190±80 BP (5230–5000 cal BC) at Caisteal nan Gillean I on Oronsay, to

4980±50 BP (3900–3700 cal BC) from Carding Mill Bay near Oban (Connock *et al.* 1993). A date of 7810±90 BP (6700–6470 cal BC) for a barbed bone point from Druimvargie Rockshelter (Bonsall and Smith 1989) is substantially earlier than this main cluster. However, the Druimvargie point is uniserial, while all other 'Obanian' bone and antler points are biserial; this may indicate that the Druimvargie find does not belong to the 'Obanian' (Bonsall and Smith 1990).

Interestingly, the other 'Obanian' sites producing early dates, Ulva Cave and An Corran, are also not entirely typical of the main cluster of 'Obanian' sites. Dates on shells from the basal 10 cm of the midden at Ulva Cave are 8060±50 and 8020±50 BP, adjusted to 7660±60 BP (6480–6410 cal BC) to allow for the greater age given by marine deposits (Bonsall 1992). Dates from the top 10 cm are, however, conformable with the main 'Obanian' series (adjusted age 5690±60 BP [4590–4460 cal BC]). The artefactual material recovered from Ulva includes platform cores and blades (not characteristic of 'Obanian' contexts) and a perforated cowrie shell, a typical 'Obanian' artefact. At An Corran, a date of 7590±90 BP (6470–6260 cal BC) has been obtained from a bevel-ended tool fashioned from red deer antler, typically 'Obanian', but found here with a microlithic assemblage (Saville and Miket 1994).

The long sequence of deposits at Ulva may represent a gradual transition between the narrow-blade and 'Obanian' sites, incorporating material typical of both. It is noticeable from Figure 7.3 that the earliest dates come from Druimvargie, where the barbed point is not typically 'Obanian', and from Ulva and An Corran where the lithic assemblages are narrow-blade in character. Morton B is also not an 'Obanian' site in that it does not contain the classic 'Obanian' artefact types, but is more simply to be regarded as a shell midden. The main cluster of 'Obanian' dates lies towards the end of the range for narrow-blade material, overlapping and extending that date range (Figure 7.3). It is possible that the 'Obanian' sites represent an intensification of settlement during the transformation from the Mesolithic to the Neolithic.

MESOLITHIC ECONOMICS

Subsistence

As noted above, Woodman (1989) has suggested that our understanding of the types of economy pursued by the Mesolithic inhabitants of Scotland is unclear and that this lack of clarity is matched elsewhere. Extensive research has been conducted, of course, into the type of economy represented by shell midden sites (Mellars 1987). Evidence from the Oronsay shell middens suggests a broadly based economy in so far as terrestrial animals, marine mammals and birds constitute part of the subsistence base on these marine exploitation sites. Indeed on Oronsay there is evidence to show that red deer, probably already butchered (Grigson and Mellars 1987), were brought to the island. The otolith size ranges from saithe indicate that Oronsay was occupied throughout the year, with different middens in use at different seasons (Mellars and Wilkinson 1980). Grigson and Mellars (1987) observed that the bimodal size distribution of deer may indicate their derivation from two sources. It is unlikely that the neighbouring island of Colonsay was a source of the deer bone in the Oronsay middens, and the two populations of deer probably came from Jura and the mainland. This evidence for long-distance contacts, combined with the small size of Oronsay during the early Holocene (Jardine 1987), suggests that Mesolithic occupation of this island was unlikely to be permanent, but was part of a wider seasonal round (Mithen and Finlayson 1991).

Apart from ubiquitous carbonized hazelnut shells (Mellars 1978; Affleck *et al.* 1988; Wickham-Jones 1990), there are few economic data from sites other than the 'Obanian' series. Morton is an exception (Coles 1971). Here, despite chronological problems, and aside from a small bone collection including those of red deer, roe deer, boar, aurochs and cod, evidence from the stratified midden deposits allowed the seasonality of shellfish collection to be examined (Deith 1983). The proximity of the site to the shore enhanced the exploitation of the lithic resources found there (Deith 1986). Zvelebil (1994) has produced a useful review of plant resources in the European Mesolithic, but evidence from Scottish sites is generally restricted.

In the absence of direct economic data from lithic scatter sites, attempts have been made to use the lithic evidence itself to suggest patterns of resource exploitation (Mellars 1976b; Myers 1988; Finlayson 1990a). These include assessments of the function of individual tools and of the type of hunting strategy implied, achieved by considering the time required to manufacture tools, their maintainability and adaptability to different uses. By combining these considerations with an assessment of the function of sites derived from their catchment characteristics, and their place within wider regional networks as revealed by the presence of imported lithic materials, various regional economic strategies can be suggested (Finlayson 1990a). As indicated above, settlement on Oronsay is presumed to have been part of a network that included Jura and the mainland, and therefore probably Colonsay and Islay too. In south-west Scotland it seems probable that the upland site of Starr at Loch Doon was part of a network that included the Ayrshire coast, while Smittons on the Water of Ken, just the other side of the watershed, perhaps belonged to a network that extended to the Solway Firth (Finlayson 1990a; Edwards 1996b) and both may well have been interlinked. The evidence from the Tweed Valley (Mulholland 1970) and Deeside in the North-East (Paterson and Lacaille 1936; Plate 7.4) probably also reflects such regional systems.

Studies of Mesolithic Britain have long been dominated by concepts of upland hunting in the summer, and processing activities in the lowlands in the winter, often by the sea (Mellars 1976b; Morrison 1980). Against this traditional view, Bonsall (1981) has shown that at Eskmeals, on the southern coast of the Solway Firth, all year round occupation was feasible, utilizing different resources in different seasons. The range and location of Mesolithic sites suggests that some are more likely to represent seasonal, or task-specific camps, while others may have been more permanent. The evidence taken as a whole might best intimate a complex pattern of mobility within local territories, combined with more permanent camps.

It has been suggested that the difference between the artefact collections from middens and narrow-blade sites may relate to functional differences (Bonsall and Smith 1989). Any simple argument that the midden sites reflect marine exploitation and that the narrow-blade ones represent the upland components of an annual

Plate 7.4 Terraces of the River Dee, near Banchory, Kincardineshire. Terrace soils all along lower Deeside have produced many thousands of flints. Copyright: K. J. Edwards

subsistence strategy, is inadequate. Few microliths have been recovered from secure 'Obanian' contexts other than at An Corran (Saville and Miket 1994). Moreover, the locations of many of the narrow-blade sites suggest that they were placed to exploit coastal resources. Indeed, the highest microlith counts come from coastal or near-coastal island sites such as Kinloch (Wickham-Jones 1990). A more complex functional role for the 'Obanian' may be envisaged. Both the large numbers of bevel-ended tools and the presence of disproportionate numbers of antlers and deer metapodials (Grigson and Mellars 1987) suggest that on Oronsay at least, the 'Obanian' sites were associated with manufacturing processes and not just marine exploitation (Finlayson 1995). The relationship between 'Obanian' and narrow-blade assemblages remains problematic.

Technology

Current understanding of Mesolithic technology is necessarily incomplete. It can be assumed that most equipment was probably associated with perishable materials such as wood and leather. It is clear, both from faunal evidence for deep-sea fishing and from the occupation of islands, that the Mesolithic population had seaworthy boats. Indeed, a dugout canoe of pine was recovered from beneath the carse clays at Friarton, Perthshire (Geikie 1881), and others were reported from the Forth and Clyde estuaries (Munro 1899). An extensive array of stone tools, mostly of chipped stone, in a variety of materials including flint, bloodstone from Rhum (Wickham-Jones 1990), chert (especially in the Southern Uplands; Affleck 1986), and quartz, particularly on Jura (cf. Wickham-Jones and Collins 1978,

Wickham-Jones 1986), has been recovered, as have a small number of antler and bone artefacts, mostly from 'Obanian' shell middens. The principal diagnostic tools for the period in general are microliths, but others include hammer-stones and bevel-ended pebbles (the stone equivalents of the antler and bone limpet scoops found in shell middens).

Increasingly, the function of the toolkit has become a subject for research. In the past, microliths have been associated exclusively with hunting, as representing the individual barbs and points of projectiles (possibly arrows; Myers 1987). Microwear analysis would seem to confirm the view of D.L. Clarke (1976) that microliths must have served a variety of purposes, and not simply hunting (Finlayson 1990a,b; Mithen et al. 1992; Finlayson et al. 1996). Equally, it is likely that the so-called limpet hammers and scoops of the 'Obanian' were probably associated with the dressing of skins (Finlayson 1995). Research on the Palaeolithic and Mesolithic has tended to see tools as related to food procurement, but for microliths and bevel-ended tools it is likely that the surviving tool types were not solely related to this quest.

Analysis of lithics indicates long-distance relationships. Arran pitchstone and Rhum bloodstone are both found, in small quantities, away from the islands where they outcrop, although it is likely that some of this transported material is post-Mesolithic in date (Williams Thorpe and Thorpe 1984; Clarke and Griffiths 1990). Assessment of the exploitation of other raw materials, such as flint and chert in southern Scotland, as revealed through reduction strategies and combined with the analysis of the function of individual tools, allows the reconstruction of Mesolithic regional networks (Finlayson 1990a).

MESOLITHIC SOCIETY

Postglacial hunter–gatherers

Interpretations of Postglacial societies are rooted in generalized models of hunter–gatherer behaviour. These are based largely on ethnographic analogy, sometimes with groups in extremely different environmental circumstances, such as the !Kung San in Namibia (Lee and DeVore 1968) or the Nunamiut Eskimos (Binford 1978). Some behaviour patterns are almost universally accepted as being generally applicable, such as men hunting and women gathering. The importance of these economic activities varies largely with latitude – hunting increasing in importance towards circumpolar regions. It is generally assumed that fishing and hunting would have been important activities in Scotland. It should not be inferred from these general principles that either society or economic practices were unchanging over the five millennia attributable to the Mesolithic. Hunter–gatherer societies do not have to be simple, and the epithet 'complex' may be apt (cf. Price and Brown 1985; Rowley-Conwy 1985).

Evidence of ritual or funerary practices is largely unrecognized from Scottish Mesolithic sites. In contrast to southern Scandinavia (Albrethsen and Brinch Petersen 1976; Larsson et al. 1981; Larsson 1989), there are no known burial sites, although a few human bones have been found in 'Obanian' contexts (Meiklejohn and Denston 1987). Saville and Hallén (1994) provide radiocarbon dates demonstrating much later burials within Mesolithic midden deposits. Much archaeological evidence for social structures comes from burials, both from the presence of grave goods and from human pathology and stable isotope evidence. The absence of such data does not mean that society in Scotland was primitive, but simply that a very useful data source is lacking.

The transition to a farming economy

Fresh environmental perspectives suggest that hunter–gatherers may have engaged in active woodland management practices, such as coppicing, during the late Mesolithic (Göransson 1987; Edwards 1993a). More importantly, perhaps, there is also the palynological evidence for early agriculture from *c.* 5800 BP (4690 cal BC) in Scotland (i.e. prior to elm decline times). The best evidence comes from the Arran site of Moorlands (Machrie Moor), and from Rhoin Farm on the Kintyre Peninsula (Edwards and McIntosh 1988). The data from North Mains, Perthshire (Hulme and Shirriffs 1985), are less reliable because of a possible hiatus in the pollen core. This evidence is contentious (O'Connell 1987; Edwards 1989b; Chapter 5), but if valid, it raises the possibility of either the adoption of cereal cultivation by indigenous hunter–gatherers and/or the immigration of Neolithic agriculturalists (Edwards and Hirons 1984; Zvelebil 1994). If the Neolithic is defined by economic criteria, this might indicate that, locally, the Neolithic began much earlier than generally envisaged.

   For the west of Scotland, it has been argued that many Mesolithic traits, such as settlement mobility and the exploitation of wild resources, continued into the early Neolithic, and that not until the later Neolithic at the earliest was there a major social change (Armit and Finlayson 1992). This interpretation assumes that Mesolithic society was capable of borrowing aspects of the Neolithic, such as pottery and possibly elements of agriculture (Dennell 1985), to act as new vehicles for Mesolithic symbolic expression that were required to emphasize group identity and status as the economy intensified. Thus, the broadly based economy revealed on Oronsay may be a clue to continuity into the Neolithic. Rather than the specialization that developed amongst late hunter–gatherers in the eastern Baltic and southern Scandinavia (Zvelebil 1986, 1989a), where intensification and a specialized technology resistant to change was produced, the breadth of the economy in Scotland enabled Mesolithic hunter–gatherers to adopt other elements, both economic and social. The change to the Neolithic as conventionally determined may therefore be far harder to see in the archaeological record. It is possible that environmental evidence may make the change more apparent; equally it may have occurred as a gradual transformation, where a Mesolithic and a Neolithic are identifiable at either end of the process, but where sites in the middle of the continuum cannot be neatly classified (cf. Edwards 1988). In the Dee valley of Aberdeenshire, Edwards and Ralston (1984) conjectured that the post-elm decline vegetational disturbances at Braeroddach Loch may have been a response to late Mesolithic impacts, at a time when hunter-gatherer territories were restricted by farming communities further down the valley, and a similar case has been proposed for the Cheviot Hills (Tipping 1994a). The possibility remains that both activities could be the result of either the same population or two economically different ones.

J. Thomas (1988) has observed that there is a conceptual gulf between those scholars working in the Mesolithic and those in the Neolithic, not only because of chronological interests, but also because of different approaches. Thus, it has been suggested that research in the Palaeolithic/Mesolithic 'is concerned with human behaviour in terms of adaptive responses to environmental pressures', while for the Neolithic, research 'is more likely to consider human beings as purposive subjects, acting in pursuit of socially defined goals' and that the transition is 'the boundary between two models of man' (J. Thomas 1988: 59). This brings us back to Clark's (1980) observation concerning the importance of environmental research in Mesolithic studies, and more immediately refers to models such as Rowley-Conwy's explanation of the collapse of Ertebølle society as the result of a catastrophic decline in oysters as a resource (Rowley-Conwy 1985). This conceptual gulf may lead on the one hand to the formation of models favouring continuity of Mesolithic populations wherever evidence related to hunting and gathering activities is discernible, and on the other to the attribution to incoming farmers of all traits (e.g. such as pottery, domesticated livestock and monuments) that are associated with the Neolithic. It is important to note that the two approaches do not give equal weight to subsistence systems. In their interpretations, Mesolithic scholars give the subsistence system primary significance, while Neolithic scholars tend to assume the characteristics of the farming economy, but concentrate on social behaviour. This incompatibility in approach thus radically underplays the range of intermediate subsistence strategies between hunter–gatherer systems and established agriculture, and ignores the wide range of possibilities from collaboration, intermarriage, and other peaceful socio-economic adaptations to outright hostility between groups (cf. Dennell 1985). In the mosaic of landscapes that make up western Scotland, parts of which are unlikely ever to have been suitable for sustained farming secure from crop failure, many communities may well have continued to derive their nutritional requirements from both wild and domesticated resources, and to that extent, a prolonged period of transition is envisaged.

CONCLUSION

Mesolithic studies in Scotland have come a long way in recent years (Pollard and Morrison 1996). When continuing research is published, it is likely that our understanding will change substantially. What is already clear is that the Mesolithic was a period which saw settlement, although probably not the earliest in Scotland, by groups which exploited the environment intensively, and which probably lived in communities, sufficiently robust both economically and socially, to have played a significant part in the establishment of farming in the area, as shown by the regional diversity within the Neolithic (Armit and Finlayson 1992).

For a long time the Mesolithic has been considered the Cinderella of Scottish archaeology (Edwards 1989a). Woodman (1989) suggested that research trends were worrying, with few projects under way, research concentrated on too few sites and an inadequately established chronology. Recent work encourages optimism, with a marked increase in the number of researchers working on dating programmes, large-scale excavations and regional analyses. Research to integrate technological and environmental studies is required to maximize the information retrieved from these investigations. There is no room for complacency, but the period which embraces half the time since glaciers vanished from the landscape is far from being ignored.

8 The Neolithic

GORDON J. BARCLAY

THE IDEA OF THE NEOLITHIC

The origins of the concept

The Neolithic is the period of the first farming communities and, as defined by
Childe (1925) and more recently by Zvelebil (1992), it comprised a package of
traits:

1. the introduction of new food resources (sheep or goat; cereals);
2. new technology (polished stone tools, pottery);
3. new economic practices (agriculture).

Traditionally the Neolithic population, with its different means of subsistence, was
seen as almost entirely foreign, having settled the British Isles from the near
continent. The role of immigrants in the spread of farming economies and the
package of Neolithic traits is now seen as far less important, although it cannot be
dismissed entirely (Kinnes 1994).

   The spread of farming as a way of life across central Europe was relatively rapid.
By 6000 BP (probably around 5000 cal BC) the fertile soils of the major north
European river valleys supported farming communities using long timber houses,
particular types of pottery, cultivated wheat and barley, and domestic cattle, pigs
and sheep. We cannot detect the transplantation of an identical Neolithic culture
from one side of the North Sea and the Channel to the other. It was assumed in the
past that Mesolithic populations continued to exist alongside incoming Neolithic
peoples until such time as their way of life was replaced through the vaguely defined
process of acculturation. However, the inception of the Neolithic in Britain in the
centuries immediately before 4000 cal BC (Kinnes 1985, 1988) certainly involved far
more complex processes than the replacement of one population and way of life by
another. Zvelebil and Rowley-Conwy (1986) have suggested that there are three
stages in the change:

1. the availability of agriculture to hunter–gatherers;
2. the process of substitution of one economic system by the other; and
3. the consolidation of the change, where a 'return' to hunter–gathering becomes
    impossible.

Scotland: Environment and Archaeology, 8000 BC – AD 1000. Edited by Kevin J. Edwards and Ian B. M. Ralston.
© 1997 The editors and contributors. Published in 1997 by John Wiley & Sons Ltd.

Recent trends in the discussion

A decade ago Kinnes (1985) critically appraised the available evidence for the
Neolithic period in Scotland; simultaneously Clarke *et al.* (1985) produced a com-
pelling summary. Darvill (1987) and Parker Pearson (1993) have since summarized
effectively many of the arguments about the nature and origins of the period. Kinnes
(1994) has provided a valuable, if condensed, survey. The recent debate on the
definition of the Neolithic and the processes involved in the change from Mesolithic
to Neolithic, can be characterized at its most uncompromising by the exchanges
between Zvelebil (1989b, 1992; Zvelebil and Rowley-Conwy, 1986) and Thomas
(1987, 1988, 1991). In these works, the range of possible processes of change is
explored, from the movement of people in some numbers, to an indigenous development of agriculture through contact between hunters and people practising some
form of farming. Most recently, Zvelebil (1994) has undertaken a valuable Europe-
wide survey of Mesolithic plant use and tools that might have been involved in the
exploitation of plants. He discusses the varying intensity of plant use, the extent to
which the environment was modified to favour appropriate plants, and the ways in
which these strategies might merge indistinguishably (although with the introduction
of foreign cultivars) into formal agriculture.

Thomas (1991) has since moved further towards Zvelebil's position, suggesting
that the earlier Neolithic population was not living by practising fully developed
formal agriculture, particularly arable cultivation. In his model, the early Neolithic
may be seen as an elaboration of native Mesolithic culture by the gradual adoption
of social structures, ritual practices and economic subsistence strategies. He has
dismissed the cultivation (if any) undertaken by these people as 'transient, hoe-
based horticulture' and 'rather small-scale, garden horticulture' (1991, 21).

He has made his disagreement with the traditional model explicit (Thomas 1991,
28):

> The population of Neolithic Britain:
> [1.]   did not live in major timber-framed buildings,
> [2.]   quite probably did not reside in the same place year-round,
> [3.]   did not go out to labour in great walled fields of waving corn,
> [4.]   were not smitten by over-population or soil decline, and
> [5.]   much of their day-to-day food may have been provided by wild crops.

This statement provides a convenient structure within which to examine the nature
of early farming in northern Britain.

Thomas is referring not only to the early Neolithic, as he goes on to say that
'traces of domestic agriculture are no more common in the later Neolithic' and that
only in the early to mid second millennium cal BC 'did field systems and permanent
domestic structures become the norm in the British Isles' (Thomas 1991, 28). Barrett
(1994) argued for the same pattern, using data mainly from the same part of south-
western Britain. Zvelebil (1992) has criticized Thomas' approach and interpretation
as being regionally restricted (to Orkney and Wessex for the Neolithic) and selective
in its use of rather inadequate evidence. However, Zvelebil himself generalizes,
mixing data and interpretations from across Europe, apparently seeing the Neolithic

as a definable constant across time and space, which it clearly is not (Thomas 1991, 11). Neither approach is helpful in understanding the Neolithic of Scotland, nor its regional variations.

THE NEOLITHIC IN SCOTLAND

Problems of the data

Kinnes (1985) characterized the problems of Neolithic studies in Scotland as 'a recurrent need to derive innovation from without and then to resort to the parochial for explanation and understanding', but even explanation and understanding are too easily imported. The interpretation of relatively poorly understood local data by analogy with better (although not always comparable) data from distant areas has often proved too tempting. In Britain, explanations of the Neolithic have generally relied on models erected using data from Wessex (cf. Thomas 1991), or from Yorkshire or Orkney, the three areas where most work on Neolithic sites has been undertaken. This author would argue that the understanding of archaeological material must first be sought in its regional context, through the erection of regionally valid sequences and interpretations, before drawing on sequences and interpretations developed in distant areas. It is also necessary to be wary of believing that there is a unified entity which can be identified as the 'Scottish Neolithic' (Kinnes 1985, 16); for example Armit and Finlayson (1992) have argued that in the Western Isles the evidence of a gradual transformation to a farming economy contrasts with the pattern elsewhere. Nor can the later Neolithic of lowland Scotland be interpreted uncritically by using, for example, Skara Brae (Figure 8.1) as a model. Scotland is a country of great diversity in landscape and climate, which must surely have been reflected in the variability of the first farming communities.

Far more data have been collected in the last century on Neolithic burial and ceremonial sites than from settlements, but there are problems even with this apparently well-studied material. For historical reasons it has long been perceived that the Neolithic (and indeed much of the archaeology) of Scotland is represented by stone monuments, especially those in the uplands of the North and West. However, aerial photography in the last 20 years has revealed a dense, hitherto unsuspected, distribution of Neolithic and Early Bronze Age timber, gravel, soil and turf monuments, most now ploughed down, but some remarkably well preserved, in lowland east and south-west Scotland (Barclay 1992). As this brief survey of the Neolithic is concerned more with evidence for settlement and economy, and attempts to deal evenly with upland and lowland material, discussion of the distribution and typology of chambered tombs (Henshall 1963, 1972), the archetypal Neolithic monument, has not been included (Plate 8.1). Ashmore (1996) has provided a useful summary of their typology, Kinnes (1985) has reviewed the value of typological analysis, and Barber (1988, 1996) has cast doubt on the reliability of analyses of the chamber contents.

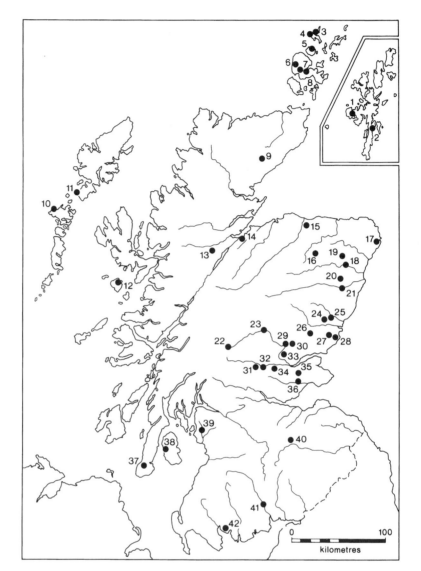

Figure 8.1 Map showing sites mentioned in the text. 1: Scord of Brouster; 2: Shurton Hill; 3: Knap of Howar; 4: Links of Noltland; 5: Rinyo; 6: Skara Brae; 7: Barnhouse; 8: Maes Howe 9: Suisgill; 10: Eilean Dhomnuill; 11: Northton; 12: Kinloch, Rhum; 13: Corrimony; 14: Raigmore; 15: Boghead; 16: Wormy Hillock; 17: Boddam Den; 18: Broomend of Crichie; 19: Castle Fraser; 20: Barmekin of Echt; 21: Balbridie; 22: Creag na Caillich; 23: Pitnacree; 24: Brown Caterthun; 25: Dalladies; 26: Kinalty; 27: Balneaves; 28: Douglasmuir; 29: Cleaven Dyke; 30: Herald Hill; 31: Bennybeg; 32: North Mains; 33: Blairhall; 34: Leadketty; 35: Kinloch, Fife; 36: Balfarg/Balbirnie; 37: Balloch Hill; 38: Machrie Moor; 39: Carwinning Hill; 40: Meldon Bridge; 41: Holywood; 42: Cairnholy

Plate 8.1 The chambered cairn known at Cairnholy I, Stewartry of Kircudbrightshire. The great interpretive value of chambered cairns, once central to Neolithic studies in Scotland, is perhaps now in doubt. Crown Copyright: Historic Scotland

The survival of evidence

As Thomas (1991) notes, little evidence has been found in southern England for Neolithic houses or arable farming. There is impatience amongst archaeologists working on the period: Bradley (1985), Thomas (1991, 8) and Barrett (1994) have all suggested that it is no longer tenable to suggest that traces of Neolithic settlement will eventually be revealed, discounting Fowler's (1981) and Bell's (1983) arguments that the evidence will survive in areas protected from intensive modern land use. The implication is that as no houses or fields have been found, few or none existed (Thomas 1991, 8–9). However, these authors take insufficient account of the evidence for domestic structures that has appeared in other parts of the British Isles and of the difficulties affecting the survival of this kind of material in intensively cultivated areas. Gibson (1992) has recently dealt convincingly with the factors which may have led to the loss of much of the evidence for Neolithic settlement in lowland Britain. The circumstances leading to the survival of the Neolithic settlement site at Lismore Fields, Derbyshire (in a field never ploughed using modern machinery) and its discovery (while searching for traces of a Roman road) amply demonstrate the problems of survival and location of domestic sites in such areas (Garton 1987). Likewise, the cropmark of the enormous earlier Neolithic building at Balbridie, Kincardineshire (Ralston 1982; Fairweather and Ralston 1993) was confidently identified as an early historic structure prior to its excavation.

Plate 8.2   The long barrow known as Herald Hill, Perthshire. Its western end points at the south-eastern terminal of the Cleaven Dyke cursus monument/bank barrow. Crown Copyright: Historic Scotland

THE MONUMENTAL NEOLITHIC

Monuments and society

Our understanding of Neolithic society is based largely on interpretations of burial and ceremonial structures and the changes in practices associated with them. Just as some have sought to push the 'agricultural transformation of the landscape' into the second millennium (e.g. Barrett 1994, 147), there have been recent challenges (e.g. Bradley 1993) to the assumption that the Neolithic and monument building began simultaneously; monuments traditionally seen as being of a sedentary, farming Neolithic, might have grown out of the needs of a hunter–gatherer, traditionally Mesolithic, population.

The end of the earlier Neolithic (in the centuries around 4300 BP; 3000 cal BC) was marked by significant changes in ceremonial and burial architecture which seem to reflect major changes in society. Suggestions of a contemporary decline in the agricultural economy, with the regeneration of scrub and woodland on previously cleared land, are dealt with below and in Chapter 5.

In most of Scotland by this time, communal mortuary structures associated mainly with long earthen mounds (long barrows; Plate 8.2), long cairns (Scott 1992), and in places, round mounds and cairns (e.g. Coles and Simpson 1965), were no longer built or used for burial. The burials of the later Neolithic (where evidence survives, e.g. Corrimony, Inverness-shire [Piggott 1956]) are more likely to be of

Plate 8.3 Professor Richard Bradley's excavations in 1995 on the north-east cairn at Balnuaran of Clava. Radiocarbon dates now suggest these monuments are constructions of the Early Bronze Age: Bradley 2000. Crown Copyright: Historic Scotland

individuals rather than of communal assemblages of, to us, anonymous bone; this has been widely interpreted as reflecting a greater capacity for the representation of individual status in death. Corrimony is a member of the Clava series of monuments, best exemplified by the group of cairns at Balnuaran of Clava, Inverness-shire (Plate 8.3).

The significant foci of ceremonial activity during the later Neolithic are no longer burial sites, but monuments known as henges (Harding 1987; Burl 1991; Plate 8.4 and Figure 8.2). Henges normally comprise a ditch with external bank, the purpose of which may have been to screen the interior from view; there are usually one or two entrances and often there are internal settings of timber or stone uprights. Stonehenge is both the best known and least typical of the class. Enclosures that can be interpreted as being henges or related to the henge tradition vary in diameter from below 10 m to almost 400 m in diameter; the smallest (below 14 m [Harding, 1987] or 30 m in diameter [Wainwright, 1969]) are often called hengiform enclosures. In the map showing the distribution of these sites (Figure 8.2) the dividing line is drawn, perhaps arbitrarily, at 20 m.

The labour input for the construction of a substantial henge is greater than for a burial monument of the earlier Neolithic. Their construction and use might imply a hierarchical society, in which there was a need for large-scale gatherings and which could have organized the large workforce necessary to build them. However, many of the Scottish henges are small, and would have required little more, or even less, effort than the construction of a long mound, and it has been suggested that the largest of the enclosures, in Wessex, may have been built in segments over a prolonged period (Barrett 1994; cf. the Cleaven Dyke, Perthshire, p. 135).

Plate 8.4 The excavated henge monument at North Mains, Strathallan, Perthshire. Crown Copyright: Historic Scotland

The process of change from burial structures of the earlier Neolithic tradition to later Neolithic ceremonial enclosures may have been detected at two sites in Scotland, at Maes Howe in Orkney and at Balfarg in Fife (4 in Figure 8.5). At the former, the tomb, a local variant of the communal burial tradition, is encircled by a ditch and bank which Sharples (1985) has compared to a henge. At Balfarg, a structure possibly used in the preparation of bodies for communal burial in the earlier Neolithic tradition, was, at the end of its use, covered by a low mound of earth and surrounded by a henge: both mound and ditch contained Grooved Ware pottery (Barclay and Russell-White 1993) and the ditch deposits were dated to *c*. 4385 BP (3275–2900 cal BC). Grooved Ware, a type of pottery with flat bases and complex decoration, appeared at the same time as (and was particularly associated with) henges and other features of the apparently changed society of the later Neolithic. While Parker Pearson's statement (1993) that Grooved Ware and henges were 'invented' in Orkney cannot be substantiated, it is clear that the radiocarbon dates for both are earlier in Scotland than in England (MacSween 1992).

Two types of monument whose date-range is still unclear are the bank barrows (exaggeratedly long mounds) and the cursus monuments (Figure 8.3 and Plate 8.6). Cursus monuments (which are probably related to long barrows and mortuary enclosures [Loveday and Petchey 1982]) appeared in England prior to henges, and, it has been suggested, in some way may have presaged their development. All Scottish

Plate 8.5 The long cairn at Auchenlaich near Callendar survives, bisected by a farm road. Its length exceeds 300m. Crown Copyright: Royal Commission on the Ancient and Historical Monuments of Scotland

examples appear as cropmarks, either of ditched cursus monuments, as at Holywood (Dumfriesshire), or – a Scottish variation – as parallel lines of pits, as at Balneaves, in Angus. However, there is one exception: the Cleaven Dyke, apparently a hybrid cursus/bank barrow, which runs for over 2 km. A substantial portion (1.75 km long) survives as an upstanding earthwork: a 9 m wide bank, standing 1–2 m high, running midway between two segmented ditches 50 m apart (Pitts and St Joseph 1985; Barclay and Maxwell 1998). The construction of the Dyke has been dated to before *c*. 3300 BC. This date is comparable with normal earlier Neolithic long barrows. A few other bank barrows and one very long cairn (Plate 8.5) are known in Scotland, but as yet they are undated.

The apparent large scale of the effort necessary in the construction of the cursus monuments may be illusory; the Cleaven Dyke at least may have been built over a prolonged period, in relatively short segments (Barclay and Maxwell 1998).

Regional variation

Sharples (1992a) has suggested that although Neolithic colonization (if that was indeed the mechanism of change) must have begun in much the same way in each region, later diversity would have been caused by environmental, social and cultural factors peculiar to those regions. He presents a number of examples of regional diversity: Orkney, the Western Isles, the Clyde and south-west Scotland.

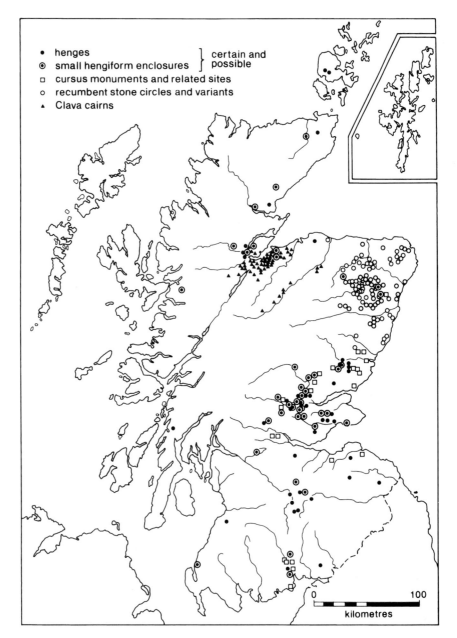

Figure 8.2 Map showing the distribution of certain and possible henges and small hengiform enclosures (less than 20 m in diameter); cursus monuments; Clava cairns (after Henshall 1963) and recumbent stone circles (after Burl 1976)

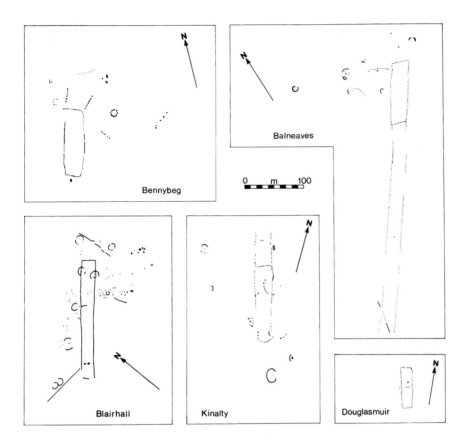

Figure 8.3 Pit-defined enclosures and cursus monuments in Scotland. From computer rectifications prepared by the Royal Commission on the Ancient and Historical Monuments of Scotland

The situation in north-east Scotland provides perhaps the clearest example of regional diversity, at least in traditions of monument building, for this period. While henges are the typical large public monuments of the later Neolithic, a very different type of site is characteristic of north-eastern Scotland: the recumbent stone circle (Plate 8.7 and Figure 8.2; Burl 1976). There are about 100 of these (and variant sites related to the tradition) in the relatively small area of Aberdeenshire, Banffshire and Kincardineshire, in an area with very few normal henges (e.g. Broomend of Crichie and Wormy Hillock, Aberdeenshire). The recumbent stone circles (now generally accepted as having their origins in the later Neolithic [Shepherd 1987]) seem to be a very different sort of ceremonial monument from the conventional henges found over much of lowland Britain at that time. This impression of regional individuality is further strengthened by the distribution of that unusual class of artefact, carved stone balls (Edmonds 1992), which is also weighted heavily towards the same area. The contrast between this area and the

Plate 8.6   An aerial view of the Neolithic enclosure at Douglasmuir, Angus. Crown Copyright: Historic Scotland

Plate 8.7   Recumbent stone circle at Loanhead of Daviot, Aberdeenshire. Crown Copyright: Historic Scotland

coastal plain and Tay and Earn valleys to the south, where probable henges abound, may suggest very different developments in ritual practice in the later Neolithic.

Astronomy, geometry and theocracy

In the late 1960s the role of astronomy and geometry in the construction and use of megalithic sites, based on the work of Thom (1967, 1971), became the subject of debate; a radical re-interpretation of the nature of late Neolithic society was proposed (MacKie 1977, 1993). Ritchie (1982, 1990) has provided characteristically thoughtful and balanced views on some of these discussions.

Most archaeologists working in the period would accept that there is clear evidence for the alignment of elements of sites, in a relatively imprecise way, on lunar rising and setting points, particularly the extremes of their ranges, and also some evidence for solar alignments on the winter solstice (Ruggles 1984), e.g. the alignment of the passage of the chambered cairn at Maes Howe, Orkney. There is general agreement in the worlds of archaeology (e.g. Burl 1980; Ritchie 1990) and of science (e.g. Norris 1988) that the interest of prehistoric peoples in the sun and moon was in ritual observation (i.e. low accuracy, within one or two degrees).

Studies of the geometry of stone circles rely on the application of precise methods of analysis upon monuments, which are often incomplete or which were altered during use, and that are constructed of stones, often rough and irregular in shape. There is no necessity to assume that the complex geometries used by proponents of these beliefs to *describe* the shape of a circle (where the 'best fit' may leave many stones off the geometrical shape) was originally used to *set out* that circle (Barber 1996); it may be that they have fallen into 'the delusions of accuracy' (Huff 1954; Moroney 1965; MacKie 1977, 13).

In 1977, MacKie, drawing on parallels with Mayan civilization, argued that Britain in the late Neolithic was a theocracy, in which an elite of 'wise men, magicians, astronomers, priests, poets, jurists and engineers with all their families, retainers and attendant craftsmen and technicians' (MacKie 1977, 186) lived in major ceremonial complexes and in other special sites (such as all the known Skara Brae type settlements on Orkney). Fed by the efforts of a peasantry living in primitive conditions, this elite undertook precise astronomical observation and set out complex ceremonial sites using advanced geometry and a standard unit of measurement. He rejected the possibility of a sophisticated and capable Neolithic society *without* an elite of priests to organize it, in contradistinction to most workers in this field. The arguments used in 1977 can now be seen to be flawed. Two examples must suffice here. First, there is no evidence that the settlements of the Orcadian Neolithic were anything other than normal settlements of the period in that area, containing buildings of complex domestic, and perhaps ceremonial and religious function, constructed and used by a sophisticated society; only MacKie still argues for the Neolithic as a society largely of primitive peasants. Secondly, there is no unequivocal evidence that the timber structures within the Durrington Walls henge in southern England were large roofed buildings, let alone that they were occupied by priests; both assertions are central to MacKie's thesis, but he largely ignores Musson's clear statement (1971, 363) that there is no evidence to prefer the interpretation that these were roofed structures over any other.

To summarize, there is no evidence for high-precision astronomy and the geometrical arguments are unconvincing; the complex structure of interpretation erected on the astronomical and geometrical arguments is flawed and, furthermore, is unnecessary to explain the data.

RESOURCES, SETTLEMENT AND LANDUSE

The exploitation of resources and the movement of artefacts

During the Neolithic, resources were systematically exploited on a considerable scale for the manufacture of artefacts (Saville 1994a). There are examples in Scotland of the production of both stone axes and flaked stone tools (Sheridan 1992). Four groups of Scottish axehead rock have so far been identified by the petrological analysis of axes (Groups XXII, XXIV, XXXII and XXXIII). Of these only the exact location of Group XXIV has yet been found, at Creag na Caillich in Perthshire (Sheridan 1992). Products of the quarry were widely distributed; one axe has been found as far away as Buckinghamshire. Radiocarbon dating places the early quarrying activity at around 4240 BP (2925–2878 cal BC).

The processes of quarrying and distribution raise many questions about the function, or range of functions, fulfilled by the axes (Bradley and Edmonds 1993). The traditional view of the process of manufacture as quasi-industrial has evolved into an appreciation of its real complexity. For example, both quarries that have been examined in detail (Creag na Caillich and Langdale, Cumbria [Bradley and Edmonds 1988]) are situated in striking locations; it has been observed that the rock is quarried from the least accessible parts of isolated outcrops, suggesting that the choice of quarrying site was not wholly pragmatic. The axes range greatly in size and in quality of finish. Many are too small to have had a function as a cutting or digging implement, or are made of special materials (such as jadeite), are very finely finished and are either unsuitable for actual work or show no signs of having been used. Axe-shaped stones may therefore be understood as both functional and symbolic objects. It is in the latter role that axes may have been distributed over considerable distances, perhaps used in formal exchanges between individuals or groups. Other goods may also have been exchanged – the most striking possibility being the carved stone balls already mentioned (Edmonds 1992). Clear evidence of large-scale late Neolithic flint extraction has been recovered by Saville from the Buchan gravels, at Boddam Den, Aberdeenshire. Here both sides of a small valley had been ravaged by hundreds of intersecting quarry pits (Saville 1994a, b). Radiocarbon dating brackets this activity in one part of the site between *c.* 4550 BP and 3800 BP (*c.* 3500–2000 cal BC).

Other stones suitable for flaking – Arran pitchstone, Rhum bloodstone, chert and quartz – were also exploited (Wickham-Jones 1986; Saville 1994a), but the processes of their distribution are even less well understood.

LAND USE, CULTIVATION AND ENVIRONMENT

The evidence for the processes by which farming became the main economic system is equivocal. The disagreements about the social aspects of the change have been outlined above, and there is also debate about the meaning of the palaeoenvironmental data. This is unfortunate, as the evidence for the changes wrought by human settlement, certainly in the earliest phases of the Neolithic, is almost exclusively palaeoenvironmental. Kinnes (1988) has been critical of the interpretation of the limited evidence. Pollen analysis may reflect in detail only a relatively small area around the sampling site and it has been suggested that clearings in woodland would not impact sufficiently to be recorded through pollen analysis. There is possible evidence for cereal pollen around 5900 BP (Edwards and Hirons 1984; Edwards 1989a) but the significance of these data is uncertain (Chapters 5 and 7).

The most widely recognized environmental event in this period is the elm decline, i.e. the marked reduction in the amount of elm pollen appearing in pollen diagrams. This was formerly explained as the direct consequence of human intervention (e.g. Pennington 1974) such as the feeding of elm leaves to animals or the felling or pollarding of trees. However, human activity alone cannot account for the vast scale of the decline and it is more likely that the reduction was caused by elm disease or a series of contributory factors which also include soil and climate change (Bell and Walker 1992, 162–163).

In the later Neolithic what has become known as the 'late Neolithic agricultural recession' is also in doubt. Whittle described the phenomenon in an influential article (1978), suggesting that there was clear evidence in Ireland and northern and southern England for a regeneration of woodland in the main between 5000 and 4500 BP, with the next advance of clearing not taking place until c. 4000 BP. Subsequently, this interpretation of the pollen data has become accepted as fact. However, there are problems with what these pollen results signify. Edwards, as early as 1979, warned that it was 'rather dangerous to talk of a general third millennium regeneration . . . unless all sites bore a relative constant and known spatial relationship with the human community or communities causing the inferred impact' (Edwards 1979a, 283). As it is difficult to determine how much of the vegetation change of the fourth and third millennia cal BC was caused by human activity (as against natural causes), and how such change would impact on the pollen rain, it cannot be argued that the apparent regeneration of woodland indicated a reduction in that activity. For example, it could be proposed that there was only a change in the pattern and size of settlements, rather than a decrease in the area of land under cultivation. Recent pronouncements from palynologists raise the possibility that the expansion of woodland pollen taxa should not necessarily be taken to signify a reduction in agriculture or population (Gransson 1987; Edwards 1993a; Chapter 5).

As noted above, Thomas (1991) rejected a particular model of Neolithic farming (large timber-framed buildings, great stone-walled fields of corn and so on) which has become familiar through the study of later prehistory. Gibson (1992, 42) has critically examined the assumptions made about the nature of Neolithic settlement – 'a nucleated, self-contained settlement of a type with which the later Bronze Age, Iron Age and Roman-British periods have made us familiar' – which has perhaps provided us with an inappropriate pattern. Thomas therefore dismissed a model of settlement organization that is irrelevant to our study, but does not replace it with one more appropriate to the interpretation of small-scale agriculture. He presents us with only two alternatives: either relatively large-scale intensive agriculture, involving the extensive use of the plough within fields of a kind we would recognize today; or a very transient existence, possibly a form of modified hunter–gathering.

Thomas' assertion that hoe and spade cultivation was incapable of supporting a substantial, fully developed Neolithic society is not sustainable. There is clear evidence in the later Neolithic and the Bronze Age for ridged plots or fields which were probably formed by hoe or spade cultivation (cf. Barclay 1989). Fenton (1974, 43; Gailey and Fenton 1970) has noted, of more recent spade cultivation in Scotland: 'Twelve men using *cas chroms* [= foot spades] could till an acre a day, and a season's work with one from Christmas till late April or mid-May could till enough ground to feed a family of seven or eight with potatoes and meal for a year.' He has also described (1974, 139), in combination with the *cas chrom*, the use of the *ristle*: 'a kind of knife or coulter mounted in a beam, was used to cut slits in the turf to ease the working of the *cas chrom*'.

The 'ploughmarks' and other marks at Links of Noltland from around 2700 cal BC (associated with a boundary ditch: Clarke and Sharples 1985) might as easily be explained by this method of working the ground as by the use of a plough (as might the later combination of 'ard marks' and 'spade marks' found at Suisgill in Sutherland (Barclay 1985, 165–167) dating from the late second millennium cal BC). It has been suggested that much, or all, ard marking resulted from ground breaking, rather than routine cultivation by ploughing (Fowler 1981).

At North Mains, Romans and Robertson (1983a) suggested that a form of cultivation leaving no plough marks had taken place in soils beneath the henge bank at the beginning of the mid third millennium cal BC. There was also later Neolithic or Bronze Age ridging under the adjacent (Bronze Age) mound. At Pitnacree (Coles and Simpson 1965), indirect evidence of cultivation was noted in the form of a very deep sub-barrow soil and the setting at an angle of potsherds and schist fragments on the surface.

The model of the Neolithic rejected by Thomas also included 'fields'. Again it must be considered whether the assumption of permanent, formally laid out fields is conditioned by modern preconceptions and by a failure to take regional differences into account. It is evident from Scotland and other parts of the British Isles that there *were* systems of land division, even in the earlier Neolithic. For example, at Shurton Hill, Shetland, a sub-peat dyke suggests that pasturage boundaries may have been in existence soon after 4750 BP (*c.* 3600 cal BC) (Whittington 1978). Caulfield's researches in the west of Ireland (Caulfield 1978) have shown just how complex systems of land division had become in the later Neolithic (by the early to mid third millennium cal BC), and the work of Whittle *et al.* (1986) at Scord of Brouster has demonstrated the existence of cleared and divided land in the late fourth and early third millennium cal BC in Shetland.

While the extensive formal division of land on a communal basis may be inferred, there are various ways of dividing land, perhaps annually, which can be almost undetectable archaeologically: for example, plots demarcated by lines of small stones (as at Suisgill, Sutherland in the Bronze Age: Barclay 1985), or plots delineated by light hurdle fences for relatively short periods and re-established on different lines (in the later Neolithic at Machrie Moor, Arran [Haggarty 1991]).

Sharples (1992a) has discussed changing patterns of land use in Orkney and the Western Isles, suggesting that less easily cultivated soils were exploited in the later Neolithic, following earlier exploitation of more easily cultivated land. He has proposed a direct relationship between the development of Maes Howe type tombs, Grooved Ware and the economic and social innovations (including the development of larger-scale settlements of the Skara Brae type) which allowed the communal effort necessary to exploit more difficult land.

What was being grown in these fields and plots? Direct evidence for cultivated cereals is limited for both the earlier and later Neolithic. Evidence for both barley and wheat was recovered from the settlement at Knap of Howar, Orkney (Ritchie 1983). At Balfarg, a carbonized barley grain was found incorporated within an earlier Neolithic pottery sherd (Barclay and Russell-White 1993); this was radiocarbon dated to around 4830 BP (3750–3520 cal BC). At Boghead, Moray, around 5200 BP (4000 cal BC) naked six-row barley (*Hordeum hexastichum*) made up 88% of the cereal grains, and emmer wheat (*Triticum dicoccum*) 11% (Maclean and Rowley-Conwy 1984). Emmer had declined to 8.4% later in the Neolithic at Skara Brae and the decline continued into the earlier Bronze Age (Maclean and Rowley-Conwy 1984). The actual size of emmer grains decreased at the same time – a sign of poor adaptation to the northern climate. Hulled barley (*Hordeum vulgare*), not represented at Boghead, was found at Skara Brae, and a gradual replacement of the naked form by the hulled has been noted generally (Maclean and Rowley-Conwy 1984). The material from the timber building at Balbridie, broadly contemporary with Boghead, has recently added considerably to our knowledge (Fairweather and Ralston 1993). Emmer wheat made up a large component (almost 80%) of the assemblage, naked barley 18% and bread wheat (*T. aestivum*) 2%. However, in one posthole the proportion of bread wheat was 76%, showing the difficulties inherent in examining limited samples of cleaned crops.

Evidence for managed pasture is preserved under the long barrow at Dalladies, Kincardineshire (Piggott 1972), which was built on long-established grassland. Indeed, 0.75 ha of this pasture was sacrificed in the turf dug to build the mound (Piggott 1972, 45–46).

There is evidence for the use of other non-food plants, e.g. cultivated flax (as at Balbridie [Fairweather and Ralston 1993]). At Balfarg Riding School, Fife, one of the larger Grooved Ware vessels contained a substance based on black henbane (*Hyoscyamus niger*; a member of the hemlock family), perhaps used as an hallucinogen (Moffat in Barclay and Russell-White 1993; Plate 8.8). At Kinloch, Rhum, evidence was found in organic residues on pottery for a cereal-based (possibly alcoholic) drink (Wickham-Jones 1990). It is interesting to note suggestions that some patterns in Neolithic decoration, including perhaps those on Grooved Ware, may originate in patterns seen in states brought on by using hallucinogens (Lewis-Williams and Dowson 1993).

Local evidence for the management of woodland is limited and indirect (e.g. in the use of timber in the construction of monuments). Nowhere as yet is there the quality and quantity of artefactual and environmental evidence found in the Somerset levels (Coles and Coles 1986).

Plate 8.8 A sherd of Grooved Ware with residues of its contents. On analysis these proved to contain black henbane, which causes hallucinations, amongst other symptoms. Crown Copyright: Historic Scotland

Few significant faunal assemblages have been published, and it is only possible to point to the presence/absence of species, and to make generalizations about the proportional representation of different species. The assemblages from individual sites are mentioned below.

There is no direct evidence of transhumance in the Neolithic. Bradley *et al.* (1993, 278) suggest that their work on the relationship of rock art to the landscapes in which it is found provides evidence of 'an essentially mobile pattern of landuse'. Simpler patterns of carvings are found in lowland areas; more complex patterns are grouped in upland areas, around basins or waterholes, or on isolated hilltops. It is possible that some of the complex patterns are related to and produced during the use of summer grazings.

HOUSES, ENCLOSURES AND ECONOMY: A SUMMARY

There is consistent, but as yet limited, evidence that the people of the earlier Neolithic in Scotland generally lived in small rectangular houses (Figures 8.4 and 8.5) (Barclay 1996). Where the evidence survives, this picture is replicated in Ireland and England. Armit's excavations at Eilean Dhomnuill, North Uist (Figure 5.3; Armit 1988, 1992a), have provided evidence of rectilinear houses measuring 6.5 m × 4 m and 4 m × 3 m internally and probably dating to the earlier fourth millennium BC. They are similar to those found at Knap of Howar (1 in Figure 8.4), measuring 7.5 m × 3 m and 10 m × 4.5 m internally (Ritchie 1983), together with

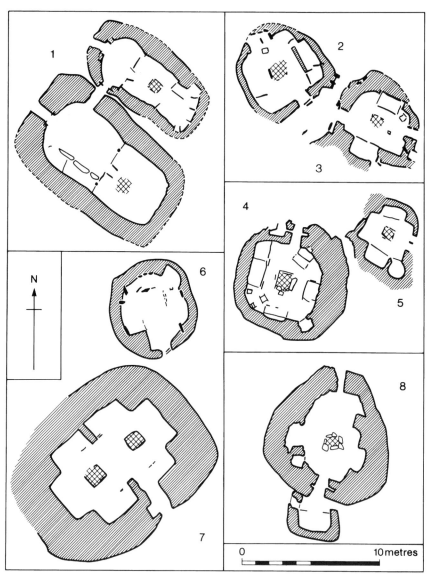

Figure 8.4 Orkney buildings. 1: Knap of Howar, buildings 1 and 2; 2: Rinyo, building A; 3: Rinyo, building G; 4: Skara Brae, building 7; 5: Skara Brae, building 9; 6: Barnhouse, building 2; 7: Barnhouse, building 3; 8: Skara Brae, building 8. The cross-hatched areas are hearths

evidence of an economy based on arable agriculture in the form of cereal grains and querns and on a wide range of wild resources. Kinnes (1985, 27) has expressed doubts about the relationship between the houses at Knap and the midden material into which they were dug (and to which the radiocarbon dates may relate), but the excavator (A. Ritchie, pers. comm., 1993) argues that these doubts have not been substantiated.

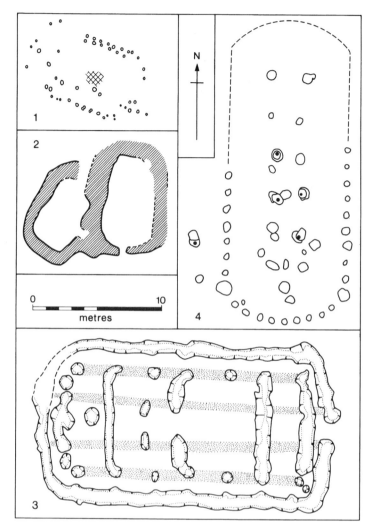

Figure 8.5 Structures on the mainland and the Western Isles. 1: Raigmore, Inverness-shire; 2: Eilean Domhnuill; 3: Balbridie, Kincardineshire; 4: Structure 2 at Balfarg, Fife – this is not a roofed building (the black dots show where one post can be proved to have replaced another). The open and filled spots are post-holes; in Balbridie the defined areas are post-holes and wall-slots and the toned areas are suggested rafter lines

These structures invite comparison with those at Ballyglass in Ireland (measuring 7.4 m × 6.4 m) (O'Nuallain 1972) and the recently excavated house at Tankardstown in Co Limerick (Gowen 1988), both of which have produced dates around 5200 BP (*c.* 4000 cal BC). The houses at Lismore Fields (Garton 1987) are of similar dimensions. These recall aspects of the Neolithic timber houses of continental Europe (Ilett 1980).

The massive building at Balbridie (2 in Figure 8.5; Plate 8.9; Ralston 1982; Fairweather and Ralston 1993) has to date no excavated parallel, either for scale

Plate 8.9   The building at Balbridie, Kincardineshire, under excavation. Crown Copyright: Aberdeen Archaeological Surveys

(24 m long and 10 m broad) or construction. Radiocarbon dating puts the building in the early/mid fourth millennium cal BC. Broadly comparable cropmark sites are not known but some are likely to be of later date or different function (e.g. the Balfarg, Fife, timber structure, 4 in Figure 8.5; Barclay and Russell-White 1993). Fairweather and Ralston (1993, 321) comment that 'the farmers of Balbridie were – in terms of their building and, it would seem, of their strategy with cereals – closer to continental European practice than has normally been identified in the British Isles'.

There is as yet no certain evidence for large-scale Neolithic enclosures to compare with the causewayed enclosures of southern Britain of the period *c.* 5100–4500 BP (*c.* 4000–3100 cal BC). The promontory enclosed by a massive palisade at Meldon Bridge, Peeblesshire, may have a domestic aspect but, in the final report, activity there is considered to have been primarily ceremonial (Speak and Burgess 1999). There are hints of enclosures at Balloch Hill, Argyll (associated with Neolithic pottery: Peltenburg 1982), and at Carwinning Hill Ayrshire (Cowie 1979) where causewayed ditches were recorded under later hillforts. The excavation of a probably domestic enclosure at Kinlock Farm, Fife (J. W. Barber 1982a), has suggested there may also be a tradition of enclosed Neolithic stettlement in eastern Scotland yet to be explored. In the cropmark record there are possible causewayed sites, such as Leadketty, Perthshire (RCAHMS 1994, 40). A number of complex multivallate hilltop enclosures, such as the earthwork element of the Barmekin of Echt, Aberdeenshire (Feachem 1966, 72–73), traditionally dated to the Iron Age, may be Neolithic in date. Their defences are pierced by many gaps in both bank and ditch, in contrast with the more normal hillforts of the area, and they bear a close resemblance in plan to the casewayed camps of southern Britain.

Anna Ritchie's (1983) excavation of the Neolithic settlement at Knap of Howar has provided a useful picture of the nature of settlement and range of resources being exploited in the later fourth millennium cal BC. There is evidence of cereal cultivation, surviving both as grains and as pollen, and of cattle and sheep or goat. There is also evidence for some pig-keeping, limited use of wild animals (deer, seal, whale and otter), and more intensive exploitation of sea birds, fish and shellfish, discussed in more detail in Chapter 6. An even greater use of wild resources is indicated at Northton, Harris, where 14 wild species were found (Simpson 1976) and at Noltland, Orkney, where 15 deer skeletons were recovered (Kinnes 1985, 30). At Knap of Howar there is indirect evidence of the collection of seaweed, perhaps as manure or food (for animals or humans).

A model that might be useful in the interpretation of the available information is crofting, as operated by communities in north and west Scotland in the recent past and, in other forms of broad spectrum, intensive resource use, by peasant agricultural communities elsewhere in Europe. While crofting was a deliberate product of changes in land tenure during the late nineteenth century AD (Hunter 1976) and involved the cultivation of the potato, this model of a small scale, intensive, subsistence economy utilizing a wide range of resources may be more helpful than comparisons with later prehistoric agricultural systems in Wessex.

It might be suggested that the pattern of agricultural economy throughout much of Scotland from the earliest Neolithic differed from that dismissed by Thomas, in the following ways. The population of earlier Neolithic Britain:

1. lived in light timber houses (cf. Lismore Fields) (or, where timber was not the most readily available building material, stone [cf. Knap of Howar]), which should not be dismissed as impermanent; in some places (e.g. Balbridie) larger structures were in use;
2. resided in one area, probably based in permanent settlements, but possibly with some of the population moving seasonally, to summer grazings or fishings;
3. worked for part of the year in productive hoe- and spade-, if not ard-cultivated plots, perhaps of considerable extent; the organization, size and boundary structures of such plots or fields might vary widely, from permanent arrangements to plots defined by shifting hurdles or even slighter demarcations, depending on local practice and land tenure arrangements; pasture was managed and enclosed;
4. used locally-available wild resources intensively for food, manure or oil (from sea birds);
5. managed herds of cattle and sheep or goats (which could be moved to summer grazings) and pigs (which probably could not [Piggott 1981]).

Thomas (1991) suggests that there is little reality in the soil conservation problems often attributed to Neolithic farming; the evidence for this and for a late Neolithic 'agricultural recession' in Scotland, involving the abandonment of cleared land (perhaps because of the exhaustion of those soils which had been the first cleared and colonized) and regeneration of woodland on it, is as yet either absent or debatable. This issue is considered more fully by Edwards in Chapter 5.

For the later Neolithic, Orkney provides the best settlement evidence, in particular from the sites at Rinyo (2 and 3 in Figure 8.4; Clarke 1983), Skara Brae

(4 and 6 in Figure 8.4; Clarke 1976a,b), Noltland (Clarke *et al.* 1978) and Barnhouse (7 and 8 in Figure 8.4; Richards 1992). The norm seems to have been relatively large-scale, communally based settlements which were occupied for long periods by people with a rich material culture and who practised an economy that incorporated both mixed agriculture and intensive exploitation of wild resources. The major difference from the earlier Neolithic seems to lie, in some areas, in the more communal organization of settlement and agriculture, reflected for example in the arrangement of the Skara Brae houses, and the shared effort necessary to construct the field systems of Ireland in the later Neolithic. The less-nucleated later Neolithic settlement at Scord of Brouster in Shetland (Whittle *et al.* 1986) perhaps suggests that the process of nucleation was not constant. However, it might be suggested that in general the differences already visible to archaeologists in the organization of ceremonial and burial monuments can also be detected in the organization of settlement and economy in the later Neolithic. These sites all lie in a limited geographical area – the uplands and the islands. Elsewhere in Britain the accidents of preservation have revealed only limited evidence; for example two wooden buildings under later burial mounds at Trelystan, Powys (Britnell 1982). The ground plans of these buildings are strikingly similar to those at Skara Brae, although the building medium is less substantial. While similar structures may remain to be found in lowland areas, it is probable that they will only survive and be discovered by chance.

It may be suggested that the people of later Neolithic Scotland:

1. were more likely to live in larger-scale communally arranged settlements;
2. may have worked less easily cultivated but more productive soils, in a more communal arrangement of land-holding, such as the field systems of western Ireland. The extent and complexity of the communalization of land use may have varied considerably from area to area;
3. probably continued to exploit locally available wild resources to different degrees of intensity;
4. managed herds of cattle and sheep, although the increase in pig numbers might indicate a reduction in the proportion of stock suitable for transhumance.

Into the Bronze Age

The chapters on the Neolithic and the Bronze Age divide with the appearance of Beaker pottery. However, the traditions of ceremonial and burial activity continued and Beaker pottery appeared on many sites which had already been in use for over 1000 years.

9 The Bronze Age

TREVOR G. COWIE AND IAN A. G. SHEPHERD

INTRODUCTION

This review will outline selected themes of the period from approximately the mid-third millennium cal BC until roughly the eighth/seventh century cal BC. This period commences with the introduction of the earliest metalwork, in the form of copper artefacts, probably in circulation among 'Neolithic' communities in Scotland prior to the formal introduction of metallurgy itself, and closes with the adoption of iron as the principal raw material for edged weapons and tools.

Conventionally, this encompasses the 'Bronze Age', but, as has long been recognized, this term must now simply be seen as a convenient shorthand. By and large, the technological changes which it and similar terms reflect do not appear to have coincided with major changes in settlement and economy, or archaeologically detectable social and political upheavals; however, alternative terminology suggested for prehistoric periods in the British Isles has either not been widely adopted (e.g. Burgess 1980) or is inappropriate for general use (Parker Pearson 1993, 125–134: 'Age of Sacred Landscapes', 'Age of Land Divisions').

For much of the period under review, the legacy of some two centuries of discovery and excavation consists principally of artefacts and data relating to ceremonial and funerary practices, and this has coloured most accounts of the period. As a result of intensive field survey over the last two or three decades, however, the balance has begun to shift in favour of achieving a fuller understanding of Bronze Age settlement and economy. The biased and partial nature of the evidence currently precludes a review in the form of a narrative prehistory, but it is hoped that the 'snapshot' approach adopted here will provide a flavour of the period and its potential for further research (Figure 9.1).

LANDSCAPES OF THE DEAD

In the lowland zone – in effect, the present-day agricultural heartlands of the country – centuries of intensive farming have resulted in the reduction or obliteration of most of the upstanding archaeological monuments of the Bronze Age. It is primarily from such areas that the bulk of the evidence that has traditionally

Scotland: Environment and Archaeology, 8000 BC – AD 1000. Edited by Kevin J. Edwards and Ian B. M. Ralston.
© 1997 The editors and contributors. Published in 1997 by John Wiley & Sons Ltd.

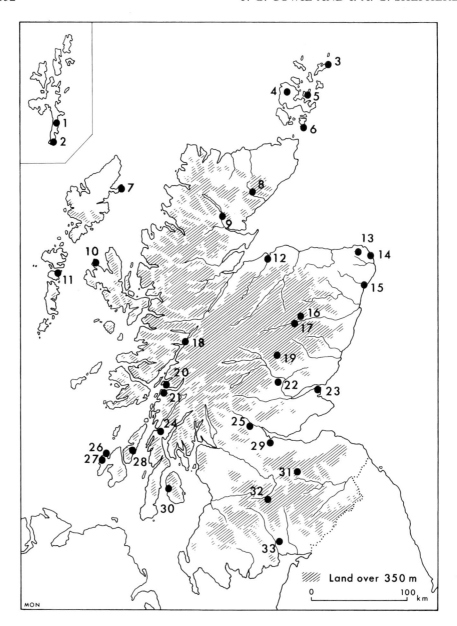

Figure 9.1 Map showing sites mentioned in text. 1: Catpund; 2: Jarlshof; 3: Tofts Ness; 4: Beaquoy; 5: White Moss, Shapinsay; 6: Liddle; 7: Sheshader; 8: Upper Suisgill; 9: Lairg; 10: Waternish; 11: Rosinish; 12: Tulloch Wood; 13: Memsie; 14: Rattray; 15: Sands of Forvie; 16: Braeroddach Loch; 17: Pass of Ballater; 18: Dail na Caraidh; 19: Balnabroich; 20: Black Moss of Achnacree; 21: Loch Nell; 22: Newmill; 23: Tentsmuir; 24: Kilmartin valley; 25: Blair Drummond; 26: Ardnave; 27: An Sithean; 28: Cùl a' Bhaile; 29: Myrehead; 30: Tormore; 31: Green Knowe; 32: Lintshie Gutter; 33: Catherinefield Farm. Drawn by Marion O'Neil

provided the stuff of the Bronze Age (e.g. the cists, urn burials, single finds and hoards of metalwork) has been recovered. Settlement evidence of the Early Bronze Age is virtually unknown from lowland Scotland apart from largely unstratified collections of material from coastal dune systems (e.g. Tentsmuir, Fife: Longworth 1967).

Aerial survey now holds the key to an understanding of patterns of Bronze Age settlement in the lowland zone (Barclay 1992). Campaigns of aerial photography since the 1970s have revealed remarkably detailed cropmark complexes which must embrace the settlement record but they remain as yet very largely uninterpreted (with the significant exception of the recent survey of south-east Perth: RCAHMS 1994). Relatively few such complexes have been the subject of systematic ground-survey and excavation, attention having focused on unitary sites subject to specific threats.

The artefactual evidence suggests that certain parts of the country may have been richer core areas during the Bronze Age than others, e.g. the Moray plain, central and north Aberdeenshire, Strathmore, Fife, the Lothian coastal plain and Mid Argyll (the importance of relatively fertile soils possibly enhanced, in the last case, by the proximity of ore sources). It is these areas that produce comparatively wealthy assemblages of prestige grave goods, e.g. bronze daggers or ornaments such as bronze bracelets or jet necklaces. Some high status objects appear to represent the provincial counterparts of types known from the rich Wessex graves of southern England (Clarke *et al.* 1985, 157–158). The paradox, for which there is as yet no satisafactory answer, is that the settlement record is virtually unknown in these areas. The answer may simply be that, in what have always been Scotland's farming heartlands, the traces of timber-built structures have been obliterated by intensive ploughing, over the last two centuries if not since the Medieval period; but until a representative sample of cropmark complexes is explored, the character of lowland Bronze Age settlement will remain unresolved.

Bradley (1984, 70) has suggested, admittedly with southern British evidence in mind, that the earlier Bronze Age settlement record is hazy precisely because the built environment of contemporary communities consisted primarily of structures asociated with the dead. Striking examples of ritual landscapes still survive, as in the Kilmartin valley, Argyll (RCAHMS 1988). In the lowlands, such survival is rare; here, antiquarian accounts of the destruction of some large burial mounds and cairns hint at what has been lost (e.g. Memsie, Aberdeenshire: Wilson 1851, 434). It has also been suggested that if the subsistence economy was geared towards pastoralism in the earlier Bronze Age, many or most settlements may have been relatively impermanent. The Scottish evidence is currently too restricted to bear out such interpretation.

Prehistoric societies were more static than previously believed (e.g. Childe 1935) but that does not mean they were stagnant or lacking all outside contact. While waves of immigrant Beaker folk and other colonists are no longer countenanced, some movements of people (e.g. early Beaker users) may have been a feature of the opening phase of the Bronze Age, a phase of immigration bringing new technology, new ideas and perhaps new animal bloodstock.

Even if the mechanisms for the procurement of goods and raw materials are uncertain, contacts between communities appear to have been widespread. Long-

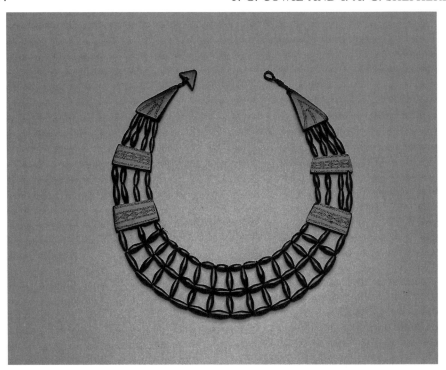

Plate 9.1 Jet necklace from Pitkennedy, Angus. The finest specimens of such necklaces were imported from eastern Yorkshire. ©: The Trustees of the National Museums of Scotland 1997

distance contacts may have had to be established, or periodically re-established, in order to procure finished artefacts or the raw materials required for various crafts (e.g. metallurgy, owing to the limited distribution of ore sources particularly tin: cf. Scott 1951). Current work on the jet necklaces found in some of the richer Bronze Age graves is proving particularly revealing, showing how the finest specimens are made from jet from eastern Yorkshire, and probably represent actual imports from that region, while locally available raw materials were subsequently used to repair or replace broken components or in imitations (Shepherd in Clarke *et al.* 1985, 204–216; Davis and Sheridan 1993; Plate 9.1). The utilization of Shetland's steatite sources (principally Catpund) for massive cinerary urns found in the Orcadian Bronze Age provides an illuminating example of contact driven by ritual requirements.

Burials are of course also the primary source of information about physical aspects of the population, ranging from issues of gender and life expectancy to physique (Plates 9.2 and 9.3) and life-style (see Bruce in Shepherd (1986, 17–22) for discussion of possible indications of archery and horse-riding). In the past, skull type was invoked as evidence of a 'Beaker folk', characterized by brachycephalic or 'round-headed' skulls in contrast to an indigenous 'long-headed' Neolithic population. The significance of these differences is not entirely clear but would not automatically now be attributed to the arrival of an immigrant population. The

Plate 9.2 Bronze Age burial from Cnip Headland, Uig, Lewis. The grave contained the flexed remains of the skeleton, with a plain pottery vessel by the skull. Mid-second millennium BC. Crown Copyright: reproduced by permission of Historic Scotland

Scottish data set is one of the best available, with the qualification that the data derive from that almost certainly limited section of the population accorded formal burial. In passing, it may be noted that palynological investigations have thrown intriguing light on the funerary rituals associated with certain cist burials; there is convincing evidence that, during the flowering season at least, floral tributes, particularly meadowsweet, were placed with the deceased (Whittington 1993; Tipping 1994b).

Long-distance travel is likely to have been waterborne or on foot. Evidence from southern Britain ranges from wreck-sites in the English Channel to the remains of various types of craft (Muckelroy 1981; McGrail 1993), but as yet there is little direct evidence for water transport from Scotland during the period under review apart from a log-boat from Catherinefield Farm, Locharbriggs, near Dumfries,

Plate 9.3 Reconstruction of the facial features of the adult male, aged about 35–40 years, whose skeleton was found in the cist burial shown in Plate 9.2. Copyright: T. Cowie and I. MacLeod

which has been dated to 3754±125 BP, (SRR-326) 2400–1970 cal BC (Jardine and Masters 1977). However, maritime and riverine activity can safely be inferred, not least because certain artefact types do indicate links that must have involved sea crossings. An early beaker from Newmill, Perthshire, is virtually identical to examples from the Netherlands (Watkins and Shepherd 1980), while many of the later beakers from the North-East bear close resemblances to Dutch Veluwe beakers (Lanting and van der Waals 1972, 31).

As a result of a programme of direct radiocarbon dating of organic artefacts, there is now evidence that wheeled transport was known by the end of the second millennium BC. Three wooden disc-wheels were found in the nineteenth century at Blair Drummond Moss in the Forth Valley: a sample from the surviving one has been dated to 2810±85 BP (OxA-3538), 1255–815 cal BC (calibrated after Hedges *et al.* 1993, 156) making this currently the earliest evidence for wheeled transport in

Plate 9.4   The remains of the wooden disc wheel found at Blair Drummond Moss in the Forth Valley. With a radiocarbon date of 1255–815 cal BC, this is currently the earliest direct evidence for wheeled transport in Britain and Ireland. ©: The Trustees of the National Museums of Scotland 1997

the British Isles. The wheel (made from ash wood) is an example of a type well-known on the continent, and is likely to derive from a cumbersome cart or waggon (Plate 9.4). Although there is no way of knowing how common they were, and such vehicles are unikely to have been suited to all types of terrain, the find evokes an aspect of the Scottish prehistoric agrarian scene that would not immediately come to mind. Antiquarian accounts also indicate the discovery of lengths of wooden track within Blair Drummond Moss; these are of uncertain date but inevitably recall the trackways recorded in the Somerset Levels (Coles and Coles 1986) and in Ireland (Raftery 1990).

LANDSCAPES OF THE LIVING: SETTLEMENT AND ECONOMY

The loss of sites in the lowlands is partially compensated for by the remarkable survival of monuments in what is today agriculturally marginal land (Stevenson

1975). The term 'uplands' embraces generally marginal zones from about 180–245 m up to 600 m; it is used here as a simple 'broad brush' term and includes much topographical variation at both local and regional levels. Contraction of settlement in recent centuries has frequently resulted in the survival of entire ancient landscapes – particularly settlement evidence principally relating to the second/first millennia BC.

The fieldwork that has revolutionized our understanding of these areas ranges from strategic survey undertaken by the RCAHMS in Perthshire (RCAHMS 1990; see also Halliday and Stevenson 1991) to survey in advance of the afforestation which continues to pose the main long-term threat to this resource (Proudfoot 1989). The surviving evidence is very uneven, reflecting modern patterns of fieldwork and site detection rather than the prehistoric picture. The recent study of lowland south-east Perth (RCAHMS 1994) provides a vital contrast to that of upland north-east Perth (RCAHMS 1990).

Individual monuments (e.g. hut circles, clearance cairns) have been recognized for a long time but the potential for recovery of complete areas of preserved landscape has been underestimated; then too, much had in the past been attributed to a later prehistoric, generally Iron Age, date. The unitary monuments can now be seen to be integrated with extensive remains of managed landscapes (Plate 9.5 and Figure 9.2), and with this has come recognition of the interlocking nature of so much of the evidence: such landscapes have to be analysed as a whole through time (as at Lairg, Sutherland, e.g. McCullagh 1992a,b; see also RCAHMS Afforestable Land Surveys, e.g. Waternish, Skye and Kildonan, Sutherland: RCAHMS 1993a,b).

In general, many hut circle groups, integrated with field systems, may now be presumed to straddle the second and first millennia BC. In most areas, the remains of houses are represented in the field by 'hut circles' – a somewhat misleading term as it carries with it with connotations of squalor and impermanence, whereas excavation has revealed complex structures fully intended for occupation. A number of regional varieties are known.

It is also always necessary to bear in mind that features visible on the surface may mask a range of evidence only detectable on excavation. Excavations at Lairg, Sutherland, for example, uncovered earlier timber structures (Plate 9.6 and Figure 9.3), while investigation of hut circles at Cùl a'Bhaile on Jura (Stevenson 1984) and at Tormore on Arran (J. W. Barber 1982b) revealed that both had undergone several phases of remodelling or rebuilding. The impression of nucleation at some hut circle complexes is therefore potentially misleading: not all structures visible on the surface need have been occupied at the same time, and, much like modern farmsteadings, may reflect complex processes of growth and decay, and changes of function and abandonment.

Unenclosed platform settlements – the remains of groups of house stances terraced into slopes – had for some time been thought likely to fill a gap in the settlement record: excavations at Green Knowe, Peeblesshire (Jobey 1980) confirmed this, with a series of dates suggesting a *floruit* for that site in the later second millennium BC. Recent excavations at Lintshie Gutter in Lanarkshire (Terry 1991) indicate that such sites (which in any case reflect a response to local topographical conditions rather than a specific architectural form) may be as early as the first half of the millennium.

Plate 9.5 Balnabroich, Strathardle, Perthshire. Hut-circles, field systems and small cairns showing from the air under snow cover. The vulnerability of such relict landscapes is clear from the adjacent coniferous plantations. Crown Copyright: Royal Commission on the Ancient and Historical Monuments of Scotland

A more clustered settlement type is found at Jarlshof, Shetland (Hamilton 1956), which is a multi-period site including remains of a late Bronze Age metalworker's workshop. Also in the Northern Isles, burnt mounds are a very common type of field monument whose existence has for long been known, but it is only relatively recently that the type has been scientifically dated. Now recognized as being of second millennium date, these fill a gap in the settlement record in many areas. The principal excavated sites are Liddle Farm, Orkney and Beaquoy, Orkney (Hedges 1975). Well over 800 have now been recorded in Scotland (Halliday 1990) and the type is now known widely distributed over the British Isles, although their date range is not necessarily restricted to the Bronze Age everywhere (Buckley 1990).

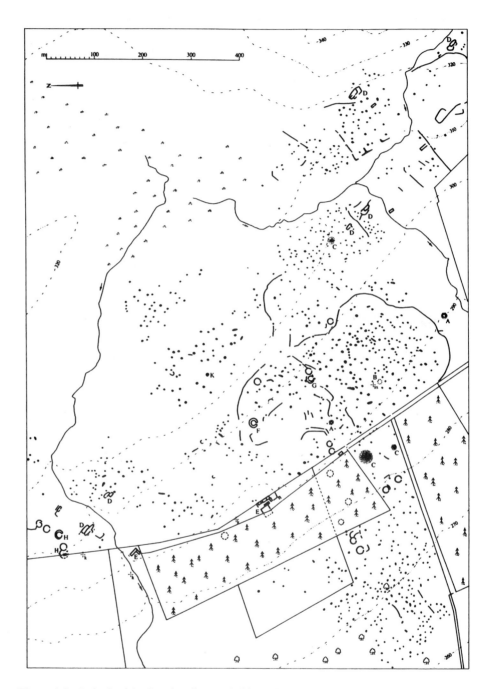

Figure 9.2 Balnabroich, Strathardle, Perthshire. The complex nature of the archaeological landscape as revealed by detailed survey. Burial cairns, ritual monuments, hut-circles, field systems and field clearance cairns, and later rectilinear buildings indicate patterns of landuse and settlement of considerable chronological depth. Crown Copyright: Royal Commission on the Ancient and Historical Monuments of Scotland

Plate 9.6 Large roundhouse in the course of excavation, Lairg, Sutherland. Crown Copyright: reproduced by permission of Historic Scotland

The grassy swards of alkaline-rich shell-sand along the western and northern seaboards of Scotland, collectively known as machair, have always proved attractive for settlement. Such areas of soft coastline are inherently vulnerable to erosion, resulting in a biased picture of the settlement pattern. Erosion has frequently favoured site- or artefact-recognition, but against this has to be set the possible loss of sites through geomorphological change in the past (either as a result of deflation or marine transgression: Ritchie 1979). A number of small domestic sites of second millennium BC date have been investigated around the western and northern seaboards of Scotland, ranging from Ardnave on Islay (Ritchie and Welfare 1983) to Tofts Ness on Sanday, Orkney (Dockrill 1987). In mainland Scotland, too, coastal sand-dune systems have also produced plentiful evidence for early settlement, as at the Sands of Forvie, Aberdeenshire (Ralston 1980). In all such areas, human settlement involved the exploitation of inherently fragile or unstable surfaces vulnerable to sand-blow, and there is evidence that prehistoric occupation may indeed have been a contributory factor to episodes of erosion.

Although the picture is uneven, some general points can be made regarding the main trends in settlement in the second millennium/early first millennium BC. Where settlements have been investigated, they are invariably unenclosed (in the sense of undefended), and often superficially simple field monuments mask sites of

Figure 9.3 Reconstruction of the roundhouse shown in Plate 9.6. Drawn by Christina Unwin. Crown Copyright: by permission of Historic Scotland

considerable complexity and longevity. In many areas, there are no longer grounds for talking of hiatuses or gaps in the record: the problem is now simply one of archaeological visibility. In the first millennium, by contrast, the settlement record is characterized by a trend towards enclosed settlement, with the appearance of stockaded enclosures and early hillforts with timber-laced defences by the eighth/ seventh centuries cal BC.

SUBSISTENCE ECONOMY

The body of evidence for prehistoric agriculture in Scotland is growing (see reviews by Halliday *et al.* 1981; Halliday 1993). It ranges from surviving field systems and enclosures to actual traces of cultivation, from plant and animal remains to cultivating implements. Just as surface remains of houses or hut circles may mask a far more complex picture, so the evidence of field systems may be much more complicated than surface indications would suggest. Careful field survey can disentangle the main strands and set up models for testing by excavation. The need for caution, however, was demonstrated by excavations at An Sithean, Islay, where a range of types of evidence (including stratification, radiocarbon dating, pedo-genesis and pollen analysis) showed that elements of what was superficially a single field system were all of different dates, reflecting phases of refurbishment and reuse

Plate 9.7 Beaker period ard marks revealed in the course of excavation at Rosinish, Benbecula. Copyright: Ian Shepherd

of a restricted area of cultivable land from the Late Bronze Age to the post-Medieval period (Barber and Brown 1984).

The reasons for enclosure include demarcation, protection and organization of land use. The scale of field walls revealed by survey or excavation often seems unsuitable as barriers to the movement of stock, and it has been suggested that they may have been augmented by hedges (Barber and Brown 1984, 186). There is little direct evidence of the use of organic materials but a salutary reminder is provided by the discovery of a stretch of burnt hurdling at Rattray, Aberdeenshire (Murray *et al.* 1993); and of course, control of stock can be achieved by herding or direct supervision. While there is no evidence of the massive organization of the landscape on the scale of the Dartmoor reave systems (Fleming 1988), some land division may have involved a greater degree of communal planning or control (e.g. the so-called 'treb dykes' in Orkney (Lamb 1980, 9), or the rare example of a co-axial field system at Tulloch Wood, Forres, Moray (Carter 1993).

The principal cultivating implements appear to have been simple wooden ploughs of the type known as ards, often tipped with stone points to penetrate the soil and prolong the life of the implement. Actual traces of cultivation have been recovered at an increasing number of sites: published examples include Rosinish (Shepherd and Tuckwell 1977; Plate 9.7), Cùl a'Bhaile, Jura (Stevenson 1984) and Rattray, Aberdeenshire (Murray *et al.* 1993). Evidence of manuring is widely attested, a practice that is at least as early as the Late Neolithic. At Rosinish on Benbecula, midden deposits may have been applied to cultivated areas in an attempt to consolidate surfaces undergoing wind erosion (Shepherd and Tuckwell 1977; Shepherd

Plate 9.8   This ox yoke was found in a bog at Loch Nell, Argyll. Radiocarbon dated to *c.* 1950–1525 cal BC, it is currently the earliest known example from Britain and Ireland. ©: The Trustees of the National Museums of Scotland 1997

1981). The remains of cultivating tools and implements have been recovered from a number of sites: the fullest evidence comes from the Northern Isles where stone shares have been recovered in large quantities and studied in detail by Rees (1979, 1981).

Animal traction was almost certainly employed during ploughing: a wooden ox yoke found late last century in a bog at Loch Nell, Argyll has recently been radiocarbon dated to 3430±85 BP (OxA-3541), *c.* 1950–1525 cal BC, making this currently the earliest dated yoke from Britain and Ireland (Hedges *et al.* 1993, 156; Plate 9.8). A further example of a wooden yoke, found in peat at White Moss, Shapinsay, Orkney, has been dated to *c.* 1516–1253 cal BC. This has been interpreted as a short yoke (or *skammjok*) or as a swingle-tree and could have been used for either ploughing or wheeled transport (Hedges *et al.* 1993). On lighter soils especially, such as the machairs of western Scotland, human traction may also have been employed after the manner of the highland *cas chrom*. Although no actual examples of spades of Bronze Age date are known from Scotland, their use can safely be inferred from the nature of cultivation ridges on a number of sites and from distinctive soil-marks recoverable by careful excavation (Shepherd 1976, 214, plate 11.VI).

The cereal evidence is dominated by barley; although occasional grains of emmer wheat are recovered (e.g. at Rosinish, Benbecula), by the second millennium this crop may have been at the very limits of its environmental range (Maclean and Rowley-Conwy 1984). Analysis of macro-plant remains has thrown light on crop-husbandry and processing techniques (e.g. Milles in Whittle *et al.* 1986, 119–124). Palaeobotanical analyses have also provided evidence for crop pests: these include an early record of ergot among the cereal remains from the Late Bronze Age settlement at Myrehead, West Lothian (Barclay 1983b; Barclay and Fairweather 1984).

The usual range of domesticated animals appears to have been reared, including cattle, sheep, goats and pigs. It has been suggested that some enclosure systems (e.g. Black Moss of Achnacree, Argyll) may have been intended for stock management rather than boundaries to arable fields (Ritchie *et al.* 1974). Besides their practical aspects, the symbolic importance of dung and midden deposits should not be overlooked (cf. Barrett 1989).

Patterns of farming doubtless varied from region to region but the evidence is too partial to permit any distinctions to be drawn; the favoured survival of organic remains in the calcareous soils of the Northern and Western Isles contrasts with the relatively poor organic preservation in the lowlands. However, the sum of the evidence would point to a mixed subsistence economy, although farming strategies may well have varied from region to region. The traditional view of a Highland Zone with a subsistence economy dominated by pastoralism is no longer tenable (Topping 1989). In coastal locations the full range of marine resources was certainly exploited, as, for example, at Ardnave, Islay (Ritchie and Welfare 1983).

Of course the scale of arable farming carried out by individual farmers or communities cannot be thought of in terms of modern agriculture. Over most of Scotland, the face of the countryside has been changed beyond recognition by agricultural improvements over the last two centuries, and it is perhaps necessary to look further afield to areas such as the west of Ireland or Brittany to obtain a flavour of a farming landscape composed of small enclosed plots associated with small farmsteads.

THE ENVIRONMENTAL DIMENSION OVER THE PERIOD AS A WHOLE

The causes and rate of expansion of settlement into the upland areas are far from clear. Construction of large ceremonial monuments might indicate sizeable human populations and even population pressure. Another factor may be increasing territoriality. By the Late Neolithic, farming settlements appear to have been universally established across the length and breadth of the country, so that the only scope for further expansion of settlement may have been inland and upland, particularly if areas of the country were assuming a more clearly defined regional identity, as seems to have been the case by the later third millennium BC.

In any case it is perhaps misleading to think in terms of an explosive expansion of settlement rather than a cumulative process of land intake which may have affected different areas at differing rates and at different times. In north-eastern Perthshire, for example, the distribution of monuments such as ring cairns suggests that upland areas may have been cleared, if not farmed and managed, by the Late Neolithic (RCAHMS 1990). By contrast, pollen analyses from Lairg, Sutherland, and Braeroddach Loch, Deeside, Aberdeenshire, give differing dates for the most intensive periods of clearance and settlement: second millennium BC at the former, and first millennium BC at the latter (McCullagh 1992a,b; Edwards 1979b; 1993b, 20; see also Chapter 5).

Locally then, exploitation of the uplands may have been in train during the Neolithic, but what does seem clear is that in many areas the second millennium BC saw considerable expansion of the limits of settlement and infilling of the settlement pattern. The process is differentiated from the Neolithic by the evidence for much greater organization and control of the landscape. By the second millennium BC, landscapes were managed in all respects, from formal physical demarcation to possible periodic burning of moorland to improve coarse grazing (cf. Sheshader, Lewis: Newell 1988). It has been suggested that some deliberately placed deposits of artefacts, particularly metalwork, may symbolically mark the limits of the contemporary settled landscape, or the frontier between 'domesticated' and 'wild' land (e.g. Early Bronze Age axes from the Pass of Ballater, Deeside, Aberdeenshire: Ralston 1984, 73; Cowie 1988, 21; or Dail na Caraidh, Inverlochy, Inverness-shire: Gourlay and Barrett 1984).

The expansion of settlement during the second millennium involved occupation of areas that were inherently vulnerable to even minor fluctuations in climatic conditions and to human mismanagement in the form of soil impoverishment resulting from clearance and grazing. The precarious nature of Scottish hill-farming is well known today; in an era without subsidies, the vulnerability of marginal areas can be readily appreciated. The fact that such landscapes survive for study and analysis today is a stark reflection of the fact that they eventually became untenable for settlement. The catalysts for their abandonment may have been several and neither palaeoenvironmental nor archaeological evidence necessarily allows all the factors to be distinguished. The proximate causes for phases of abandonment have occasionally been suggested by excavation (e.g. destruction of a building by conflagration at Tormore, Arran: J. W. Barber 1982b; massive gravel inwashes at Upper Suisgill, Sutherland: Barclay 1985), but such catastrophic events may have been the exception rather than the rule.

In recent years, work on tree-rings and tephrochronology has begun to suggest that major volcanic eruptions may have caused climatic upset at various times in the past on a world-wide or hemispheric basis. When such major eruptions occurred, the climate of the northern hemisphere may have been altered for several years. In particular, attention has been drawn to the possibility that a phase of depopulation and contraction of settlement may have been initiated by the effects of an eruption of the Icelandic volcano Hekla in 1159 BC (Hekla 3) (Baillie 1989). It is as yet unclear just how sudden or severe such environmental effects may have been: it has been suggested that much of northern Britain may have been rendered uninhabitable by an environmental catastrophe resembling the 'nuclear winter' that some scientists believe would follow an atomic war (Burgess 1989) but the severity and long-term effects of such volcanic events are now being seriously queried by palaeoenvironmentalists (cf. Bell and Walker 1992, 135–136; Grattan and Gilbertson 1994).

Monocausal explanation oversimplifies a complex situation and fails to distinguish long-term from proximate causes. We should also recognize that the situation may have differed from region to region. A moderate view might see such episodes as one of a set of variables combining to tip the scales between success and failure in inherently marginal areas, where the immediate or proximate cause for abandonment of a particular site or areas may have been social or economic rather than environmental. In any case, the field evidence varies from region to region; in the Borders, for example, the case for widespread abandonment of upland areas has not gone unchallenged (Halliday 1993, 77; Mercer and Tipping 1994), though reiterated recently by Burgess (1995). It has been suggested that a small number of hilltop settlements found at relatively high altitudes may in fact date to around the turn of the second/first millennia, at a time when uplands are claimed to have been deserted (Ralston and Smith 1983; Mercer 1991).

What is clear, however, is that there was an overall trend away from open, unenclosed settlements to enclosed farmsteads and settlements – hinting perhaps at greater territoriality – and certainly by the middle centuries of the first millennium,

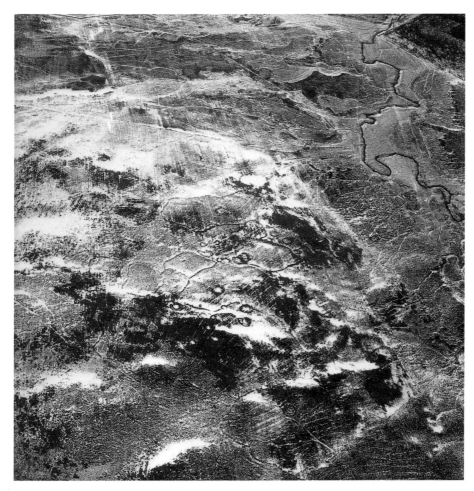

Plate 9.9 Traces of a relict prehistoric landscape showing under light snow: hut-circles and field system at Drumturn Burn, Alyth, Perthshire. Crown Copyright: Royal Commission on the Ancient and Historical Monuments of Scotland

there is a clear hierarchy in settlement types and scales, although not necessarily of functions.

CONCLUSION

While the major phase of destruction of archaeological field evidence has taken place since the agricultural improvements of the eighteenth and ninteenth centuries AD, fieldwork and excavation show clearly how monument survival is the product of an interplay that has gone on since the Neolithic (RCAHMS 1990, 1). Although the loss of sites in the lowlands is partially compensated for by the remarkable survival of relict landscapes (Plate 9.9) in what is today agriculturally marginal land, our present picture of Bronze Age settlement in Scotland remains a particularly fragmented one, the product of diverse patterns of survival, monument recognition and discovery.

The immense scope for integrated archaeological fieldwork and palaeoenvironmental work has been well demonstrated by projects such as those at Lairg in Sutherland (McCullagh 1992a,b) or the Bowmont Valley in the Cheviots (Mercer and Tipping 1994). The challenge now must be to investigate long-term change across broader transects of country that encompass both lowland and upland terrain, study areas where the respective strengths and limitations of both zones can be examined, and the changing nature of their interrelationships investigated (cf. Barclay 1992, 119–123).

10   The Iron Age

IAN ARMIT AND IAN B. M. RALSTON

INTRODUCTION

General

The Iron Age is a label used, rather unsatisfactorily, for the period between the eighth century BC and the arrival of the Roman military (Ralston 1979). This closing date is of limited significance since the presence of the Roman army is not reflected by significant changes in the archaeological record of indigenous communiuties for the country as a whole. The starting date is dependent on the appearance of the first iron artefacts in hoards deposited in keeping with Late Bronze Age practices, and on indications of smithing and smelting on some settlement sites. How widely the new material was worked at an early date is debatable, as is its importance in defining a major change in the development of previously bronze-using societies. Radiocarbon dates make it clear that some of the architectural developments discussed in this chapter occurred first in the context of bronze-using societies.

Throughout this period, Iron Age societies in Scotland shared traits with southern Britain, Ireland and, in some cases, continental Europe. Such traits included a preponderance of roundhouses in domestic architecture, and, at least intermittently, a preference for enclosed and indeed fortified settlement units – a characteristic shared with much of temperate Europe. Some technological innovations, such as the manufacture of rotary querns, were taken up as early in Scotland as elsewhere in north-west Europe: others, however, including wheel-thrown pottery and the production of coins, were not adopted at all. High status objects, decorated in styles termed 'Early Celtic Art', also bear witness to external contacts, doubtless of varying kinds (MacGregor 1976; Megaw and Megaw 1986). In sum, there is no reason to doubt that the societies of Iron Age Scotland were aware of contemporary communities elsewhere; but the material culture they have left illustrates that developments within Scotland could be independent.

Landform and province

The study of the Scottish Iron Age continues to operate within a regionalized framework. Stuart Piggott's model (1966) of four geographical provinces remains a

Scotland: Environment and Archaeology, 8000 BC – AD 1000. Edited by Kevin J. Edwards and Ian B. M. Ralston.
© 1997 The editors and contributors. Published in 1997 by John Wiley & Sons Ltd.

pervasive influence underpinning the structure of recent research (Figure 10.1). The diversity of the Scottish physical landscape promotes a feeling that Iron Age settlement archaeology ought to be divisible by landscape units, e.g. by contrasting the fragmented western coasts with the lowland plains of the North-East. Even though Piggott was essentially importing a model developed for southern Britain, it seemed to accord both with the topographic diversity of Scotland and with the long-perceived variety of Iron Age monuments.

Why did this idea of Provinces appear so attractive to students of the Iron Age rather than, say, the Neolithic or the Bronze Age, where no such formalized pattern won acceptance? The country's topography was no less varied in other periods. Part of the answer lies in the embarrassment of riches that is Scotland's Iron Age settlement record. Brochs, duns, hillforts, crannogs, enclosures, souterrains, hut circles, wheelhouses and so on, and the broad regional patterning in their distributions which was identified from an early stage, made a geographical framework very appealing (Figures 10.2 and 10.3, Plate 10.1). Studies could be focused on particular areas and monument types rather than attempt to assimilate large volumes of data from a range of site types. This perspective was enhanced by the uniqueness of some Scottish monument types, of which brochs are the classic example.

Regionalism underpinned many of the contributions to the main post-war overview of the Scottish Iron Age (Rivet 1966). This zonal focus to research has only recently been tackled anew (cf. Hingley 1992). In general, studies of the Atlantic Iron Age, and of the Scottish part of Piggott's Tyne–Forth Province in the South-East have been pursued separately, by different researchers: attempts at the integration of methodologies or results have been few. The South-East saw a burst of research c. 1980 which produced notable works of regional interpretation and a significant body of excavation evidence, the latter still largely awaiting publication (summarized in D. Harding 1982). Atlantic Scotland has seen more sustained activity developing from a series of important excavations in the late 1970s and 1980s. Contrastingly, the South-West, and to a lesser extent the zone between the Tay estuary and the Moray Firth, have been neglected in terms of works of synthesis and interpretation, although data collection has continued albeit on a relatively limited scale (Ralston 1996).

Settlement studies

Settlement archaeology has long dominated the interpretation of the Scottish Iron Age. Artefact studies have played a minor role in all but the Atlantic regions, and even there they have been marginal to considerations of the settlement evidence. Burial and specialized ritual sites remain difficult to detect: the overwhelming impression is that, for most of the Iron Age over much of Scotland, domestic settlement, whether enclosed or open, was the principal forum for social interaction which is detectable by archaeological means.

The existence of abundant, often elaborate settlement sites is a factor which unites the Iron Age in many parts of Scotland and distinguishes the archaeological record of this period from that of earlier prehistory. The attention paid by the

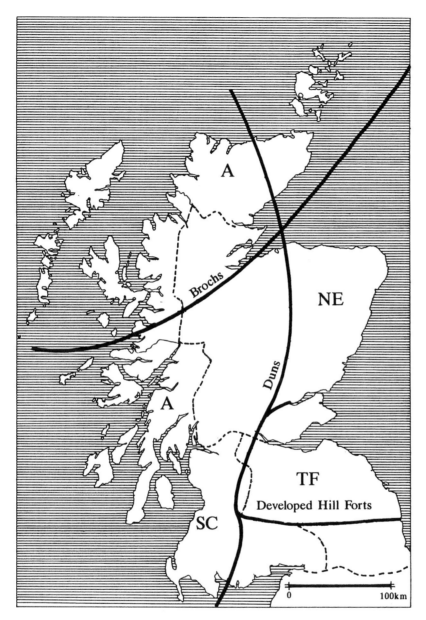

Figure 10.1 The conventional scheme for the subdivision of the Scottish Iron Age, as developed by Piggott (1966), and the principal distributions of some of the regionally-distinctive settlement types, after Cunliffe (1983). A: Atlantic; NE: North East; SC: Solway Clyde; TF: Tyne–Forth

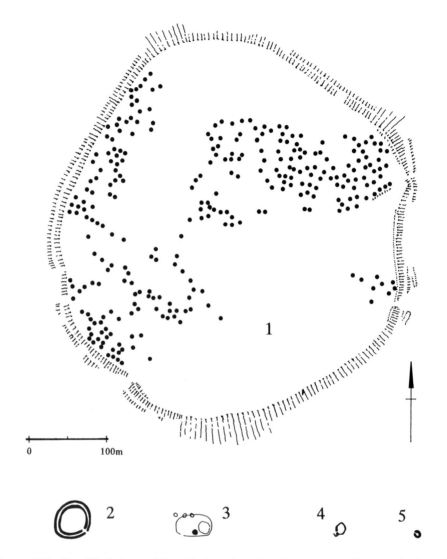

0   100m

Figure 10.2 Simplified plans of Scottish Iron Age sites drawn to a uniform scale. 1: major hillfort at Eildon Hill North, Roxburghshire; 2: hillfort at Cairnmore, Aberdeenshire; 3: palisaded site at Dryburn Bridge, East Lothian; 4: blockhouse fort at Loch of Huxter, Shetland; 5: broch, Dun Carloway, Lewis. Drawn by Gordon Thomas

builders to the physical appearance of settlements suggests that comparable processes of social development were under way across much of the country. Rather than using the variability within settlement architecture to restate the regional perspective traditionally offered, it seems more appropriate to analyse in outline similarities in settlement and, by extension, social development, thus promoting a more broadly based study of the Scottish Iron Age with which environmental evidence can be integrated.

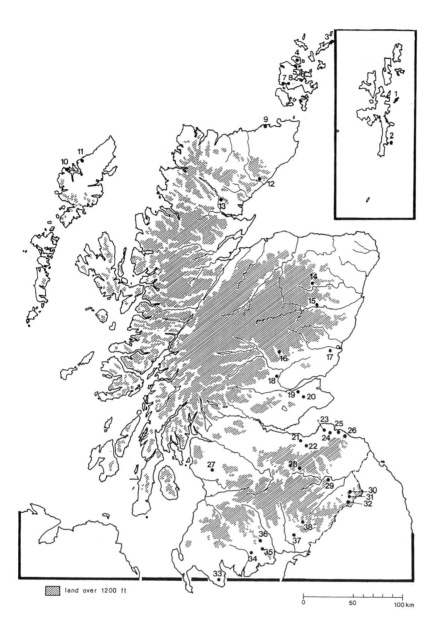

Figure 10.3 Sites mentioned in the text. 1: Loch of Huxter; 2: Mousa; 3: Tofts Ness; 4:
Midhowe; 5: Gurness; 6: Quanterness; 7: Bu; 8: Howe; 9: Crosskirk; 10: Cnip; 11. Dun
Carloway; 12. Kilphedir; 13: Achany Glen, Lairg; 14: Cairnmore; 15: Howe of Cromar; 16:
Dalrulzion; 17: Douglasmuir; 18: Newmill; 19: Black Loch; 20. Scotstarvit; 21: Castlesteads/
Newton; 22. Crichton Mains; 23. Chesters, Drem; 24: Traprain Law; 25: Broxmouth; 26:
Dryburn Bridge; 27: Bloak Moss; 28: White Meldon; 29: Eildon Hill North; 30: Hownam
Law; 31. Hownam Rings; 32: Hut Knowe; 33: Rispain; 34: Torrs; 35: Milton Loch; 36:
McNaughton's Fort; 37: Burnswark; 38: Boonies

Plate 10.1   Hillfort, Hownam Law, Roxburghshire, with hut platforms – the former stances of timber roundhouses – highlighted by snow. Copyright: D. W. Harding

Society and change

Recent work has concentrated on the evolution of settlement over time, and has offered interpretations on the significance of, for example, enclosure, the elaboration of settlement layouts and the monumentality of individual buildings. The development of social complexity and the transformation, on a longer time-scale, from tribal to state-level organizations (however fragile these may have been) have re-emerged as important themes. These trends, marked by an increasing readiness to speculate on the social meaning of archaeological data, are particularly apparent in Atlantic Scotland (e.g. Armit 1990a,b; Barrett and Foster 1991), but can also be discerned in more general works (cf. Hingley 1992).

The Iron Age is increasingly seen as one chronological slice within a longer period stretching from the Later Bronze Age to the emergence of the Pictish and Scottish states, and the Viking incursions around AD 800. Already apparent in studies undertaken in the 1970s (e.g. Thoms 1980), this perspective has been reinforced by the recent concentration of research on Atlantic Scotland: the absence of a significant Roman interlude in the North and West has buttressed the adoption of this wider time-frame. Although terminology to replace the classic chronological sequence (Iron Age–Roman–Early Historic) has yet to be developed for other parts of Scotland, there is little dissent among present students of Iron Age Scotland that the Roman incursions may have been less disruptive to the development of native societies than was once propounded (cf. Whittington and Edwards 1993).

Plate 10.2 Circular cropmarks (overlain by indications of rig-and-furrow cultivation), corresponding to the former positions of the houses of an unenclosed settlement in the North Esk valley, Angus. Crown copyright: Aberdeen Archaeological Surveys

EASTERN SCOTLAND

Settlement

Most modern work on eastern Scotland, other than the reporting of individual excavations, has been concentrated on the Borders and East Lothian. Fife and the area between the Tay and the Moray Firth have been less intensively treated, although there are numerous pointers (e.g. in the burgeoning aerial photographic record) that areas to the north of the Forth display many of the settlement characteristics of zones further south, although with different proportional emphases (Plate 10.2). This discussion will focus primarily on areas south of the Forth.

The approach to settlement in eastern Scotland is dominated by consideration of the relationship between enclosed and unenclosed settlement, the latter including some large timber-built houses of imposing proportions (reaching 20 m in diameter). Hingley (1992) terms such structures 'substantial houses', irrespective of their architectural detail. The geographical and chronological relationships between open and enclosed settlements remain far from clear-cut, inhibiting the development of general models of settlement and social development.

Substantial houses, often unenclosed, are a conspicuous feature of the settlement record of the early first millennium BC. These include many ring-ditch buildings, large timber roundhouses with concentrically defined, internal areas, best known from excavation in East Lothian and at Douglasmuir in lowland Angus (Hill

1982a,b). Halliday (1985) has compared these houses with substantial double-walled hut circles, named after Dalrulzion in Perthshire. Structures of similar scale, once considered as being of advanced design, occur elsewhere, as at Scotstarvit in Fife. Such buildings imply the existence of large individual domestic units and/or the coexistence of various economic and social activities within single structures. Reynolds (1982) proposed that cattle may have been kept in ring-ditch houses, but some of the more extensive groups of such buildings are located rather far from surface water, essential for cattle-keeping, and other types of livestock are perhaps more likely. It can also be argued that such buildings were two-storeyed, thereby offering additional internal space for a range of activities. The restricted excavation evidence does not contradict such hypotheses.

During this half-millennium, substantial houses are a recurrent if discontinuous feature of the archaeological record from southern England to the Northern Isles. The subsequent history of this tradition of building imposing individual structures varies widely. In Atlantic Scotland, substantial houses continued to be erected until at least the end of the first millennium BC, but in other areas the tradition was abandoned several centuries earlier. In southern Scotland, a clear trend in favour of the erection of smaller structures has been identified towards the end of the millennium. These, 'Votadinian' (Hill 1982a) houses were defined by low stone walls, but generally enclosed distinctly reduced areas compared with the majority of the ring-ditch series. Their smaller sizes suggest that some domestic buildings were now used in different ways. None the less, further north, substantial houses like that enveloping the entrance to a souterrain at Newmill (Perthshire: Watkins 1980) may have continued in use into the first millennium AD, although the proposed reconstruction of this building has been disputed (Halliday 1985, 247). Datable artefacts associated with the scatter of complex stone roundhouses in the Atlantic tradition known between the Tay and the Tweed confirm that at least a small number of such substantial buildings were erected in eastern Scotland in the early centuries AD (cf. Macinnes 1984a).

The hillforts and other enclosed settlements characteristic of much of the southern Scottish Iron Age have become less straightforward to interpret as a result of recent excavations. The Hownam sequence, developed by C. M. Piggott (1948) from her excavations at Hownam Rings, Roxburghshire, presented an internally coherent picture of settlement development well-integrated with post-war views on chronology and diffusion. This model proposed a sequence of settlement forms through time: from unenclosed to palisaded sites, through univallate to multivallate forts; and finally back to unenclosed settlements (in this case comprising round-houses with stone footings) in the Roman period. This perspective implied increasingly aggressive and competitive societies, marked by progressively stronger fortifications, until these became redundant with the apparent stability and peace resulting from the imposition of the *pax romana*.

The generalized application of the Hownam model on a wider geographical scale has been considerably undermined by the results of recent excavations and also by the removal of the framework of compressed chronology and repeated invasion favoured during the 1950s and 1960s. Interim accounts of Hill's excavations at Broxmouth, East Lothian, illustrate how elaborate occupation and defensive sequences can be (Hill 1982c; Figure 10.4). This site varied in character over several

HOWNAM RINGS

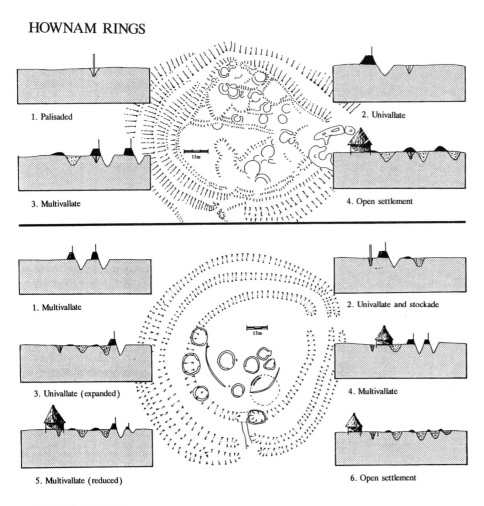

BROXMOUTH

Figure 10.4 Schematic representation of the principal phases of enclosure represented at Hownam Rings, Roxburghshire and Broxmouth, East Lothian, based on information in Piggott (1948) and Hill (1982c). Drawn by Gordon Thomas

centuries, with several lines of defence in use singly or together at different times; there was, however, no indication of the increasing defensive elaboration predicted by the Hownam model. At Dryburn Bridge, also in East Lothian, the sequence included the replacement of a palisaded settlement by unenclosed ring-ditch houses, further demonstrating the difficulties in applying a simple evolutionary model to settlement development at a regional scale (Triscott 1982).

Some structural features used in the Hownam model are now recognized as of little value as chronological indicators. Palisaded enclosures, for example, have now been dated from the later Bronze Age through to the Early Historic period (Hill 1982b). Clearly, the unilinear model of settlement development is less appropriate

Plate 10.3 Remnants of an experimental timber-laced wall, after it had been ignited. The vitrified forts of Scotland are considered to be the product of the destruction by fire of walls containing internal structural timberwork. Copyright: J. Livingston

now than was thought to be the case two decades ago. None the less, the Hownam sequence probably remains valid at the local scale though over a longer timespan than was initially supposed: it is possible that re-examination of the type-site would demonstrate that its enclosure sequence was more complicated than was initially recognized. Palisades, for example, appear generally to predate earth ramparts and/ or stone walls on sites at which these more substantial forms of enclosure are represented. Palisade construction may have been appropriate to newly established settlements, where surrounding land was initially cleared thereby making quantities of straight lengths of timber available for construction (Reynolds 1982), although this enclosure style does not appear to be particularly consumptive of wood.

Defensive types as chronological markers have also fallen from favour in the North-East. Timber-laced forts (including vitrified examples) have a longer period of construction and use than previously thought (Plate 10.3). Although the various dating techniques produce somewhat contrary results, construction and use extend at least through the Iron Age into the Early Historic period (Gentles 1993). As elsewhere in eastern Scotland, cropmark aerial photography is expanding the settlement record, previously dominated by visible enclosed sites in the upland zone of survival, although testing of such discoveries by excavation remains rare.

Macinnes used the basic elements of the Hownam sequence, within the extended chronology provided by radiocarbon dating, in proposing sequences of economic and settlement development in sub-areas of eastern Scotland, particularly East Lothian and lowland Angus (Macinnes 1982, 1984b). Differences between settlement patterns recognizable in East Lothian and those elsewhere were identified.

Macinnes envisaged the later Bronze Age settlement pattern as comprising mostly open settlement supported by a mixed economy and concentrated in low-lying areas. Although individual communities appear to have been largely self-sufficient, exotic metalwork may have been redistributed from sites like Traprain Law, East Lothian (Jobey 1976), which may also have been centres of production. The first enclosed settlements, represented by palisaded sites and the earliest hillforts, seem to mark the extension of settlement into the uplands (Macinnes 1984b, 181). By this stage, there may have been more economic specialization, with lower-lying agricultural settlements complemented by upland pastoral farming. Such patterns may have enhanced the role of redistribution centres such as Traprain by providing an impetus for the economic integration of increasingly specialized sites.

There is now consensus that open settlements of 'Votadinian' houses, common in much of the Tyne–Forth area, were first constructed before the Roman invasion in the AD 80s. The dense pattern of unfortified farming settlement that these houses represent appears, therefore, to belong to a period of stability and relatively high population that originated before and thus independent of Roman influence. It has even suggested that this settlement tissue was 'severely disrupted during the Roman occupation', leaving 'tableaux of desertion' comparable to the Highland Clearances of recent centuries (Hill 1982b, 9). Some examples of this type of settlement seem however to have endured for several centuries, and, as noted previously, the view that the Roman incursions were not especially disruptive seems currently to be increasing in favour.

Little progress has been made in the study of the distinctive rectilinear, non-defensively sited enclosures identified in East Lothian. Comparanda from north-eastern England suggest that these probably belong to the Roman period, but independent dating evidence is lacking for the Scottish examples. Their dense distribution within East Lothian, far from any Roman installations but clustering around Traprain Law, suggests that examination of these sites may be of the highest importance in determining the relationship between Rome and the local tribe, the Votadini, and perhaps also in identifying agricultural innovations in Lothian such as are identifiable in particular areas of north-eastern England (see below).

The role of the series of fortifications enclosing the major fort on Traprain Law itself can be questioned (Plate 10.4). Substantial (by the standards of the time) excavations in the early part of this century yielded, *inter alia*, a range of agricultural ironwork and an impressive hoard of late Roman material. The settlement on this volcanic plug is conventionally described as the capital of the Votadini. Traprain is often presumed both to have been in continuous occupation since the later Bronze Age and to have fulfilled the central functions ascribed to *oppida* of the later pre-Roman Iron Age and Roman periods further south. The early excavations, concentrated on the western slopes of the hill, however, produced little secure artefactual evidence for pre-Roman Iron Age occupation (Jobey 1976), although the largely undated defensive sequence may imply otherwise. There is insufficient structural evidence to indicate dense settlement, another trait believed typical of *oppida* (cf. Feachem 1966, 77–82). The comparably-sized site of Eildon Hill North (Roxburghshire), proposed as a minor *oppidum* of the Selgovae and containing several hundred house-platforms, has also failed, on recent small-scale excavation,

Plate 10.4   Traprain Law is a conspicuous volcanic dome standing proud of the agricultural lands of East Lothian. The combination of snow and low sun is highlighting part of the enclosing circuit on its western flank. Copyright: Ian Ralston

to produce convincing evidence of substantial pre-Roman Iron Age occupation (Owen 1992; Rideout (with Owen) 1992). It remains possible that these sites were not densely occupied during the Iron Age. Their *floruit* may have been in the Bronze Age; Roman Iron Age activity may have included periodic visitations for ritual practices which are detectable archaeologically (cf. Hill 1987, for Traprain Law). A sequence in which early hillforts are generally larger than their successors, whilst not squaring with the conventional wisdom of expansion and increasing centralization, is not unknown elsewhere: the eastern zone of Germany offers a parallel (Figure 58 in Audouze and Büchsenschtz 1992).

The whole question of the functions of enclosed settlement, including hillforts, has been a matter of debate in recent years (Figure 10.5). Formerly, when cultural change was viewed as the product of (often violent) population movement, such sites were regarded as essentially defensive. The recognition of the symbolic

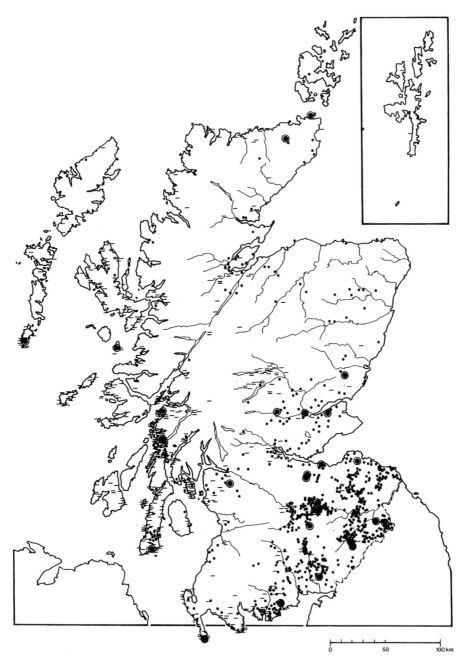

Figure 10.5 Distribution of Scottish hillforts and duns, based on the corpus assembled by Hogg (1979). The large symbol indicates hillforts over 2.5 ha in area, and the horizontal line is used for duns, some of which may have been roofed buildings rather than enclosures. Complex Atlantic roundhouses (brochs) are not included. Drawn by Gordon Thomas

qualities of enclosure has been significant in realigning perceptions of social relations and cultural change in Iron Age societies. Recent work has played down the practical defensive functions of hillforts and stressed their social implications: the command and mobilization of labour, the territoriality and authority implicit in their scale and locations. Warfare clearly played a part in Iron Age society (Sharples 1992b) but there is no *a priori* reason to link it with hillfort construction. Whilst substantial defences, suitable for military purposes, are found on some Scottish sites, this is far from universally the case. The diminutive scale of the enclosing banks on some major hillforts (e.g. White Meldon in Peeblesshire) almost certainly precludes a serious defensive intent. Hillforts, as well as containing settlement, may well have fulfilled social roles not dissimiliar to those claimed for stone circles and henges in Neolithic times.

The periodic remodelling of enclosing works, notably at Broxmouth, allows the inference that labour provision may have been a significant component of social relations. Such activity might represent the marshalling of client labour for an elite residence area, thus symbolizing the authority of those living on the enclosed site. Alternatively, the deployment of communal labour may have served to bond community members who were the site's inhabitants, reinforcing their separation from the wider society. The absence of well-defined chronological spans for the various types and scales of settlements still bedevils attempts to establish definitively whether there was a clear settlement hierarchy at any time during the pre-Roman Iron Age; social, economic, regional and chronological variations remain difficult to disentangle and the ambiguities of the data continue to foster a wide range of interpretations.

Native–Roman interaction

The first Roman invasion of Scotland in the later first century AD serves conventionally to define the end of the Iron Age although, as has been mentioned, this division is unhelpful. None the less, this perspective has shaped research, particularly in southern Scotland, with different specialists treating the Iron Age, Roman and Early Historic periods. Topics such as the development of native societies contemporary with the Roman incursions, and the social transformations which led from Iron Age communities to the Early Historic kingdoms, have suffered as a result. Whilst Chapters 11 and 12 consider the Roman period and the Roman legacy to Early Historic Scotland, it is useful to discuss briefly the Roman incursions from the native Iron Age perspective.

In so far as radiocarbon dates allow definition, the building of hillfort defences does not seem to have characterized the later pre-Roman Iron Age (Alcock 1987; cf. Ralston 1996). Although not straightforwardly to be correlated with less stressful times, the absence of direct evidence for fortification building at this time may be juxtaposed with other indications of relatively tranquil conditions. Settlement in the South-East at the time of the first Roman incursions seems to have been dominated by small unenclosed farming settlements of stone-walled Votadinian houses. The status of larger centres such as Traprain Law is uncertain: it is not clear that social differentiation was manifested in a formal settlement hierarchy at this period. The apparent growth in population represented by the numbers of settlements with

Votadinian houses, and the indications of their expansion, suggests that stable political and economic conditions prevailed.

The impact of Roman contact is difficult to gauge in Scotland, given its peripheral location on the margin of the Empire: even the army's presence was intermittent. Military works on the scale of the Antonine Wall and the networks of roads, forts and lesser installations must have profoundly affected the way in which native people saw their world. But, compared with areas beyond the *limes* in Germany (Hedeager 1992), Roman material reaching native hands seems restricted both quantitatively and qualitatively. Indigenous societies had been manipulating environments and erecting monumental constructions for several thousand years and it is easy to draw too great a distinction between the 'civilized' and the 'barbaric' because of the enhanced scale of planning implicit in Roman structural programmes.

Roman influence may have helped to catalyse the transformation from tribal organizations to kingdoms, by providing a model for the Votadini and other tribes to emulate. Imperial power may have impacted sufficiently on native psyches that it provided a model for displays of power well into the post-Roman period: thus the incorporation of dressed Roman masonry into the walls of a souterrain at Crichton Mains, Midlothian (and other structures in southern Scotland) can be proposed as more than simply the reuse of convenient stone blocks.

In an alternative perspective, the Roman presence may actually have destabilized existing power structures. In this hypothesis, the pre-Roman transformation from hillfort-dominated landscapes to open settlements of Votadinian houses may be advanced as an indication of the centralization of power. The Votadinian elites may have been bolstered by a Roman presence near at hand, but the Army's subsequent withdrawal may have contributed to the collapse of such arangements, leaving a state of political fragmentation not experienced (to judge from the settlement evidence) since the middle of the previous millennium. The abandonment of Votadinian house settlements after the second century AD proposed by Hill (1982b) may support such a hypothesis, as might the refortification of Traprain Law in the late or sub-Roman period (Hill 1987).

THE ATLANTIC REGIONS

Architecture and social change

Prior to the 1980s, the study of the Atlantic Iron Age was firmly diffusionist in approach. The origins and development of broch towers were basic concerns: discussion was based primarily on structural typology and conditioned by a belief that the impetus for construction came from southern English immigrants at the time of the Roman invasion. As a result of major excavations during the late 1970s and early 1980s at Bu and Howe in Orkney (Hedges 1987; B. Smith 1994), research directions altered radically and attention shifted to the social context of broch architecture. The concentration on broch towers and related heavy-walled structures has meant that the significant scatter of hill and promontory forts which extend

Plate 10.5   The surviving portion of the broch at Dun Carloway, Lewis, remains a conspicuous feature in the landscape. Copyright: Ian Ralston

throughout Atlantic Scotland, has been largely neglected. This focus on monuments in its distinctive architectural traditions has probably exacerbated the lack of integration between archaeological studies of this region and those of other areas of Scotland.

The principal factor underpinning new interpretations of social change in the Atlantic Iron Age of Scotland has been the extended chronology for elaborate drystone circular buildings (cf. Armit 1991). The earliest substantial stone roundhouses, termed simple Atlantic roundhouses (Armit 1990a) appear to have developed in the early to mid first millennium BC. Drystone roundhouses of this type, best demonstrated on Orkney, as at Bu (Hedges 1987), Tofts Ness and Quanterness (Renfrew 1979), lack distinctive architectural details characteristic of broch architecture (Armit 1991). Later in the first millennium BC, roundhouses of greater architectural complexity were built at sites like Crosskirk (Caithness) and the early phases at Howe in Orkney. Whilst these display additional architectural details also recorded on broch sites (e.g. intra-mural cells), these buildings were apparently not built on the scale of the broch towers.

The culmination of this architectural tradition was the erection of the broch towers of the later centuries BC. Classic examples, albeit unevenly preserved, include Gurness and Midhowe in Orkney, Mousa in Shetland, and Dun Carloway, Lewis (Plate 10.5). These melded the innovatory technique of hollow-walled construction with the long-established roundhouse form, thereby attaining stability and wall heights in excess of 10 m. Broch towers still dominate the landscapes of parts of Atlantic Scotland, providing some impression of their initial visual impact.

In restricted areas within the North and West, notably on Orkney, the last centuries of the first millennium BC were marked by the appearance of village settlements of drystone structures clustered in the immediate lee of certain broch towers, as at Gurness and Midhowe. Such settlements are generally interpreted as the homes of subsidiary households contemporary with the broch tower itself, which is hypothesized as the elite residence (e.g. Foster 1989; Armit 1990a) forming the nucleus of the settlement. This pattern is not widely replicated, the distinctive combination of densely packed but small-scale settlement units being rare in many areas of temperate Europe.

The early, simple roundhouses may be proposed as the homes of households of varying status, a view which the small finds from the excavated examples does not contradict. These structures were generally isolated. In some instances, their functions may have included display to make apparent their inhabitants' power and territorial control, although on a very localized level, akin to that proposed for earlier prehistoric funerary monuments: the Quanterness house is indeed built into the cairn of a much earlier, but still monumental, chambered cairn (Renfrew 1979). The subsequent architectural elaboration of houses may be proposed to have translated into stone the power and status of the households that inhabited them; it may be surmised that the holding of power and land by the elite was made to seem both permanent and legitimate by the imposing characteristics of their dwellings. The development of monumental domestic architecture as evidenced by these increasingly complex Atlantic roundhouses parallels that of ring-ditch houses in the South at approximately the same period, and may intimate similar concerns with social, political and territorial status.

During the first millennium BC in the North and West, a number of developments in architecture can be noted. Individual structures became more elaborate; enclosures were built around some complex roundhouses; and settlements grew in size. These features have been considered to betoken increasing centralization marked by control by fewer, more powerful households. None the less, compared with hillfort-dominated landscapes, or even some of the cropmark palimpsests of the eastern lowlands, the overall size of Atlantic roundhouse settlements remained small, never exceeding one hectare.

These parallel developments of architectural elaboration, enclosure and increasing settlement size were not, however, common to all areas where Atlantic roundhouses were constructed. In the Western Isles, isolated broch towers and other complex roundhouses appear to represent the standard domestic settlement. Their distribution and numbers demonstrate that they cannot have belonged only to elite households (Armit 1992b). The absence of nucleated settlements focused on broch towers here (and on Shetland) contrasts with the Orcadian pattern. In the Hebrides and possibly Shetland, brochs were being abandoned and replaced by wheelhouses at the end of the first millennium BC. The fashion for wheelhouses (an architectural form much less ostentatious than broch towers when viewed externally; and with radial internal partitions) suggests a radical change in the social significance of buildings. Contrastingly, at this period – and still within the Atlantic ambit – the Orcadian broch tower settlements may have enjoyed their *floruit*. The absence of wheelhouses within Orkney may also imply a different social development on that island group (Figure 10.6).

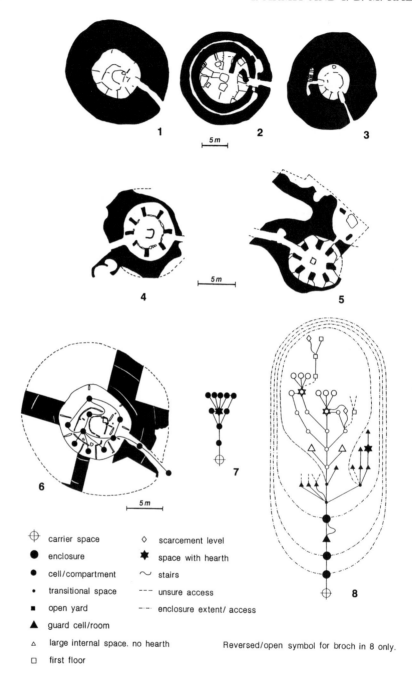

Figure 10.6 Plans, and access maps indicating differential complexity, of complex Atlantic roundhouses and wheelhouses. 1: Bu, Orkney; 2: Gurness, Orkney; 3: Howe, Orkney; 4: Clettraval, North Uist, wheelhouse; 5 Cnip, Lewis, phase 1; 6: Bu, unjustified access map; 7: Bu, justified access map; 8: Midhowe broch and external features, Orkney, justified access map. After Armit (1990a) and Foster (1989)

By the end of the first millennium BC, Shetland and the Western Isles may have been integrated into more extensive power structures centred on Orkney and Caithness (Armit 1990c). It can be argued that a distribution of power based on a pattern of numerous, largely autonomous, Atlantic roundhouse settlements, typical of the mid-first millennium BC in the northern Atlantic Province, was replaced by authorities whose influence encompassed larger regions. During this later phase, broch architecture appears to have been restricted to a few elite centres, primarily on Orkney and Caithness: this building tradition disappeared altogether in the first or second centuries AD. The imposing architecture of these sites may have been a potent means of demonstrating social dominance, but only at a local scale. More outward-looking elites may thus have found it an inappropriate means of conveying their aspirations. Moreover, these structures were manifestly ill-adapted to the harsh climate of northern Scotland; they must also have required quantities of substantial timbers which may not have been readily available.

Beyond the brochs

There is no reason to suppose that the Roman incursions in the south had a very profound effect on the communities of Atlantic Scotland, despite some evidence for contacts between Orcadian elites and Rome (cf. Fitzpatrick 1989). Similarly the withdrawals need not have destabilized these northern societies. As discussed above, the cessation in broch building can be envisaged as representing the end of the need to display power at the local scale (cf. Armit 1990a,b). There are indications that, in the post-Roman period, control over craft production and subsequently support for Christianity and, more tentatively, the deployment of literacy, replaced monumental architecture by providing alternative discourses for the framing of social and political relationships.

THE SOUTH-WEST

Relatively little work has been published on Piggott's Solway-Clyde province in recent years. Aerial survey has made less impact here than in the eastern lowlands. Elements of the settlement tissue noted previously are known however. Substantial houses include that on Milton Loch crannog, near Castle Douglas (Stewartry of Kirkcudbright: C. M. Piggott 1953), and the excavated ring-groove example within the rectilinear enclosure at Rispain, Wigtownshire (Haggarty and Haggarty 1983).

Enclosed sites are present in considerable numbers in some areas (e.g. on the coast of Galloway), and excavated examples can also show evidence of dense internal settlement of smaller buildings, as at Boonies, Dumfriesshire. In terms of scale, many of the smaller enclosed sites fit within the RCAHMS' 'dun' range (i.e. less than 375 m^2), but the region also includes more substantial sites. Among smaller sites, some may have been entirely roofed, as has been proposed for MacNaughton's Fort, Stewartry of Kirkcudbright.

Partial excavation of the large hillfort at Burnswark in Annandale, has served to divorce its enclosure, and inferentially its occupation, entirely from the period of the Roman siege camps that surround it (Jobey 1978). There are indications (from an initial palisaded phase) that this site may have had its principal occupation at an earlier, mid-millennium date.

The lack of recent work prevents a full evaluation of the relationship between developments in this region and the two regions discussed above, although clearly the broad types of settlement are comparable. Other factors such as the appearance of elaborate metalwork decorated in the styles of Early Celtic Art (and, in some cases its deposition in watery surroundings as in the case of the pony-cap from Torrs Farm, Kelton, Stewartry of Kirkcudbright: MacGregor 1976, no. 1) late in the period suggest further points of comparison with the South and East in particular.

ENVIRONMENT AND ECONOMY

Regional variation

Issues concerning the environment and agrarian economy have become more prominent since the mid-1970s. Several contributions in a key volume on the South-East addressed the economic context of Iron Age settlement in that region in outline (e.g. Halliday 1982; Macinnes 1982; Reynolds 1982) while Fojut's work on the environmental setting of the Shetland brochs is a notable contribution for the Atlantic province (Fojut 1982). Larger-scale, integrated programmes of fieldwork which comprise palaeoenvironmental and archaeological components characterized the mid to late 1980s, including Sheffield University's programme in the Uists and Barra and McCullagh's work at Achany Glen near Lairg in Sutherland (McCullagh 1992a,b).

Iron Age economic strategies probably varied widely across Scotland, dependent on soils, topography, climate, and the inheritance from Bronze Age patterns of exploitation. There is evidence for declining economic potential in areas of Atlantic Scotland prior to and during the Iron Age, including podsolization and the extension of peat cover. Fieldwork at Lairg indicates soil deterioration leading to economic stress and depopulation (McCullagh 1992a,b), amplifying results from earlier studies, for example at Kilphedir, also in Sutherland (Fairhurst and Taylor 1971) and in projects undertaken by the former Central Excavation Unit on Arran. In North Uist, settlement became concentrated on the coastal margins where a wider range of resources could be exploited (Armit 1992b). It is likely that many communities in Highland Scotland were engaged in broad spectrum economies, with fishing, hunting and fowling being essential complements to arable and pastoral activities. The balance would have varied considerably depending on local conditions and strategies. At Cnip in Lewis, for example, red deer seem to have provided the principal source of meat (F. MacCormick, pers. comm., 1995) – a highly localized pattern of subsistence.

There is less evidence for environmentally induced economic stress in eastern and southern Scotland, where greater potential existed for economies more reliant on agriculture (although this would not necessarily hold true at higher altitudes). Preliminary consideration of the faunal assemblage from Broxmouth, on the coastal plain of East Lothian, indicates very limited exploitation of wild animals and fish (Barnetson 1982), in contrast to the likely situation further north.

Horses and ponies also seem to be rare at this site, which has produced the largest collection of Iron Age animal bones recovered in excavation on the mainland. The rarity of horses in the osteological record offers a contrast with the surviving bronzework, amongst which horsegear forms a conspicuous component (MacGregor 1976). Tacitus records chariots in use on the native side at the battle of Mons Graupius, perhaps in Aberdeenshire, in 83 AD: and the significance of horses in speeding up overland communications and allowing the direct dominance of more substantial territories cannot be discounted in estimating socio-political developments during the Iron Age.

Settlement in the Southern Uplands in the first half of the first millennium BC unsurprisingly appears often to have occurred in areas of established grassland, where livestock probably played a significant part in preventing regeneration (Halliday 1986, 584). The shallowness of soils means that evidence for the natural removal of woodland is sometimes preserved in the form of distinctive hollow-and-mound traces left by windthrown trees. Similarly, areas of artificial smoothing, where such surface irregularities are absent, identified by Halliday in the Cheviots, are believed to be indicative of intensive early land use.

For much of eastern Scotland south of the Moray Firth, the first millennium BC seems to be marked by the renewed expansion of settlement into the uplands. This is a notable contrast to the evidence for northern and western Scotland and reinforces the dangers of simplistic, environmentally deterministic models of universal deterioration and retreat from higher altitudes.

In the lowlands there is extensive evidence for highly organized patterns of landscape division, although these are inevitably difficult to date and to relate to specific classes of settlement site. Pit alignment systems such as those at Chesters, Drem (East Lothian) and Castlesteads/Newton (Midlothian) are prime examples of such complexes revealed by aerial photography (Figure 10.7) (Halliday 1982). Further north, the density of cropmark evidence around Leuchars in north-eastern Fife, in the Lunan Valley of Angus and in the Laigh of Moray testifies to the potential wealth of the intensively farmed areas of the landscape. Lowland pit alignments are complemented by more fragmentary survivals at greater altitude (Halliday *et al.* 1981), where linear earthworks may also reflect the broadly contemporary definition of larger tracts of land for livestock control.

Crops and the extent of arable

The cereal and other crops of Iron Age Scotland are not well known. As befits the northern latitude and more especially the indications of high-altitude cultivation, barley (*Hordeum vulgare*) seems to have been the dominant crop, although wheat (probably mostly *Triticum dicoccum*) occurs as a secondary cereal at a range of sites from a hillfort in Kintyre to coastal machair sites in the Hebrides (Boyd 1988). Spelt is certainly present in the Tweed–Tyne area, but seems to have been less important here than in the Tees lowlands where it became the dominant wheat in the last centuries BC (van der Veen 1992). Oats are first recorded during this period, with *Avena strigosa* seemingly representing the usual cultivated form. In comparison with the relatively small quantities of carbonized grains recovered, querns (although perhaps not exclusively used for the grinding of cereals) are more frequently found.

Figure 10.7 Landscape subdivisions marked by pitted boundaries, and settlement evidence, in the cropmark record around Castlesteads and Newton in Midlothian. After Halliday (1982)

The rotary quern is one of the earliest circular-motion technologies to be disseminated throughout Scotland, perhaps indicating that the processing of cereals into flour was of considerable significance for many settlements.

A major addition to our understanding of Iron Age economies has been the identification of patches of cord rig, which resemble miniaturized rig-and-furrow field systems and are likely to have been hand-dug (cf. Halliday 1986). Primarily a discovery of the 1980s, cord-rig systems have been identified in association with upland palisaded settlements in the Borders and with later, possibly Roman-period field systems successive to hillfort defences as at Hut Knowe in the Cheviots (Plate 10.6). The surviving evidence for cord-rig field systems in upland areas above the limits of later cultivation, suggests that arable farming must have been fairly extensive, perhaps particularly on reasonably drained, lower-lying areas where repeated use means that archaeologically-recoverable traces are no longer extant.

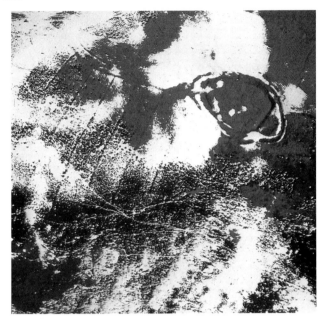

Plate 10.6 Hut Knowe, Roxburghshire: an enclosed settlement with an external trackway and bounded plots of cord rig agriculture, under partial snow cover. Copyright: D. W. Harding

Recently-obtained radiocarbon dates demonstrate that such cultivation practices survived in Highland Scotland into the present millennium (Carter 1994). In southern Scotland, valley-side cultivation terraces of similar dimensions to the tracts of cord rig may represent further vestiges of Iron Age systems. Elsewhere in the lowlands, the storage capacity of later Iron Age souterrains, particularly densely represented in Angus (cf. Maxwell 1983), may offer a further intimation of the scale of arable production.

This evidence for field and cultivation systems, coupled with the indications of cereal processing provided more particularly by quernstones, has finally banished the romantic but oversimplified notion of 'Celtic cowboys' (Piggott 1958), relying on an almost exclusively pastoral economy. What does remain different between the cereal-growing regimes practised in Scotland and those further south, and which prompted Piggott's much-quoted phrase, are the associated storage technologies: both above-ground four-poster granaries and storage pits are still very rare in the North.

Woodland and its removal

The transformation of architectural styles from ring-ditch to 'Votadinian' houses in southern Scotland, is indicative of a shift from timber to stone as a major building material, for structures with smaller diameters and stone walls required fewer major timbers. Allusion has already been made to the replacement of palisaded enclosures with stone and earth enclosures. Both trends may indicate increasing strain on the timber resources in the region, although the absence of stone structures in the later phases at Eildon Hill North suggests that the pattern was not universal, even within what may have been a single tribal territory (Owen 1992).

Palynological evidence supported by ancillary techniques and isotopic dates provides clear indications that in some tracts of Scotland, clearance during the first millennium BC was on a more substantial scale and more enduring than in previous millennia. At Black Loch in the Ochil Hills (Fife), Whittington *et al.* (1990) identified evidence for sustained agricultural activity in the catchment throughout the first millennium BC. Cereal pollen (Whittington *et al.* 1990, table 2), primarily of barley type, was identified only intermittently, but the significant expansion of grasses (to 30–50% of total land pollen) and of weeds, such as *Plantago lanceolata*, that colonize disturbed ground was maintained until *c.* 2000 BP. Before mid-millennium, the inwash of charcoal attained quantities that were not repeated until medieval times. Similarly removed from areas that have been the focus of recent archaeological excavations, the lochs within the Howe of Cromar on middle Deeside, Aberdeenshire, provided Edwards (1979a, figure 1) with palynological evidence of sustained mixed pastoral and arable activity again lasting for most of the last millennium BC, coupled with indications of soil erosion within the catchment.

Other indications of the extent of unforested land are provided by the implantation of Roman temporary camps, the enclosing earthworks defining tracts sometimes in excess of 40 ha, implying the previous existence of land devoid of tree cover. Whilst there is thus a variety of evidence for the substantial, long-term decline in woodland cover during the Iron Age, this pattern should not be considered universal. At Bloak Moss in lowland Ayrshire, for example, Turner identified an episodic pattern of woodland regeneration and clearance during the first millennium BC, suggesting repeated use of the environs of this substantial moss (Turner 1975, 1983).

The apparent population expansion from the last centuries BC, inferred from the numbers of 'Votadinian' house settlements, may have contributed to the diminution of timber resources. Increased grazing pressures and domestic requirements for heating and cooking may have been coupled with inadequate management of what remained of the southern Scottish woodlands.

Economic development and change

Just as economies varied between regions, they also varied through time. There are few secure indications of such developments for much of the country except for the indirect evidence of changing settlement densities and juxtapositions between different categories of field evidence. Thus Halliday has interpreted the apparent absence of cord rig around Cheviot hillforts as evidence for a more pastoral economy successive to the mixed agrarian regime represented by palisaded sites and ring-ditch houses (Halliday 1986, 584). This in turn was possibly succeeded by a period of 'considerable arable expansion' in the later Iron Age and Roman period (Halliday 1986, 585). Although based on a small sample area, this interpretation has interesting implications for the economic as opposed to the social role of upland hillfort-dominated landscapes. Such a perspective also undermines any unidirectional model of progressive environmental decline, although the potential significance of changes in crop regimes is presently unknown (cf. van der Veen (1992) for northern English comparanda). The apparent arable expansion in southern Scotland is suggestive of an increased demand for cereals broadly contemporary with the Roman incursions in the early first millennium AD.

ECONOMY AND SOCIETY

In recent years different approaches and research priorities have created rather divergent perspectives on the economic and social aspects of the Scottish Iron Age. For example, the interpretation of architecture in social terms has been promoted most strongly for the Atlantic area and this zone has also provided the core data underpinning most wider political interpretations. Economy and society are clearly, however, inseparable throughout the Scottish Iron Age and there is a pressing need further to integrate studies of the Atlantic province with those in the rest of the country.

In spite of the interest in economic aspects shown in work on southern Scotland, interpretation of the role of hillforts and other enclosures remains notably insecure in the absence of much relevant data. Is the shift to enclosed settlement, for example, associated with an increasing emphasis on pastoral farming brought about by climatic deterioration, or is it predominantly the result of social change? Furthermore, the agricultural cycle has been proposed as providing the central symbolic metaphor for social relations in Scotland during this period (Hingley 1992, 37–39). Whether or not such a view is accepted, questions of economic organization and patterns of land use and tenure must remain major concerns if we are to advance our understanding of social organization and the emergence and maintenance of complex social systems during the Iron Age.

11 The Roman Presence: Brief Interludes

WILLIAM S. HANSON

INTRODUCTION

The Roman occupation of Scotland may be characterized as little more than a series of brief military interludes in a wider pattern of development, though there is a long tradition of seeing it as the cause of, or stimulus to, various changes within indigenous society. The aim of this chapter is to outline the nature and extent of the Roman presence and to consider its impact on environment and society in Scotland, both in the immediate term and in relation to the longer term trends outlined in preceding and succeeding chapters.

THE CHRONOLOGY OF CONQUEST

The first direct Roman involvement with Scotland occurred in the late first century AD. Credit for overcoming tribal resistance and completing the conquest of the area is assigned to Julius Agricola, the action occurring over some five years of his governorship between AD 79/80 and 83/84. However, the process seems to have begun slightly earlier than has generally been assumed, the first contact with the area probably coming in the governorship of one of his predecessors, Petillius Cerialis, early in the AD 70s (Hanson 1991a). Agricola's campaigns culminated in the famous battle of Mons Graupius, somewhere in the North-East of Scotland, at which the final resistance of the Caledonians was successfully crushed. Yet by AD 87 Roman occupation had been withdrawn to the Southern Uplands, and by shortly after the turn of the century did not extend north of the Tyne–Solway isthmus (Hobley 1989).

The establishment of Hadrian's Wall in the early AD 120s saw Roman outpost forts located on the fringes of the Lowlands (Breeze and Dobson 1987). The Romans did not return to occupy Scotland until the reign of Antoninus Pius (Figure 11.1). Re-conquest began probably in AD 139, shortly after the accession of the new emperor. The limited objective, the reoccupation of the Lowlands, seems to have been completed by AD 142, culminating in the construction of a linear barrier,

Scotland: Environment and Archaeology, 8000 BC – AD 1000. Edited by Kevin J. Edwards and Ian B. M. Ralston.
© 1997 The editors and contributors. Published in 1997 by John Wiley & Sons Ltd.

196 W. S. HANSON

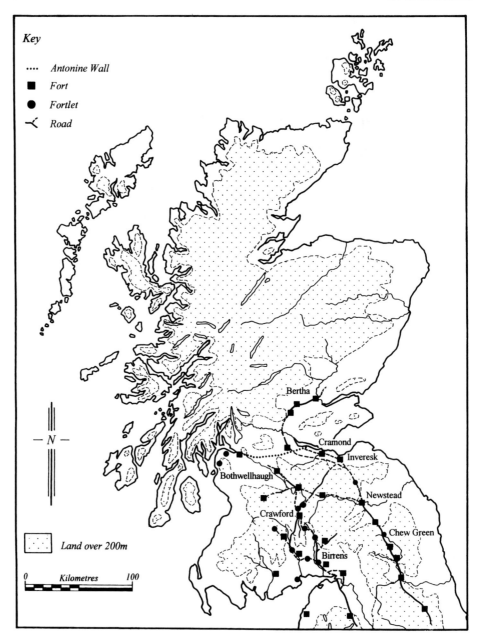

Figure 11.1 The Roman occupation of Scotland in the early Antonine period (*c.* AD 142–158)

Plate 11.1   The line of the Antonine Wall ditch and upcast mound across Croy Hill from the air. Crown Copyright: by permission of Historic Scotland

the Antonine Wall, across the Forth–Clyde isthmus. Little survives of its turf rampart, but the massive ditch and the mound created by the excavated spoil thrown to the north are still notable landscape features across various parts of the central belt (Plate 11.1). With some minor fluctuations, occupation continued until the mid-160s AD (Hanson and Maxwell 1986), though there are still some adherents to a longer chronology (e.g. Mann 1988).

The abandonment of the Antonine Wall and the return to the frontier on the Tyne–Solway isthmus did not see the complete cessation of Roman contact with Scotland, though the Roman presence thereafter was restricted to its southern periphery. In the West, outpost forts were maintained immediately north of Hadrian's Wall as they had been earlier, though in the East they extended as far north as Newstead, near Melrose, Roxburghshire, for a further 20 years or so (Breeze and Dobson 1987). From the mid-180s AD, though outpost forts continued to be maintained north of Hadrian's Wall, there were none sufficiently far beyond it to impinge on what is now Scottish soil (Hanson and Maxwell 1986; Breeze and Dobson 1987).

The third and final Roman occupation of Scotland was even more short-lived. Between AD 208 and 210, the emperor Septimius Severus conducted major campaigns in Scotland either personally or through his elder son Caracalla (Reed 1976; Breeze 1982). Only two garrison posts are known definitely to have been occupied at this time, at Cramond on the Forth and Carpow on the Tay, but neither seem to have continued in use for more than a year or two after Severus' death in AD 211. Thereafter the Roman frontier reverted to Hadrian's Wall and its outposts in northern England (Hanson and Maxwell 1986; Breeze and Dobson 1987), though some would still argue that Rome maintained political control through the supervision of tribal meeting places within the Scottish Lowlands (Mann 1992). Although the Romans did campaign into Scotland on several occasions thereafter, these were solely punitive exercises and did not result in any further attempt to occupy territory (Hanson 1978a; Hanson and Breeze 1991).

The total period of Roman occupation of any substantial part of Scotland was, thus, limited to some 40 years. Even if those areas in the southern Lowlands which were intermittently controlled as outposts of a more southerly frontier line are included, the figure does no more than double to 80 years. This does not, however, take into account any influence exerted or contacts maintained beyond any formal boundaries of Roman territory (e.g. Hanson forthcoming a). None the less, the time-scale is very short.

THE GEOGRAPHY OF CONQUEST

Moreover, even the Roman presence in its fullest form in the first century AD did not involve occupation extending to half of Scotland's land mass (Figure 11.2). Though Roman armies traversed north-east Scotland as far as Moray, perhaps marching back through the Highlands, and the fleet circumnavigated the island (Hanson 1991a), no Roman forts are known to have been constructed north of the Howe of the Mearns. Despite assertions to the contrary, particularly in relation to the excavations at Easter Gallcantry (Cawdor), none of the postulated sites discovered by aerial survey in Moray and Nairn over recent years has the distinctive morphological characteristics of a Roman fort (*contra* Jones 1986). All the known fort sites are located south of the Highland Boundary Fault. In the second century there was no attempt to extend the occupation even this far. The most northerly garrison was located at Bertha (Perthshire) on the Tay, serving as one of the outposts forts to the main frontier line drawn across the Forth–Clyde isthmus. Finally, in the brief third century presence the only permanent bases attested did not extend beyond the geographical limits of those areas occupied in the second century, and may have been constructed primarily to protect seaborne shipments from the South being brought in to supply troops on campaign (Breeze 1982) (Figure 11.3).

The influence of geography on the nature of the Roman occupation is becoming increasingly apparent as our knowledge of its location and extent is refined. The attraction of the Forth–Clyde and Tyne–Solway isthmuses as transverse lines of

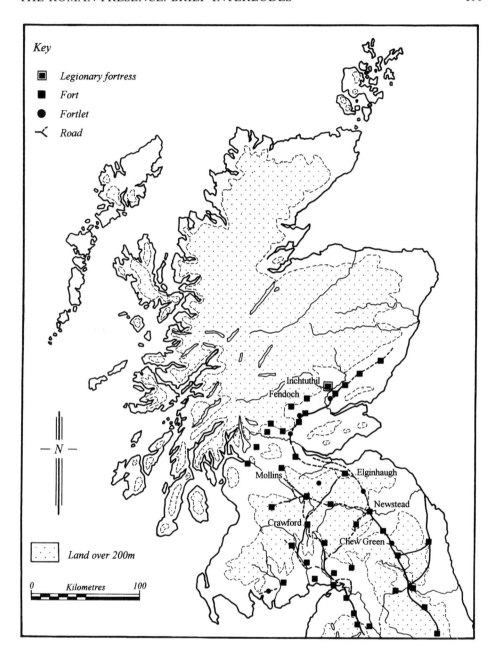

Figure 11.2 The Roman occupation of Scotland at its furthest extent in the Flavian period (*c*. AD 84–87)

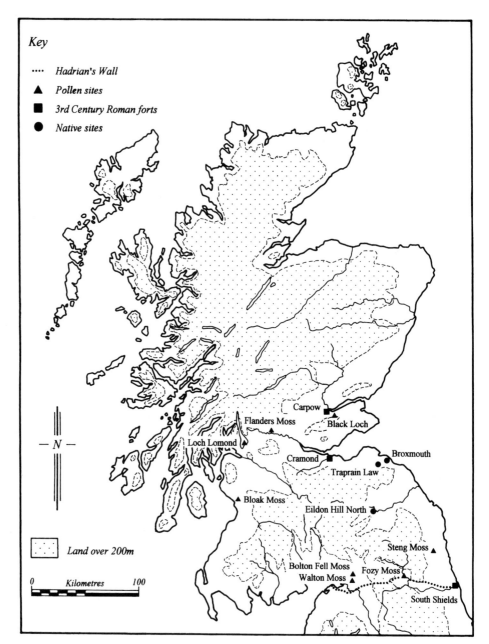

Figure 11.3 Hadrian's Wall, third-century Roman forts and non-Roman sites mentioned in the text

Plate 11.2 The line of the road, picked out by its quarry pits, and the adjacent timber watchtower on the Gask Ridge at Westerton showing as cropmarks from the air. Copyright: Colin Martin

demarcation and communication has long been obvious. Accordingly, their use by the Romans as frontiers is unsurprising and their importance is evident in their continued employment as major lines of east–west communication within the modern landscape. Less obvious are the geographical factors, both topographical and human, which influenced other strategic dispositions. The utilization of the Forth–Clyde line was accompanied by the maintenance of a series of outpost forts along a road running north through the Stirling gap and then eastwards through Strathallan and Strathearn to the Tay. Again this follows another natural routeway which is broadly mirrored by the modern communications system. This pattern occurs not just in the second century as an accompaniment to the Antonine Wall (Figure 11.1), but also during the first century under Agricola when a temporary halt was called to the advance (*Agricola* 23). Moreover, in the latter case the line of forts along the road was augmented by fortlets and timber watchtowers, best known along the Gask Ridge to the south-west of Perth (Plate 11.2), and was clearly conceived as demarcating a frontier line which served to include the Fife peninsula to the south (Figure 11.2) (Hanson 1991b). In terms of political geography, these dispositions imply closer links between Fife and the central Lowlands than is normally assumed and add support to the suggestion that the territory of the Dumnonii extended beyond the Forth–Clyde isthmus as far as eastern Perthshire (Hanson and Maxwell 1986).

That the furthest limit of Roman occupation in the first century closely followed the Highland Boundary Fault has already been noted, and again this is suggestive of the way that topography seems to have influenced the nature of the Roman

Plate 11.3 The Antonine auxiliary fort and parts of two adjacent temporary camps at Glenlochar showing as cropmarks from the air. Copyright: W. S. Hanson

occupation. What is less frequently remarked upon, however, is that such a disposition is the minimum necessary to ensure the inclusion of all agricultural land in Scotland considered to be of first class quality according to the Land Utilization Survey (cf. Coppock 1976). This would seem to indicate a concern to ensure the control of only the wealthier areas which had the ability, or at least the potential, to sustain Roman garrisons.

The Roman presence in Scotland was primarily, if not exclusively, military in nature. All the sites known are associated with the army. The most abundant evidence of the Roman presence comes in the form of temporary works, which vary in size from less than 5 acres (2 ha) to over 160 acres (65 ha), and in use from the defensive enclosures of armies on campaign to the housing of construction parties. The main manifestations of the Roman occupation are the forts and fortlets of the auxiliaries, varying in size from fractions of an acre to some 10 acres (4 ha). Not infrequently, camps cluster around forts, representing the repeated mustering of troops at particular strategic locations, as for example at Glenlochar (Stewartry of Kirkcudbright) (Plate 11.3). These, along with the rare, larger legionary bases, housed the garrisons, from a few dozen men to over 5000, whose presence ensured the maintenance of Roman control. The only non-military personnel would have been those resident in the civilian settlements, or *vici*, situated outside the forts. The existence of such settlements is generally assumed to have been the norm (Sommer 1984), though they are rarely attested at Scottish sites, particularly in the Flavian period, and have been examined even more rarely. The best-known example, outside the fort at Inveresk, seems to have been quite extensive (Thomas 1988), though this may relate to its probable role as the port of supply for the garrisons of the Antonine Wall.

Though the location of forts must have been linked to political geography at a strategic level in order to control the territory of recently conquered tribal groups, it does not seem to have been dictated primarily by population density at a tactical level, even along the line of the Antonine Wall (Breeze 1985). Lowland areas of Scotland, such as Fife and East Lothian, which show the greatest density of native occupation based on cropmark evidence, do not appear to have been provided with Roman garrisons, though this may in part be explained by the possible pro-Roman attitude of the tribal groupings concerned (Hanson 1991a). There are, on the other hand, military posts in locations which seem to bear little relationship to local population density. One need only consider the remote fortlet at Chew Green near the present Anglo-Scottish border to appreciate this. The overriding criterion evident in the general disposition of Roman installations in Scotland seems to be the control of movement. Forts are invariably located at regular intervals along natural lines of communication, usually following river valleys. They are often positioned at river crossings, such as Glenlochar on the Dee (Plate 11.3), confluences, such as at Bertha where the Almond meets the Tay, or other key topographical positions, such as at the mouths of glens, best exemplified at Fendoch in Perthshire.

ECONOMIC DEMANDS OF THE ARMY

Any attempt to consider the impact of the Roman presence on the environment of Scotland must consider not only its extent but also its size. This varies considerably over the three major periods of occupation outlined above and may also differ between phases of active military campaigning and subsequent, more permanent occupation. The size of campaign armies can be difficult to determine. Estimates of the size of Agricola's army at Mons Graupius, based on the account of Tacitus, vary between 17 000 and 30 000 (Mann 1985; Hanson 1991a); the second century re-conquest may have involved no more than 16 500 troops on the basis of the size of the largest postulated Antonine temporary camps known in the Lowlands and an estimate of troop density not exceeding 300 men to the acre (740 per ha); while the size of the better attested Severan camps would indicate a force of between 40 000 and 50 000. Knowledge of the size of the forces of occupation depends entirely upon the number of posts constructed and estimates of their individual garrisons, though such figures can provide only a general guide. In no case is there confidence that we have the complete pattern of Roman dispositions; nor is it certain that all known installations of a particular period were in fact occupied contemporaneously; and finally, in few cases is enough known of any single fort plan to estimate the size or nature of the individual garrison with confidence.

Calculations based on current knowledge of the number of forts probably occupied simultaneously in the Flavian period at the furthest extent of Roman occupation suggest that the number of troops involved was just over 22 000 (Table 11.1). Since it is likely on the grounds of topography and spacing that there are at the very least five other forts which remain undiscovered, mainly in south-western

Table 11.1 Estimated Roman garrisons in Scotland

Period	Fortress	Forts			Fortlet	Garrison
		Large	Medium	Small		
Flavian	1	9	18	3	9	22 280
Antonine I	–	11	20	7	12	21 040
Antonine II	–	8	19	6	6	17 440
Severan	1	1	–	–	–	3 300

Notes
1. Figures are based on the maximum extent of occupation in the period concerned and include all sites north of the Tyne-Solway isthmus thought to have been occupied contemporaneously.
2. Figures are calculated on the following criteria and assumptions: Flavian fortress = 5000 men; Severan fortress = 2500 men; large fort (6 acres/2.4ha and above) = 800 men; medium fort (3–5.5 acres/1.2–2.2 ha) = 480 men; small fort (1–2.5 acres/0.4–1 ha) = 240 men; fortlet (less than 1 acre/0.4ha) = 80 men.
3. Where fort sizes are not known they have been assumed to be of medium size.
4. No allowance has been made for garrisons below strength, but neither has allowance been made for larger auxiliary units or the possibility of different units brigaded together. It is assumed that these two factors will effectively balance each other.
5. No allowance has been made for the possibility of forts being larger in order to house cavalry as it is difficult to confirm the regular presence of stables within forts.

Scotland, an overall figure of 25 000 seems not unreasonable. This, understandably, is the period of the largest garrison, though the likely total in the first Antonine period may have fallen not far short, assuming a similar number of missing forts. Given the lesser area occupied, this may seem surprising, but it is accounted for by the greater intensity of provision required for the control of a linear frontier. Indeed, this seems to have proved more expensive than the Romans originally intended and clearly put pressure on the available resources of manpower (Hanson and Maxwell 1986). The real reduction in the second Antonine period was probably even greater than is implied by Table 11.1, for the simple model employed for the calculation does not allow for garrison reductions unless reflected in major structural changes to the fort enclosure.

The primary requirement for any garrison is food, of which the bulkiest element is likely to have been grain (mainly wheat and barley). The military grain ration has been variously calculated, but using the weight to volume ratio for wheat quoted by McConnell (1968), the figures provided by Polybius (*Nat. Hist.* 6, 39, 12–14), the original Greek source on which most calculations seem to be based, convert to between 59 lb and 75 lb per person/month, which also seems to comply with the size of the corn dole in Rome (Rickman 1980). On this basis, a garrison of occupation of 25 000 would require between 8000 and 10 000 tons of wheat per annum. To this would need to be added the cereal feed requirement of the cavalry horses. Polybius also indicated a ration which converts to between 400 and 560 lb of barley provided to each cavalryman each month, though it seems likely that these figures include feed for remounts, which were generally not provided in the imperial army. Assuming that they relate to one and two remounts respectively (Walker 1973), these figures need to be reduced, the smaller by one-half or the larger by two-thirds. Taking the ratio of cavalry to infantry as approximately 1:2, this would suggest a further annual requirement of between 8350 and 8930 tons of barley. Thus the annual grain provision for the army in Scotland at its maximum in the first century is likely to have been of the order of 16 000–19 000 tons.

The military diet could be extremely varied, as both literary and archaeological evidence attests (Davies 1971): meat was certainly eaten, but may not have formed as significant a part of the diet as has been assumed (Dickson 1989). Indeed, the role of meat in the diet has proved extremely difficult to quantify, a situation not helped in Scotland by the generally poor preservation of bone evidence. If a sixth century AD papyrus from Egypt may be considered an appropriate source of information for the imperial army (Jones 1964), the daily ration might have been as much as 1.4 lb per man, though this figure seems very high. At the other extreme a recent calculation, based on the analysis of well-preserved faunal remains from the waterlogged auxiliary fort site at Valkenburg in the Netherlands, has suggested a figure of only 0.13 lb (Groenman-van Waateringe forthcoming), which is seen to comply with the calculated daily distribution of meat in Rome in the late third century AD (*Theodosian Code* 14.4; Sirks 1991). Though the figure seems low, in combination with the estimated wheat ration it would have provided sufficient sustenance for a physically active male (Groenman-van Waateringe forthcoming), and would give an annual total for the putative maximum garrison of only 537 tons. Taking the average proportions of the different species found on native sites as a reflection of the likely availability of animals at the time of the Roman invasion of Scotland, and applying widely used published estimates of total meat yield per animal (Carter *et al.* 1965), the annual military requirement would translate into 670 cattle, 1200 pigs and 2890 sheep. But the archaeological evidence from Britain indicates a preference for beef and pork amongst the military (approximately 70% and 20% respectively), with a much higher proportion of lamb/goat (approximately 50%) eaten by the native population (King 1984). If this military dietary preference was met, the first figure would increase to 935 and the last decrease to 960. For an average auxiliary garrison the corresponding figures are surprisingly small (13 or 19 cattle, 24 pigs and 57 or 19 sheep per annum). Similar calculations based on the estimated dressed weight of carcasses (Chaplin 1971) would give figures of 2000–2800 cattle, 4800 pigs and 4800–14 400 sheep for the total annual requirement, and 40–55 cattle, 95 pigs and 95–285 sheep for an average auxiliary garrison.

Cattle were also required for their hides, since the army was a heavy consumer of leather for shoes, clothing and other equipment, notably tents. It has been calculated that the hides of some 2000 calves would have been needed to provide tents for a single auxiliary unit (Breeze 1984). The rate at which such items needed replacement, however, is more difficult to estimate, but excavation at waterlogged sites invariably produces soles of shoes and pieces of tent leather indicating something of the regularity of repair or replacement. One unusual statistic is the requirement of animals for sacrifice at festivals in the military calendar: some 43 animals per unit per annum (Breeze 1984). For the putative average first-century Scottish garrison this would have represented some 1500 animals each year. Presumably a steady supply of cavalry horses and pack animals would also have been required to maintain an efficient level of operation for the units, though again the size of the annual demand is difficult to quantify. Assuming a working life of 12 years for cavalry horses (Dixon and Southern 1992), the army in Scotland in the first century would have required some 520 new mounts per annum in optimum conditions simply to allow for natural retirement. This makes no allowance for additional needs as a result of disease or losses in battle.

As excavation makes clear, the Roman army were great consumers of material goods. Most prolific within the archaeological record are the broken pots which necessitated replacement, thereby generating a regular and probably quite large demand. But more personal items of equipment, such as bronze brooches or buckles, iron axes or spearheads, or even imported lava quernstones, are not infrequently recovered (e.g. Curle 1911; Pitts and St Joseph 1985). Though this may reflect the disposal of surplus or damaged equipment on evacuation of a fort, or an element of votive deposition, it none the less highlights the need for regular repair or replacement.

After sherds of broken pottery, the next most common finds recovered from the excavation of a Roman fort are iron nails. These are generally little reported upon unless well preserved or found in particularly large quantities, of which the most famous example is the 10 tons buried on abandonment of the legionary fortress at Inchtuthil, Perthshire (Manning 1985). This serves to remind us of the considerable quantites of building materials needed for the construction of all the garrison posts to house the army of occupation. The major requirements were timber, turf and stone, with greatest emphasis on the first of these. Stone buildings were not a regular feature of Roman forts until the second century occupation, and even then they were often restricted to the central range of administrative and storage buildings, many of which may have been provided with stone only in their footings. Most fort ramparts were of turf or earth. The provision of stone ramparts was the exception rather than the rule, even in the second century occupation. Only two of the 13 forts on the Antonine Wall for which evidence is available had stone-built defences. The building of the Antonine Wall alone represented the stripping of between 800 and 950 acres (325–385 ha) of turf (Hanson and Maxwell 1986), while an average Roman auxiliary fort might involve the removal of some 5 acres (2 ha). The timber requirement of the same fort would amount to some 22 000 cubic feet (632 m^3) and for all the first-century forts in Scotland some 1 million cubic feet (28 315 m^3) (Hanson 1978b; Hanson and Macinnes 1980), with figures for the second century occupation perhaps 20–30% less in recognition of the greater use of stone.

MEETING THE DEMANDS OF THE MILITARY

The impact of these various demands upon the Scottish landscape depends in the first instance on the extent to which they were met locally. The literary and documentary evidence from the wider empire makes clear that the process of supplying the army was both complex and variable (Breeze 1984). On the one hand, there is clear archaeological evidence of the provision of military supplies from long distances. Thus, even at the most remote military establishment, wine and olive oil are regularly attested by the presence of distinctive amphorae; the availability of other Mediterranean foodstuffs is sometimes indicated by botanical evidence; and imported fine pottery, particularly samian ware from Gaul, is ubiquitous. On the other hand, it has long been recognized that in the ancient world, where overland transportation was slow and expensive, supplies would have been obtained locally whenever possible (*Theodosian Code* 7.4.15). Thus there is growing evidence that the army manufactured its own coarse pottery in Scotland (Breeze 1986). The quarrying of building stone and burning of lime to make mortar by the military are also well attested (Hanson forthcoming b); while recent work at Elginhaugh has indicated that it may have served as a military collection centre for animals, though not until it had ceased to operate as a fort (Hanson forthcoming a). Clearly turf was used for building precisely because it could be obtained locally. What remains disputed, however, is the extent of the local provision of the two major requirements, grain and timber, the winning of which involved potentially the greatest environmental impact.

A strong case has been argued, primarily in the context of Wales, that the army would have obtained its grain supplies locally if at all possible (Manning 1975). The difficulty lies in demonstrating to what extent it did so. There are some indications of local supply. At South Shields in north-eastern England, the presence in spelt wheat (*Triticum spelta*) of specific weed species characteristic of assemblages from the region, indicates that the grain had come from the immediate area in the third century AD (van der Veen 1992). Similarly, at Caernarvon in Wales, the weed species identified were commensurate with locally grown cereals (Nye 1993). Moreover, the presence of grain driers within the annexe of the first-century fort at Elginhaugh implies the processing of local supplies (Hanson 1997a). On the other hand, various strands of evidence combine to suggest that importation of some considerable quantities of grain would have been necessary. Tacitus implies that grain requisitioned in the civil province was sometimes sent to garrisons on the frontier (*Agricola* 19), though no specific historical context is indicated. The presence of two places in the north named *Horrea Classis* (= Granaries of the Fleet) (Rivet and Smith 1979) is strongly suggestive of the need to trans-ship grain supplies. The remodelling of the fort at South Shields at the mouth of the Tyne to contain 22 granaries has long been associated with the supply of Severus' campaigns in Scotland (Dore and Gillam 1979) and recent work has indicated that it continued to serve as a base for seaborne supplies to Hadrian's Wall throughout the third century (Bidwell and Speak 1989). Indeed, it has been argued that the widespread distribution of imported fine pottery comes about as makeweight cargo carried on the back of major grain shipments to feed the military (Fulford 1984).

Moreover, although the long held view of a pastoralist basis of the local economy in north Britain is now seen to be an oversimplification, it seems likely that the Roman dietary preference for wheat would not have been readily met given the local emphasis on barley growth indicated in the record of macrofossil plant remains (Boyd 1988). Indeed, a consideration of recent pollen analyses from Roman fort sites in central Scotland would seem to support the view that locally grown grain would not have been readily available regardless of species. The consistent pattern recorded is one of extensive grazed pasture land, though at least that should mean that the supply of cattle and other animals would have been assured (e.g. Boyd 1985a,b). At best there are only hints of arable cultivation, usually of barley (e.g. Butler 1989; Dickson forthcoming).

It seems likely, however, as Boyd (1984) himself recognized, that the role of arable agriculture is under-represented for three reasons: the well-established rapid fall-off rate in the dispersal of cereal pollen; a tendency for samples to be taken from rampart turves thus biasing the sample towards pastoral species; and the difficulty sometimes encountered of differentiating cereal pollen from that of some wild grasses. Archaeological evidence of later prehistoric arable cultivation continues to grow. Ploughmarks are regularly discovered preserved beneath the ramparts of Roman forts, as for example at Cramond, Midlothian (Goodburn 1978). Remains of narrow or cord-rig cultivation in later prehistoric contexts have been widely recorded, particularly in the Borders, even in what are now environmentally unsuitable locations, though attaching precise dates to these field remains is still a problem (Halliday 1982; Topping 1989; Carter 1994). Quernstones are common finds on Iron age and Romano-British settlement sites in north Britain (e.g. Hill 1982c; Jobey 1988), though they have been surprisingly little studied. Additionally, dated regional pollen diagrams from eastern Scotland suggest the presence of mixed farming involving barley cultivation in the pre-Roman Iron age (Whittington and Edwards 1993), though wheat is attested as a relatively minor part of the plant assemblage from some sites further south in northern Northumberland (e.g. van der Veen 1992) and cereal pollen, specified as wheat in one case, is attested in Iron age levels in two pollen diagrams from Cumbria (Dumayne and Barber 1994). Thus, the weight of evidence seems at present to favour importation of perhaps a large part of the wheat requirements of the Roman army, though the availability of barley has probably been underestimated and it seems reasonable to assume that as much as possible would have been obtained locally.

By contrast, the long-established view has been that timber supplies were not of local origin. It was on the basis of results from excavations of the Roman fort at Fendoch in Perthshire, that a case was argued for the stockpiling of prefabricated timber by the Roman army and its subsequent shipment north to supply military fort-building needs (Richmond and McIntyre 1939). But detailed consideration of the evidence does not support the existence of such stockpiles: there is no consistent record of standard sizes of timbers; the seasoning of timber, thought to be an essential prerequisite for building, is an unnecessary extravagance; and the army regularly employed species which were not ideal for building purposes, such as alder, which implies the use of whatever was locally available (Hanson 1978b). The probability must be, therefore, that the Roman army would have tried to obtain locally all of its timber requirements for fort building, even if its preferred timber species were not always immediately available. This necessarily leads on to the question of whether or not such demands were feasible.

The extent of the natural forest cover of Scotland and the date at which it was extensively cleared has long been a subject of interest and research, and remains a matter of debate and topical concern (e.g. Dickson 1992; Dumayne 1993a, 1994; Whittington and Edwards 1993; Hanson in press). Recently the role of the Roman army in the clearance process has been strongly re-asserted (Dumayne 1993a, 1994; Dumayne and Barber 1994). However, this represents an oversimplification of the evidence (Hanson in press). Radiocarbon dates are not sufficiently precise to allow links to specific historical events. At best they will indicate the probability of Roman period rather than Iron Age clearance. Moreover, most of the published analyses suggest either that extensive woodland clearance was already well under way before the Roman conquest, as a subsequent paper by Dumayne seems to

Table 11.2 Dated pollen diagrams and the onset of major forest clearance (Figure 11.3)

Site	Onset of major clearance[1,2]	Closest estimated date range[2,3]	References
Black Loch, Fife	3035±75 BP cal 1392–1138 BC	cal 1392–1199 BC	Whittington and Edwards (1993)
Bloak Moss, Ayrshire	1535±90 BP cal AD 439–608	cal AD 439–608	Turner (1965)
Flanders Moss, Stirlingshire	1860±110 BP cal AD 27–327	cal AD 58–261	Turner (1965)
Loch Lomond, Dunbartonshire	1730±59 BP cal AD 251–393	cal AD 251–393	Dickson et al. (1978)
Steng Moss, Northumberland	1970±20 BP cal AD 21–69	cal AD 21–69	Davies and Turner (1979)
Fozy Moss, Northumberland	1820±45 BP cal AD 139–311	cal AD 139–249	Dumayne and Barber (1994)
Bolton Fell Moss, Cumbria	1860±60 BP cal AD 86–235	cal AD 116–235	K. E. Barber (1981); K. E. Barber et al. (1994a)
Walton Moss, Cumbria	2000±40 BP cal 32 BC–AD 62	cal 8 BC–AD 62	Dumayne and Barber (1994); K.E. Barber et al. (1994a)

[1] The calibrated figures given are for the full range at one standard deviation.
[2] All calibrations are derived from the radiocarbon calibration and statistical analysis program produced by the Research Laboratory for Archaeology, Oxford.
[3] The dates indicated here are statistically the most probable, at 80% confidence or higher, within the range of one standard deviation.

acknowledge (1993b), or that it occurred later in the Roman period (Table 11.2). Even that from Fozy Moss, Northumberland, just to the north of Hadrian's Wall, where extensive Roman period clearance is indicated and which seems to have prompted the recent reassessment, represents a process which seems to have lasted over 200 years, and the high grass pollen figures were not attained until towards the end of the phase of activity involved (Dumayne 1993a, 1994). Thus, this clearance is more likely to be linked to the long-term expansion of settlement and agriculture than any short-term needs of the Roman military. Moreover, other evidence does not support a picture of rapid forest clearance as a result of Roman demands. Site-based pollen analyses in central Scotland consistently indicate a largely cleared landscape at the time of the Roman arrival, though the difficulty with such analyses is the extent of the area from which the pollen is derived and the precise extent of clearance. There is also increasing archaeological evidence of arable agriculture even in the more remote uplands by the pre-Roman Iron Age (see above and Chapter 10). Furthermore, the rapid creation and short-term use, as well as the large size, of many temporary camps constructed by the Roman army while on campaign in the Lowlands of Scotland, such as that at Kirkbuddo, Angus (Plate 11.4), is likely to have precluded any substantive forest clearance during their construction. This provides further support for the existence of extensive areas of open land before the Roman conquest, though only occasionally can this be demonstrated directly (e.g. Welfare and Swan 1995).

Plate 11.4   The 63 acre (25 ha) Severan camp at Kirkbuddo showing as cropmarks from the air. Copyright: W. S. Hanson

But even if the pollen analyses, whether regional or site-based, show a significant reduction in tree pollen, there is still uncertainty about exactly how this relates to actual forest clearance. Examination of the pollen diagrams indicates that in most cases tree species still constituted some 45% or more of the pollen recorded (e.g. Dumayne 1993a), so that woodland must still have been an integral part of that landscape, though there is no direct correlation between the percentage of pollen recorded and the area of tree cover. Although Tipping (1992) argues, on the basis of radiocarbon-dated soil erosion in river valleys, that pollen diagrams tend to overemphasize the amount of tree cover and that clearance in some places may have been extremely extensive, this indication in the pollen record of continued woodland cover is supported by the presence of macroscopic remains of wood at Roman sites, in the form of bark, leaves, fruits or roundwood of small diameter, which are unlikely to have derived from non-local timber. These remains serve also to confirm the presence of a wider range of species than may be indicated by the pollen analyses (Table 11.3). Further support is provided by the contemporary or near-contemporary written sources. Though these can present problems, given the tendency of classical authors to rely less on historical accuracy than on dramatic effect and literary style when dealing with such detail, the presence of woodland within the landscape is a consistent feature of the historical accounts of Roman campaigns in the North. Even more convincing are the numerous references to the Caledonian forest. This seems to have been sufficiently extensive to impress those who first came into contact with it during the conquest in the later first century and thus to fix the literary stereotype so that it came to be synonymous with remoteness

Table 11.3 Tree species from Roman forts represented by macrofossil evidence

Fort site	Alder (*Alnus glutinosa*)	Ash (*Fraxinus excelsior*)	Birch (*Betula*)	Elm (*Ulmus*)	Hawthorn (*Crataegus*)	Hazel (*Corylus avellana*)	Oak (*Quercus*)	Rowan (*Sorbus aucparia*)	Willow (*Salix*)
Bar Hill	x	x	x	x	x	x	x	x	x
Bearsden	x		x			x	x		x
Birrens	x		x						x
Bothwellhaugh							x		x
Crawford			x			x			x
Elginhaugh	x					x	x	x	
Mollins			x				x		
Newstead	x		x			x	x	x	x

and difficulty of progress. The location and extent of this forest is more difficult to determine. The name suggests that it ought to have been focused on the Highlands, or at least north of the Forth, but knowledge of the forest before the campaigns of Agricola implies that it may have extended into the Southern Uplands (Hanson and Macinnes 1980).

IMPACT ON THE LOCAL ENVIRONMENT

Assessing the impact of these military acquisitions is more speculative. When the timber requirements of the army for fort building are translated into areas of woodland, the scale of the potential impact seems small. Calculations based on average modern yields for all productive woodland suggest that the timber required to build a typical auxiliary fort could have been obtained from clear-felling between 17 and 30 acres (6.9–12.1 ha). If the area in which it was constructed was still heavily wooded, much of this timber would presumably have been obtained from clearing the 14 acres (5.7 ha) or more necessary to provide sufficient room for the fort with its defensive ditches and an open area beyond. However, in areas that had already been extensively cleared of their woodland for agriculture, the impact of such a requirement would have been commensurately greater, particularly in respect of the larger and therefore older trees. Clearly, felling for building purposes immediately after the Roman conquest at the end of the first century AD was a single event after which woodland could have regenerated, but its effects may still be evident in the second-century occupation when the more massive squared timbers which had been commonly used in the earlier period for the construction of gateways and towers seem to have been less readily available, and greater use was made of roundwood generally (Hanson 1982).

The locally-provided food requirements of the army were, of course, continuous as long as the military presence was maintained. Estimating the impact of obtaining locally the full barley requirement calculated above, requires that it be translated into a percentage of local production figures, which in turn requires knowledge of both the annual yield and the area under cultivation. The latter figures can at best only be guessed at and would in any case vary according to local circumstances. But assuming yield figures of only two-thirds of those attested in eastern Scotland in the 1950s (SSSES 1951), the area necessary to produce the barley required would have been of the order of 8500 acres (3440 ha). This represents some 570 acres (230 ha) for each of the estimated 15 forts or fortlets likely to have been occupied in the first century by an element of cavalry. This in turn represents the product of only 7% of the 8040 acres (3254 ha) of land within a 2 mile (3.2 km) radius of any single site. The figure is reduced to less than 2.5% if all the forts occupied at this time are included in the calculation. The annual requisition or purchase of the product of such an area does not seem excessive for the more fertile parts of eastern Scotland and may have served to stimulate the local economy to produce a surplus (Breeze 1989), were they not already doing so. It may have also encouraged them to take advantage of the Roman market, perhaps by attempting to grow more wheat, as appears to have occurred in the Netherlands (Groenman-van Waateringe 1989). This might be reflected in the archaeological evidence in a number of ways: by the

Plate 11.5 Ditched field systems outside the *vicus* at Inveresk showing as cropmarks from the air. Copyright: W. S. Hanson

establishment of new field systems, such as is indicated at various sites in northern Northumberland (Gates 1982) or near the second-century Roman fort at Inveresk (Leslie 1990) (Plate 11.5), assuming the latter is not part of the military *territorium* around the fort (see below); by the presence of quantities of Roman traded goods on native sites, perhaps best attested in the case of Traprain Law, East Lothian (Jobey 1976) and the lowland brochs (Macinnes 1984a); and by an increase in the production of wheat indicated by samples recovered during excavation on native sites or from pollen analysis. Only the latter cannot currently be substantiated to some extent, though it should be stressed that few native sites of the Roman period have been examined in recent years and dated regional pollen cores with identifiable cereal pollen from lowland contexts are rare.

In less fertile areas, however, with both lower yield figures and less land suitable for cereal growth, the effect may have been more dramatic, as Higham (1989) has argued was the case in northern England. In areas that were not capable of reacting to the stimulus of demand, whether because of more marginal environmental conditions or the absence of a socio-economic structure that could readily adapt to the potential of a market economy, the Roman presence is likely to have had a depressing effect in terms of the subsistence economy. This might be reflected by the abandonment or contraction of settlements, and a consequent reduction in agricultural activity which ought to be visible in the pollen record. On present evidence, at least, neither circumstance can be substantiated. Pollen evidence for forest clearance, as noted above, indicates that agricultural expansion continued through the Roman period, while examination of native settlements in the more marginal areas of northern Northumberland and the Scottish Borders suggests an increase rather than a decrease in population during the Roman period (Jobey 1974). Though in the present context the latter may seem more likely to reflect the generation of a surplus from pastoral rather than arable farming, attention has been drawn to the possible correlation between settlements showing signs of expansion and those associated with probable arable field systems (Gates 1982). It must be stressed, however, that the chronology of these sites is not sufficiently precise to assert categorically that such developments are the result of the Roman occupation.

A more immediate and dramatic impact on the environment as a result of the Roman presence has been claimed recently. A marked decline in levels of cultivation and a regeneration of woodland has been noted in a number of pollen diagrams from eastern and north-eastern Scotland dating in broad terms to the early first millennium AD (Whittington and Edwards 1993). The suggested explanation for these phenomena is that they are a direct result of the ravaging action of the Roman army while on campaign through hostile territory. Clearly there are problems with the precision of dating based solely on radiocarbon assay, but this is an attractive hypothesis concerning an area within which Roman activity was primarily antagonistic. However, the suggested impact seems out of proportion to the size and scope of the military actions which are thought to have stimulated it. Certainly the Roman army lived off the land while on campaign, but the effects would have been localized and are likely to have been short-lived, unless the phenomena recorded are a reflection of the longer-term impact on the farming population of losses in battle through death or slavery (Hanson and Macinnes 1991). However, at present the concomitant suggestion of depopulation and decline cannot be supported in the archaeological record of settlement patterns in the North and East.

It has long been argued that the Roman conquest and occupation brought about major changes in native settlement patterns generally. In particular the move from defensive to enclosed but non-defensive settlement has been seen as a direct consequence of the imposition of the *pax romana* (Jobey 1966). There is growing evidence, however, that this development occurred before the Roman arrival, as indicated by excavation at Broxmouth, East Lothian (Hill 1982c) and more recently at Eildon Hill North, Roxburghshire (Owen 1992), where the defences had ceased to be maintained before the arrival of Roman material on the sites. This is not to say that no defended sites remained in use, but rather that they had already ceased to be the most numerous settlement type. Indeed, the direct effect of the Roman presence on the nature and location of native settlement has probably been overestimated. Nor can the appearance in the Lowlands of the distinctive Highland settlement form, the broch, as for example at Edin's Hall in Berwickshire (Plate 11.6), any longer be attributed to an indirect effect of the Roman departure at the end of the first century (Macinnes 1984a).

To what extent it was standard Roman practice to confiscate large areas of land around forts for military use is uncertain and the nature of the control over any such lands is much disputed (Sommer 1984). The presence of military land (*territorium* or *prata*) is certainly attested epigraphically in various parts of the Roman empire, though it is usually associated with legionary fortresses. If such confiscations were commonplace in relation to all types of military establishment,

Plate 11.6 The broch, unenclosed settlement and earlier hillfort at Edin's Hall from the air. Crown Copyright: by permission of Historic Scotland

the potential detrimental impact of the military presence on the local economy of Scotland could have been considerable, depending upon the size of the areas confiscated and whether or not the indigenous occupants were cleared from them. On the other hand, the clustering of native settlements around forts in north-western England would suggest that the local population retained use of the land even if it was in military ownership (Higham 1991).

Indeed, once the conquest had been completed it was in Rome's interest to maximize the economic return from taxation, while a general principle of minimal interference was always preferred. Neither of these requirements is compatible with any major disruption of local settlement. The distribution of Roman finds from native sites suggests that contact with the occupying forces was limited, and confined in the main to the upper elements within the local social hierarchy (Macinnes 1984b, 1989). Nor is this distribution entirely commensurate with the supposed stimulatory effect of the imposition of monetary taxation and a market economy (Breeze 1989; Hanson and Macinnes 1991). As has recently been suggested in a different geographical context, taxation in kind may have been more widespread in the early Empire than is normally assumed (Braund 1991). This would simplify the relationship between local taxation and the food requirements of the garrison of occupation. It would also serve to reduce the impact of such requirements on the indigenous population if they were simply paying tribute to a different master. The longer-term effect of this process on native social structure, within which exchange systems would previously have been embedded, is more difficult to calculate. Finally, though we see political unification taking place in Scotland during the later third and fourth centuries AD, with kingdoms beginning to grow out of the previous tribal-based social structure, linking this process directly with the Roman presence remains speculative (Mann 1974).

For many years it has been almost axiomatic in studies of the period that the Roman conquest must have had some major medium- or long-term impact on Scotland. On present evidence this cannot be substantiated either in terms of environment, economy or, indeed, society. The impact appears to have been very limited. The general picture remains one of broad continuity, not of disruption. Though some sections of society may have been affected to a greater extent than others, the core remained largely untouched. It is always more difficult to explain why something did not happen, but in the context of acculturation and change much depends upon the potential cultural homogeneity between the native population and the invaders, since Rome relied heavily upon the conquered peoples to romanize themselves. The short timespan of the occupation was no doubt also an important factor, for it should not be forgotten that the Roman presence in Scotland was little more than a series of brief interludes within a longer continuum of indigenous development.

ACKNOWLEDGEMENTS

I am grateful to Dr L. Macinnes, Dr J. Dickson and Mrs C. Dickson for their helpful comments on a draft of this chapter, and to Prof. W. Groenman-van Waateringe for the opportunity to read her paper on the diet of Roman soldiers in advance of its publication.

12   The Early Historic Period: An Archaeological Perspective

IAN B. M. RALSTON AND IAN ARMIT

INTRODUCTION

General

In the first millennium AD the archaeology of Scotland becomes, for the first time, substantially 'text-aided'. The existence of near-contemporary documents has largely determined the agenda archaeologists have set themselves. So much so in fact that, until recently, the principal use of the evidence of material culture was to illustrate historically derived perspectives. This focus has affected both broad-scale treatments of the first millennium AD archaeological evidence, as well as more detailed considerations, in which correlations between archaeological and literary evidence may perhaps more pertinently be made. For example, the writings of the Venerable Bede furnish a range of expressions for settlement types that should, in theory at least, be matchable from archaeological evidence (Alcock 1988a).

Over the last 20 years or so our knowledge of the range of sites and monuments, particularly settlements and burials, relating to this period has grown considerably. The following review highlights the range of archaeological data now available and indicates the main directions of current research.

The legacy of Rome

After the campaigns of the Emperor Septimius Severus, which died with him in AD 211, Roman intervention in Scotland seems to have been slight, until such later fourth-century events as the shadowy Pictish wars (Maxwell 1987, table 1). During the first half of the millennium, there are hints in various Classical sources of the progressive amalgamation of previously independent tribes in Scotland (Breeze 1982, 1994). This trend, duplicated on the Continent amongst the Germanic tribes, has been seen as a response to Roman proximity, and proposed as an essential step on the route towards state formation. In Free Germany, nearness to the Empire was also marked by the movement in considerable quantity of high-status Roman goods and weapons substantial distances beyond the imperial frontier, some in due course to be deposited in princely graves (Hedeager 1992). Although high-quality Roman

Scotland: Environment and Archaeology, 8000 BC – AD 1000. Edited by Kevin J. Edwards and Ian B. M. Ralston.
© 1997 The editors and contributors. Published in 1997 by John Wiley & Sons Ltd.

material of third and fourth century date has been found in Scotland, sometimes, as in the case of the collection of copper alloy containers from Helmsdale, Sutherland, mixed with earlier items (Spearman 1990), neither its quantity nor its contexts suggest that its impact was especially significant. Hoards of third-century coins, especially from eastern Scotland north of the Tay, may indicate the payment of the Roman forerunner of danegeld. The significance of such evidence is sometimes wholly ambiguous: thus the hoard of silver treasure, recovered within the hillfort at Traprain Law, East Lothian, might have originated as either a diplomatic gift to a local potentate or the by-product of successful raiding.

Continuities from the pre-Roman period

The abbreviated tenure of the Roman military over Scotland, and the failure to consolidate this with civil rule, has already been noted (Chapter 11). It is thus unsurprising that the archaeological record fails to provide straightforwardly readable evidence for the impact of Roman activity on the indigenous communities.

Significant changes in the settlement record previously thought to reflect the *Pax romana* have lost their assumed Roman associations with the advent of radiocarbon dating. Undefended settlements of stone-footed roundhouses, sometimes overlying earlier forts, can now be seen to begin well before the Roman horizon, as at Broxmouth, East Lothian (Chapter 10), although how long they continued in use is less certain. At The Dod, Roxburghshire, for example, such circular structures were overlain by rectilinear ones during the course of the first millennium AD. Contrastingly, in Strathmore and adjacent areas, souterrains, with their implications of agricultural surplus, were abandoned in the early centuries AD. The decline of these numerous storeplaces may relate, however, to increasing centralization in the control of resources amongst the emergent Picts, or perhaps to the adoption of different storage technologies, rather than to the influence of the Roman army. Equally, the construction of complex Atlantic roundhouses in Scotland south of the Forth seems to date primarily to the decades between the Flavian and Antonine occupations, but equivalent sites continue to be occupied later further north, as at Leckie, Stirlingshire.

In sum, no direct correlations between Roman activity and the characteristics of the native settlement record can be maintained. Furthermore, site types conventionally ascribed to the pre-Roman Iron Age, notably hillforts and promontory forts, continued to be occupied and indeed to be constructed *de novo* during the first millennium AD (Alcock 1987, figure 4). This emphasis on continuity has led some scholars, especially those working in Atlantic Scotland, to define the entirety of this span as the Iron Age – such that terms like 'middle Iron Age' now have a radically different chronological meaning in Atlantic Scotland from that employed in the south of Britain.

PROTOHISTORY

Archaeology, documents and history

Interpretation of the written sources for first-millennium AD Scotland is fraught with difficulties. For example, the King Lists include mythical reigns, particularly amongst the earliest entries, representing attempts by the literate servants of sub-sequent royal houses to bolster pedigrees. Many historians are sceptical of the value of much detail recorded for periods before about the seventh century: Smyth (1984, 36) has described the absence of reliable records as likely to have provoked 'immense distortion' in our understanding. Other sources, notably the *Senchus Fer nAlban*, provide enumerated detail of considerable potential for making archaeological correlations (Nieke 1983), in this instance of the households of the Dalriadic Scots, but the surviving text is several centuries more recent than the circumstances it describes (Bannerman 1974). The most sustained recent effort to link archaeological remains in detail with the historical record has been the Alcocks' examination (1981 *et seq.*) of sites referred to by name in the Annals and like sources; such sites are inevitably associated with either the religious or secular elites and form the corpus of 'Early Historic' sites in the narrow sense.

Amongst place-names, one may be singled out as an indication of the positive potential of this set of evidence (Figure 12.1). The initial element 'Pit' (roughly meaning a piece of land) is common in place-names between the Moray Firth, Drumalban and the Forth Estuary (Whittington 1975). This prefix is Pictish, but such names usually have specifics of Gaelic/Scottic origin (Nicolaisen 1995). Thus, the naming horizon marked by these mixed names is conventionally attributed to the ninth century, and to the eastward expansion of the Scottic aristocracy. More contentious, however, is the interpretation of what these names may imply in landholding terms – whether the transference of ownership of established units to incoming Gaelic speakers, new creations for new proprietors, or simply a change in the dominant language. Whichever view is preferred, this episode must coincide with a social transformation of some magnitude. 'Pit-' names are associated with good-quality land, at low altitude and generally at some distance from the coast: the case for seeing these as the cores of estates, themselves components of larger units (cf. thanages), has been argued by Driscoll (1991), more especially using evidence from Strathearn.

The peoples of Scotland

Historical sources and place-name studies demonstrate the complex political and cultural mix present in Scotland during the first millennium AD. Indeed 'it may be doubted whether any country of comparable size anywhere in Europe had to contend with so many different ethnic groups' (Cowan 1984, 135). By the middle of the millennium, these comprised the Britons in Clydesdale and south-west Scotland, the Gododdin (the successors of the Votadini; and also British) in the South-East and the Picts north of the Central Lowlands. These groups appear to have spoken Celtic languages and are envisaged as the successors of Iron Age tribes listed on Ptolemy's map. The Picts also may have employed another, non-Celtic, language, but the extent to which this is simply represented by loanwords is unclear (Forsyth 1995). The Scots appear in Argyll by around AD 500. Another Celtic-speaking group, the dynastic movement from Ulster that marks their historical emergence probably consolidated much longer contacts across the North Channel. During the seventh century AD, Germanic-speaking Angles penetrated into south-east Scotland, as a northward expansion of Bernicia, initially at the expense of the Gododdin. The

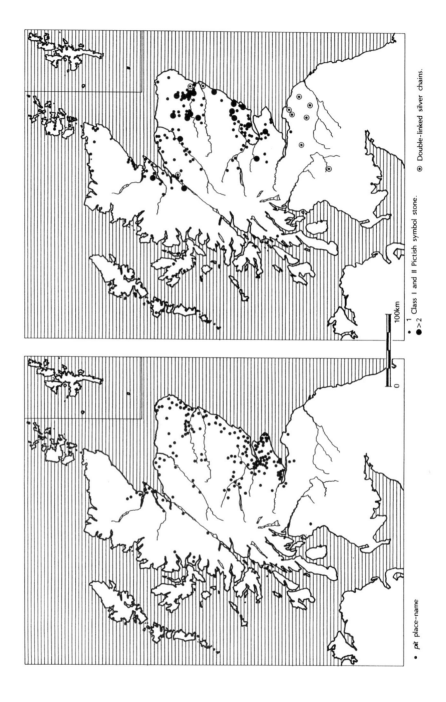

Figure 12.1 Distribution of 'Pit-' names, Pictish symbol stones and Pictish silver chains. After Whittington (1975) and Henderson (1979)

existence of these groups has long been accepted as well established, although it is not at all clear from archaeological evidence that their identities and ethnic reality were uniformly recognized by the layers of society below the ruling elites.

Archaeologists have long attempted to associate these groups with specific forms of material culture. This procedure has met with understandably limited success, there being no necessary coincidence between material culture and ethnic affiliation (Alcock 1987, 1993a); it is sometimes underpinned by dangerously circular arguments. While symbol stones coincide in their distribution more or less with historically known Pictish territories and are thus accepted as a Pictish product, heavy silver chains, two of which are inscribed with symbols cognate with those found on the stones, tend to be found outwith Pictland (Henderson 1979: Figure 12.1); their identification as of Pictish manufacture may be less secure. In other instances, whilst both findspot and the cultural tradition of the craftsman who made it may be clear (e.g. the ninth-century Anglo-Saxon silver blast-horn fitment from Burghead fort in Moray), the significance of the object in the context from which it was recovered may elude us. In general, it remains difficult to correlate the Early Historic kingdoms of Scotland straightforwardly with the archaeological evidence of structures and of more mundane material culture. Such evidence potentially provides a different perspective from the themes of dynastic rivalry, territorial aggrandizement and church–state relations, which were major concerns in the monastic *scriptoria* in which the surviving historical record was composed.

Warfare

Politically and militarily, however, the distinctions between named cultural groups seem to have been of major significance. The tenor of the historical record, and the characteristics of the enclosed settlement sites, are eloquent of 'nations . . . organized for war' (Alcock 1988b, 328), arguably in a systematic fashion distinct from the intermittent raising of war-bands. The fluctuating fortunes of their horse-borne leaders mean that it is not possible here to recount the frequently changing territorial holdings of the various dynasties; thus, although by the mid ninth century the Scottic leadership had extended its hold over the eastern lowlands north of Forth, a century earlier they had been at the mercy of Pictish assaults on the heartland of Dalriada. That warfare was not solely land-based is made plain by accounts like that in the Annals for AD 729, recording the loss of 150 Pictish ships perhaps on the southern shore of the Moray Firth (Anderson 1922, 226). The importance of seaways is also critical to trade and other external linkages during the first millennium, and significant for the location of many of the major settlement centres.

Centralization

Towards the end of the first millennium AD, written sources intimate the increasing coalescence of elites within Scotland, most famously marked by the accession of Kenneth MacAlpin in the mid ninth century to the kingdom of the Picts as well as the Scots. By this time, historical records point to repeated raids by Vikings (Chapter 13) into the heartland of Scotland. This new threat may have stimulated considerable reorganization within native elites, including dynastic marriages to underpin alliances. Attacks also weakened certain states, e.g. the Britons of Strathclyde and Anglian Northumbria, allowing the Scottic leadership subsequently to enlarge its territories. Success at the battle of Carham (AD 1018) meant that the Scots, at least nominally, had extended their lands to the Tweed. As in more recent times, however, the level of control exercisable by lowland magnates over lands beyond the Highland line was limited, and Moray for example often pursued its own interests (Ralston and Inglis 1984; Foster 1992, 1996; Shepherd 1993). In sum, it may be contended that during the first millennium AD, the conduct of territorially aggressive kings and their retinues led to the identification of peoples, rather than the contrary.

At a different scale, Driscoll (1991) has attempted to trace the archaeological correlates of the historical thanage or small shire which comprised the fundamental land units of the Pictish heartlands. The formation of such holdings will have provided the building blocks for the establishment of extensive kingdoms in the mid-first millennium AD, as the assumed kinship-based patterns of Iron Age landholding were increasingly formalized and overlain by relations based on clientage. This organic model of indigenous development has been proposed as an alternative to a perspective in which early state formation in Scotland is envisaged either as a faint imitation of practice further south in Europe (Driscoll 1991), or as a direct reaction to the nearness of Roman armies. Thus, too, the hypothesis that the leaders of the immediately post-Roman kingdoms of southern Scotland were pro-Roman kings of a series of buffer states, on account of their latinate names, has equally been challenged (Breeze 1994).

SETTLEMENT EVIDENCE

Fieldwork over recent decades has greatly increased the variety and numbers of first-millennium settlement sites (Figure 12.2). This has particularly been the case with secular examples, although religious centres, such as monasteries, should not be treated wholly apart as they could and did (e.g. in the cases of Iona and Whithorn in Galloway) play a significant role in political, economic and social

Figure 12.2 (*opposite*) Sites mentioned in the text. 1: Pool, Sanday; 2:Buckquoy; 3: Brough of Birsay; 4: Howe; 5: Cnip and Loch na Berie 6: Ackergill; 7: Wag of Forse; 8: Helmsdale; 9: Eilean Olabhat; 10: Coileagean an Udail (Udal); 11: Allt na Fearna Mór, Lairg; 12: Burghead; 13: Sculptor's Cave, Covesea; 14: Green Castle, Portknockie; 15: Gaulcross; 16: Rhynie and Tap o' Noth; 17: Monboddo; 18: Iona; 19: Dunollie; 20: Pitcarmick; 21: Restenneth; 22: Boysack Mills; 23: Carlungie; 24: Dundurn; 25: North Mains, Strathallan; 26: Forteviot and Dupplin; 27: Scone; 28: Clatchard Craig; 29: Easter Kinnear; 30: Tentsmuir; 31: Lathrisk; 32: St Andrews and Hallow Hill; 33: Lundin Links; 34: Dunadd; 35: Loch Glashan; 36: Dumbarton Rock; 37: Leckie; 38: Cairnpapple; 39: Ratho; 40: Castle Rock, Edinburgh; 41: Traprain Law; 42: Whitekirk; 43: Dunbar; 44: Doon Hill; 45: Courthill, Dalry; 46: Buiston; 47: Deil's Dyke; 48: The Dod; 49: Sprouston; 50: Mull of Galloway; 51: Whithorn; 52: Ardwall Island; 53: Hoddom

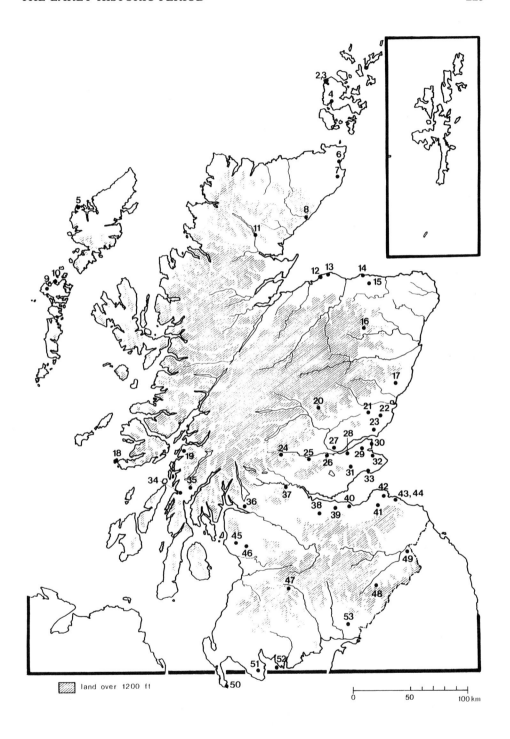

land over 1200 ft

0   50   100 km

Plate 12.1   Aerial view of the hillfort at Dundurn, Perthshire; the external cultivation terraces are also visible. Copyright: Ian Ralston

changes. Many of the categories of site are not restricted to the territory ascribed to a single, named people.

Forts

Early Historic defended sites include an important class of high-status settlements conventionally known as nuclear or nucleated forts, e.g. Dundurn, Perthshire (Alcock *et al.* 1989; Plate 12.1) and Dunadd, Argyll (Campbell and Lane 1993, figure 6.2). The topography of the craggy hills they occupy permitted the development – often over time – of a hierarchically linked set of enclosures. The configuration of other sites recorded in historical sources (e.g. Castle Rock, Edinburgh) is less certain owing to medieval and later rebuildings. Excavations at Clatchard Craig, Fife (Close-Brooks 1986), moreover, demonstrated that hillforts, of seemingly conventional Iron Age appearance, could also be of first millennium AD date, for here, despite the recovery of Iron Age pottery, three defensive lines (two timber-laced and the third incorporating reused Roman masonry) are datable solely to the post-Roman centuries. Some of these enclosed sites formed foci for the military and political power struggles during the second half of the millennium: references to sieges at this time imply that some at least held sufficient supplies for this activity to be necessary.

Forts, generally less than one hectare in extent, have been identified in both hilltop and promontory locations and were defended by either drystone or timber-laced walls, or on occasion initially by palisades. An important series of excavations by Alcock has recently increased our knowledge of their nature and chronology.

Plate 12.2   The timberlaced rampart (built of reused oak) at Green Castle, Portknockie, is dated by radiocarbon to a period when Viking activity in the Moray Firth is likely. Crown Copyright: Ian Ralston

The identification of a nailed timber framework at Dundurn, along with that previously recorded at Burghead, Moray, demonstrates the conspicuous consumption of iron in certain constructions. Such seats of power are significant both at the local scale and in terms of the wider geopolitical make-up of Scotland. Relatively numerous, they coincide with peripatetic monarchies (Alcock 1988c) and senior aristocracies, who were required to travel through their territories to secure their powerbase and to obtain and consume the resources that underscored their status. Such forts were complemented by other enclosed sites of lesser scale – some of which may have been simply sturdily built houses – including duns in Argyll and ring-forts in upland Perthshire (Taylor 1990).

From the limited evidence available, it seems that nucleated sites and their equivalents became less important as kingly centres late in the millennium, when undefended lowland locations, as at Forteviot in Strathearn, Perthshire (Alcock and Alcock 1992, illustration 11), were increasingly favoured. Forts continuing in use include the coastal promontories edging the southern Moray Firth, notably Burghead, exceptionally large amongst known first-millennium AD fortifications, and Green Castle, Portknockie, Banffshire (Plate 12.2); these may have acted as land bases for Pictish naval units facing the Norse across the Firth.

Cellular buildings in the North and West

While the construction of visually impressive fortified structures continued throughout much of Scotland, in the Northern and Western Isles the broch towers and related roundhouses had ceased to be built long before mid-millennium. Wheel-houses too seem not to have been constructed after the early centuries AD. Such monumental buildings gave way to a tradition of less imposing cellular architecture, often occupying former broch (as at Howe, Orkney: Smith 1994) or wheelhouse sites with no apparent hiatus.

Cellular buildings, frequently semi-subterranean or built into midden mounds, had a long history in Atlantic Scotland, as for example at Skara Brae (Chapter 8). This structural tradition survived during the 'broch' period, when spatial arrangements within structures clustered around Orcadian towers such as Gurness, and even the towers themselves, mirrored those of earlier buildings (Armit 1990b).

First-millennium AD cellular structures include examples ranging from the first century AD at Cnip, Lewis (Harding and Armit 1990) to the apparently eighth-century buildings at Buckquoy, Orkney (Ritchie 1977) and Loch na Berie, Lewis (Harding and Armit 1990) (Figure 12.3). These stone-footed buildings display considerable diversity in their forms, excavations at Pool on Sanday, Orkney (Hunter 1990), for example, increasing the known range. The Buckquoy and Berie structures, however, suggest the appearance of a more formalized figure-of-eight plan in the later pre-Norse period in peripheral Pictland. That such cellular structures remain absent in more easterly Pictish areas is wholly unexceptional, and emphasizes the fact that mundane domestic architecture is unlikely to be consistent in form over large and geographically disparate territories.

Crannogs

The exploitation of Scotland's rich heritage of settlement sites located within its lochs remains underdeveloped. Whilst work on the Hebrides and in Loch Tay has shown that this niche was used for the construction of artificial islets for settlement from the Neolithic, and especially during later prehistory, pioneering excavations by Robert Munro in the nineteenth century, supplemented by more recent work (Barber and Crone 1993; Crone 1993b), has demonstrated that this setting continued to be used during the first millennium AD. Classic instances include the Ayrshire site of Buiston, where dendrochronological and other evidence indicates a complex history for a site that eventually contained a timber roundhouse and surrounding features which, from their scale and the associated artefacts, again betoken a first-millennium elite residence (Crone 1991). Earlier rescue work at an example in Loch Glashan (Argyll) produced a brooch perhaps manufactured at Dunadd and a considerable range of wooden artefacts of this period (Earwood 1990).

Timber halls

Excavated examples of substantial timber buildings of post-Roman date otherwise continue to be relatively few. On the shoulder of Doon Hill, East Lothian, two successive rectilinear timber halls were excavated within a polygonal palisaded enclosure (Figure 12.4). The later building (Hall B), incorporating an annexe at one end, conformed to a type examined at Yeavering, Northumberland, and there attributed to the seventh century (Hope-Taylor 1977, 1980; Reynolds 1980; cf. Scull

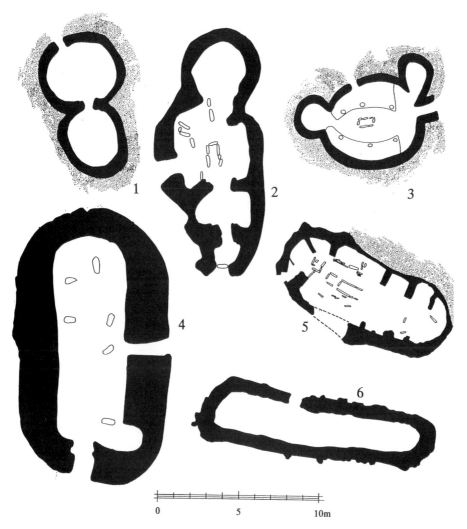

Figure 12.3 Pictish buildings, mainly from peripheral Pictland, in a variety of architectural styles. 1: Carlungie, Angus; 2. Buckquoy, Orkney; 3: Coileagean an Udail (Udal), North Uist; 4: Wag of Forse, Caithness; 5: Howe, Orkney; 6: Pitcarmick, Perthshire. After Alcock (1984), Smith (1994) and Stevenson (1991)

1991 for a fuller discussion of chronology). In south-east Scotland, the Hall B type has thus been correlated with the Anglian expansion of the mid-seventh century. Accepting this historically derived date, it follows that the preceding Hall A must have been erected by an earlier community, presumably of the Gododdin; the coincidence between its position and that of its successor seems too exact to argue for reuse of the site of a much older structure (Hope-Taylor 1980; cf. Smith 1991).

Further substantial rectilinear timber buildings dating to and before the seventh century may thus be anticipated, although major halls continued to be built into the present millennium, as the reinterpretation of the Courthill at Dalry, Ayrshire,

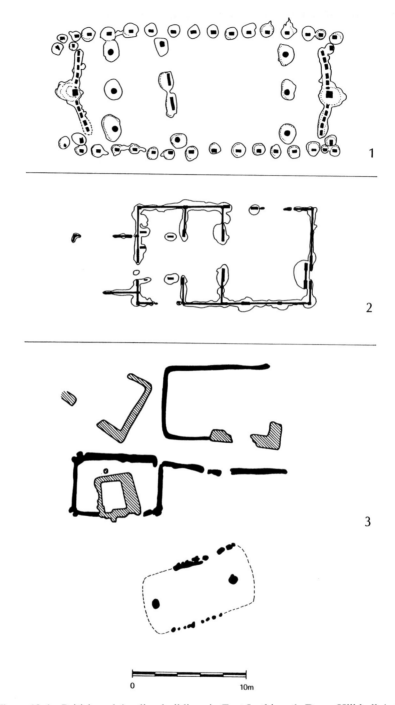

Figure 12.4   British and Anglian buildings in East Lothian. 1: Doon Hill hall A; 2: Doon Hill hall B; 3: structures of periods 3 (timber-built: solid line) and 4 (stone-built, hachured) at Castle Park, Dunbar; the *Grübenhaus* is the southernmost structure. After Reynolds (1980) and Holdsworth (1993)

demonstrates (Scott 1989). While some of these halls were presumably occupied by secular nobles, other examples, as at Hoddom, Dumfriesshire, dated to *c.* cal AD 650–790 (1340±70 BP; GU-3030) occur in an ecclesiastical context (Lowe *et al.* 1991, figure 2). All known unexcavated instances are represented by cropmarks, those on the terrace above the Tweed at Sprouston, Roxburghshire, lying within a palimpsest of considerable archaeological potential (Smith 1991), while an apparently smaller and simpler cluster has been identified at Whitekirk, East Lothian (Brown 1983). Further examples have been found as far north as the Moray Firth, but the only excavated hall north of the Mounth at Balbridie is Neolithic (Chapter 8). The cropmark evidence suggests considerable variety amongst these structures, there being significant differences between the presumed buildings at Lathrisk, Fife (Maxwell 1987, figure 2) and that at Monboddo in Kincardineshire (Ralston 1984; Foster 1996, figure 35), but it is likely that, isolated or grouped, imposing rectangular buildings will form a component of aristocratic architecture in much of Scotland away from the Atlantic coastlands.

Grübenhäuser

In recent years, small, semi-subterranean rectilinear structures in the *Grübenhaus* tradition, previously recognized as far north as Northumberland (Gates and O'Brien 1988; Scull 1991), have been recorded from Scotland. Primarily associated with the Anglian expansion, the most northerly excavated examples are at Dunbar, East Lothian (Holdsworth 1991, 1993; Figure 12.4) and, set within a palisaded enclosure, at Ratho, Midlothian (Smith 1993). As elsewhere in Anglo-Saxon areas, *Grübenhäuser* seem often to have served as workshops; loomweights from both Dunbar and Ratho indicate weaving.

Another semi-subterranean structure has been identified within the ecclesiastical complex at Hoddom in Dumfriesshire (Lowe *et al.* 1991) and there interpreted as a smoke-house for curing meat. Sunken-floored circular buildings with drystone revetting walls have also been found at Easter Kinnear in Fife, seemingly dating to the mid-first millennium AD (Selkirk 1992) and there replaced by a set of rectangular timber buildings. These structures lie firmly in Pictish territory and do not represent the same cultural tradition as *Grübenhäuser*; rather it is postulated that below-ground cellarage within these buildings fulfilled the role previously carried out by souterrains.

Pitcarmick houses

Survey by RCAHMS (1990) in north-east Perthshire identified another building type which may relate to rural, lower-status Early Historic settlement, on the upland margin of the southern Pictish heartland (Stevenson 1991). Pitcarmick houses are approximately rectilinear or slightly trapezoidal stone-footed buildings varying in length from about 10 m to, exceptionally, 30 m (Figure 12.3). They occur both in clusters and singly, often in association with extensive field systems. Many appear to include sunken internal areas possibly representing partial use as byres. Such buildings postdate later prehistoric hut circle groups stratigraphically; they also differ from buildings in pre-Improvement settlements. Radiocarbon determinations from recent excavations at North Pitcarmick, Perthshire (Barrett and Downes 1994, in prep.) indicate that buildings of this type were in use between *c.* 600 and 1000 cal AD.

Unenclosed centres

Certain unenclosed settlements seem to stand apart from the remainder and suggest centres of some importance. The constellation of cropmarks north of an apparently Neolithic complex at Forteviot in Strathearn, although unexcavated, seems to correspond to one of a series of historically documented royal centres of the later first millennium AD (Alcock and Alcock 1992; Foster 1996, figure 29 and colour plate 3), others including Scone and St Andrews. Their juxtaposition with what were probably still then imposing remains of earlier date is a feature noted at other first millennium sites (Inglis 1987) and, although the cropmarks at Forteviot do not meantime include a feasting hall, there is evidence there for an important church.

At the Brough of Birsay, Orkney, the density of remains of Pictish date (Hunter 1986), the quantity of debris of metalworking (Curle 1982) and the subsequent importance of this tidal islet in Norse times, led Hunter to suggest that the site may have been 'proto-urban' in Pictish times, but in general archaeological indications of such complexity are few and tantalizing (Spearman 1988).

ECONOMY AND MANUFACTURE

Documentary sources imply that elite associations can be assumed for timber halls and some at least of the forts, although status can be harder to identify from the artefactual and other evidence recovered. Imported materials, primarily ceramics and glass, and the on-site working of copper alloys or precious metals have often been taken as indications of high status. Metalworking is regularly identified on Alcock's 'Early Historic' sites, and on sites cognate with them, such as Clatchard Craig, but is not, however, either universally present (small-scale examination of Dundurn produced no evidence of metal-working: Alcock *et al.* 1989) or exclusive to them. In the Western Isles, for instance, precious metals were in use, and handpins and penannular brooches made, in a ruined small stone building at Eilean Olabhat, North Uist (Armit 1990c, 1996), indicating that direct control of the metalworker by an aristrocratic patron is here unlikely.

Although Alcock (1987) has noted that numbers of sites which fulfil his criteria for 'Early Historic' status are located close to good-quality agricultural land, this would be expected of all sites which relied on their immediate hinterland for their subsistence needs. Indeed, the peripatetic character of royal progresses might argue against the need for the concentration of resource wealth in the immediate vicinity of such sites, as their function was to draw in consumable resources from wider territories. It was thus perhaps less essential for these to be favoured agricultural locations in their own right than to be both impressive in their settings and readily accessible to wider productive hinterlands.

Stock-raising, agriculture and wild resources

Direct evidence for agriculture and stock-raising attributable to this period remains quantitatively slight. In bone collections, cattle generally predominate in percentage terms and wild animals are markedly rare: this pattern is recognizable at Dundurn, for instance, where pigs were numerically the second domesticate and where the cattle seem to have been particularly small (Alcock *et al.* 1989, 222). Here, there was no evidence to support the importance of dairying, argued from the cattle populations of some contemporary Irish sites (McCormick 1983, 1992). A similar pattern was recorded from a smaller collection of bones from Dunollie, Argyll (Alcock and Alcock 1987), where cattle again formed the principal domesticate and wild species were rare – red deer, for example, only being recorded by the presence of antler.

It is only from Orkney that there are substantial published collections of animal bone. Buckquoy's Pictish (and also its Norse) horizons produced an important assemblage, extending to birds, fish and shellfish. For most of the Pictish horizons, cattle represented about 50% of the assemblage, with sheep second in importance (Noddle in Ritchie 1977, table 1); goat and horse were securely identified in small numbers. Poultry bones are extremely rare, but a range of wild birds are represented (Bramwell in Ritchie 1977, table 10). Fishbones, less significant than in Norse deposits, nevertheless indicate fishing from boats at least in inshore waters as well as from the shore (Wheeler in Ritchie 1977). Shellfish, dominantly limpets and winkles, may have been a foodstuff, but may also indicate bait (Evans and Spencer in Ritchie 1977). The bone collections (Seller, Colley, Jones and Turner in Hunter 1986, Appendix 4) from the settlement on the offshore islet of the Brough of Birsay (Orkney) must have resulted from bringing in livestock (probably already slaughtered) from sites such as Buckquoy on the adjacent mainland (Plate 12.3), although pigs and sheep could have been kept on the Brough in small numbers. Cattle are again the dominant species represented, with pigs being quantitatively unimportant. As at Dunollie, many bones had been shattered to extract marrow. Horse is, perhaps unsurprisingly in view of the location, absent. Although shellfish (principally limpets, perhaps for baiting lines) and fish species that require to be caught from boats are represented, neither is believed to have made a major contribution to diet. The small farmstead represented by the phase 8 deposits at Howe has also furnished interesting data on livestock: here both sheep (generally kept for longer than in previous phases) and pigs outnumbered cattle in terms of minimum numbers of individuals present, although cattle will still have provided the most substantial quantity of meat. Horse – the size of a small pit pony – was also present, and was eventually butchered. Hunting is represented by red deer (although these were in dramatic decline in proportional terms compared to earlier periods in the site's history), and antlers may have been imported (C. Smith *et al.* 1994, 143). Domestic cat, as well as dog, is recorded. The former was exploited for its pelt, as were foxes and otters. Seabirds as well as red grouse in some numbers are represented; the latter may have been taken by hawking, as kestrel and peregrine falcon, as well as goshawk – the last mentioned perhaps a prestige possession (Alcock 1993b) – are represented. Poultry comprise domestic fowl, goose and (rarer) ducks, which first appear in phase 7 (Bramwell 1994). The fish identified

Plate 12.3 The tidal islet of the Brough of Birsay. In the foreground, excavations are taking place at the Pictish and Norse settlement of Buckquoy. Crown Copyright: Ian Ralston

at Howe phase 8 show the same preponderance of species that could be taken from the shore, with bottom feeders such as cod that would require to have been caught offshore less numerous; the fishing strategy of the inhabitants is described as 'opportunistic' (Locker 1994, 159). A range of shellfish was also recovered, with common mussel being more frequently encountered in phase 8 deposits than previously.

In general, cattle of small stature seem to have been the main domesticated animal used as a food source. Whilst it would not be reasonable on present evidence to equate stalled buildings with the practice of keeping livestock indoors – the stalled building at Howe, for example, has an internal hearth (Figure 12.3) – one of the structures at Pool on Sanday has feeding bins along either side, and the more tentative evidence provided by the Pitcarmick-type houses has also been noted. Such structures may indicate that special facilities for sheltering livestock were becoming more prevalent. The importance of cattle more generally is underscored by textual references: Adomnan's *Life of St Columba*, for example, records the Saint's ability to cause cattle to increase in number to the population's benefit (Foster 1996), or to suffer when owned by wrongdoers (Anderson 1922, 58).

Crop plants are best indicated by Dickson's (1994) work on the phase 8 assemblage from Howe, Orkney, which amplifies the earlier summary offered by Boyd (1988). Numbers of recovered grains are very low, and are dominated by barley, principally *Hordeum vulgare var. nudum* (naked six-row barley), although the hulled form, *H. vulgare var. vulgare* (bere barley), appears during phase 7. Association with chickweed intimates that barley fields were manured (Dickson

1994, 135). Oats (*Avena* sp.) are sufficiently frequent to suggest, although not prove, that the cultivated form was present from phase 8; and flax (*Linum usitatissimum*) – to become much more prevalent in Norse deposits – is also recorded. Emmer wheat (*Triticum dicoccum*) appears meantime to be represented essentially in south-western Scotland (Boyd 1988).

It seems likely that animal products were of primary importance as commodities to exchange with the exotica (such as amphorae and casks of wine, imported ceramics and glass) that arrived along the western seaways. In a recent discussion, woollen cloth, cattle-hides, kidskins and vellum, as well as products from the wild have been proposed: the latter may have especially included white furs such as ermine and those of seal-pups, but also river-pearls, birdfeathers and down (Alcock and Alcock 1990, 127–128).

Terracing of hillslopes and valley sides, found particularly in south-east Scotland, but now known as far north as Strathbogie, Aberdeenshire, has traditionally been attributed to this period, but there seems little reason to restrict this technique for stabilizing steeper slopes for agricultural use to a narrow chronological band. Such terraces, for example, edge the summit on which Dundurn sits (Alcock *et al.* 1989; Plate 12.1). The wider geographical occurrence of this phenomenon certainly precludes its direct relation with the Anglian expansion, in contrast with Graham's (1939) view. Radiocarbon dates also make it clear that the narrow rig cultivation plots, some at least dug with hand tools, continued in use during the first millennium AD (Carter 1994), those at Allt na Fearna Mór, near Lairg, Sutherland, being overwhelmed by peat at a variety of dates between the fourth and thirteen centuries cal AD. The principal technological development in relation to cultivation seems to be the beginnings of ploughing with mouldboard ploughs, as opposed to the simpler ards known in Scotland from the Neolithic period onwards. Early evidence for such ploughing comes from Whithorn in Galloway, where Hill and Kucharski (1990) have identified plough pebbles in two horizons, the earlier attributable to about the seventh century AD. In contrast, little is known of the management of grassland. At Tentsmuir Sands, Fife, however, Whittington and McManus (forthcoming) report a phase of heather burning to encourage pasture dated to around the sixth century cal AD and subsequently sealed by aeolian deposits.

RELIGION AND LITERACY

Christianity

In the second half of the millennium, Christianity played a major part in political developments within Scotland, as well as in cultural and ritual spheres. After the initial Ninianic episode (Thomas 1981a, 275–294), it is represented by two strands, Irish–Scottic monasticism in the West, with its major centre on Iona, and the Roman version, established in Northumbria during the seventh century AD and thereafter extending its hold northwards. The latter's expansion is illustrated by the appeal of Nechtan, king of the Picts, to Ceolfrith, abbot of Jarrow in northern England, in AD 710, for Northumbrian masons to build him a church '*in more Romanorum*', implying a stone-built and mortared construction.

The distinction between the organizational structures of Irish and Roman Christianity contributed to the rather different rates of their expansion in northern Britain. The Irish–Scottic form, based on monastic *parochia* run by abbots, was accompanied by an ascetic, eremitic tradition that seems initially to have encouraged rapid expansion, particularly in the North and West, from its major centres. An archaeological correlate of the early establishment of Christian communities in the East, as within Dalriada, may be offered by the extensive series of simple cross-carved stones catalogued by Henderson (1987). As has already been noted (Chapter 6), the spread of Christianity will have impacted on food consumption patterns and preferences.

Contrastingly, Roman Christianity was established around territorial sees, each controlled by a bishop. In the longer term, Roman Christianity conveyed distinct advantages to the new elites, providing a divine model for an increasingly hierarchical conception of society. The contrast, however, can easily be exaggerated. For example, historical evidence for Iona, the main centre of Celtic Christianity, points both to the orderliness of its internal succession of abbots and to its important connections with the royal house of Dalriada.

Literacy and symbolism

An important concomitant of early Christianity was the spread of literacy (Nieke 1988). The setting down of records provided a new means by which secular kings could legitimize their power, enable elaborate pedigrees for themselves to be established, and communicate with distant peers and subordinates. Important as it was, this partnership of religion and secular authority seems unlikely to have been uniquely a product of Christianity. North of the Forth–Clyde isthmus, and primarily on the eastern side of the country, there developed from the sixth century or earlier a vigorous school of incised carving on unshaped boulders. These Class I Pictish symbol stones carry no overt Christian symbols, although stylistically they are influenced by Christian iconography. The inclusion of half-human/half-animal figures certainly suggests a mythical component. These stones probably served to transmit messages relating to control of land or resources, or commemorated the activities of powerful groups or individuals: some seem to have been set up over graves. Amongst the symbols are depictions of elements of the fauna, drawn, as Alcock (1993b) has emphasized, from life (Figure 12.5).

The later, Class II stones meld the designs found on the earlier stones with overtly Christian symbols and celebrate the status of the aristocracy through illustrations of their favoured pursuits, including warfare and the hunt, both conducted on horseback. The keenly observed depictions of horses (Alcock 1993b) indicate specimens that befit an aristocratic milieu (Plate 12.4). New evidence for the continuing importance of monumental sculpture in conveying politically charged messages in the last part of the millennium is provided by the recognition of an inscription on the Dupplin Cross, Perthshire, from the hillslope above Forteviot, linking the elaborate iconography thereon (Driscoll 1988) specifically with royalty (Alcock and Alcock 1992, 283; Plate 12.5).

Figure 12.5   The fauna of Pictland as depicted on symbol stones. After Jackson (1984)

Death, burial and ritual

Burial sites of the first millennium AD are now better known although many are recent aerial photographic discoveries which can only be inferentially attributed to these centuries. Amongst types securely attributable to the first millennium AD are long cist cemeteries in which extended inhumations were placed in stone-lined pits (Dalland 1992), and graves covered by small circular or rectangular cairns, from Ackergill in Caithness to Lundin Links, Fife (Close-Brooks 1984).

Long cist cemeteries are found at a number of early Christian centres, as at Ardwall Island off the Galloway coast (Thomas 1967), and at Hallow Hill near St Andrews (Proudfoot 1995, figure 16), and are also associated with inscribed stones, as at the Catstane cemetery to the west of Edinburgh (Cowie 1978). They may originate in seemingly less structured arrangements in pre-Christian contexts, as in the vicinity of henges at Cairnpapple, West Lothian and North Mains, Strathallan, Perthshire.

Plate 12.4   A battle scene on the Class II slab at Aberlemno Kirkyard, Angus. The helmeted figures are believed to represent the Anglian army, defeated at the battle of Nechtansmere in AD 685. Copyright: Ian Ralston

In the lowlands north of the Forth, aerial photography in the 1970s brought to light a regional tradition of square barrow burials, sometimes arranged in cemeteries and in some instances associated with round barrows; the former display considerable variation in size and form (Plate 12.6). Excavation has been too restricted to attribute securely an Early Historic date to these, although an example at Boysack Mills, Angus, is dated to the first millennium, both from a ring-headed pin and from radiocarbon determinations associated with subsequent disturbance (Murray and Ralston forthcoming). It may be suggested that the appearance of conspicuous funerary arrangments was a reaction to Christian practices. In some areas, there is also evidence for renewed interest in earlier ritual sites. At Gaulcross in Banffshire, for example, a hoard of silver items was deposited within an earlier stone circle, whereas the Sculptor's Cave, Covesea, on the Moray coast, its dark interior strewn with human bone deposited during the Late Bronze Age, had a series of Pictish symbols carved on its walls and at least one deposit of counterfeit late Roman coins made within it (Shepherd 1993).

The evidence from ecclesiastical sites

In recent years the quantity and range of archaeological evidence from some of the major centres of early Christianity has increased significantly as a result of new fieldwork. Alongside continuing work at Iona (RCAHMS 1982), of particular importance is the recognition of wooden churches, for example the recovery of an arcaded Anglian example, datable from the later eighth century (Hill 1991), in the

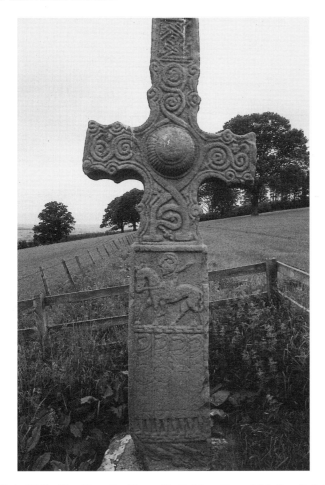

Plate 12.5 The Dupplin Cross, Perthshire. Copyright: Ian Ralston

impressive sequence of deposits at Whithorn. Contrastingly, stone church architecture of the period remains elusive, the lower part of the tower at Restenneth (Angus), for example, having been reassessed as of more recent date (Fernie 1986). A sculptured arch, datable to the second half of the ninth century, from the Water of May near Forteviot (Alcock and Alcock 1992, illustration 6) provides strong evidence for the former existence of a major stone church there.

Evidence of the range of activities carried out at monastic sites has been enhanced by finds including worked wood and leather from the enclosing ditch at Iona (J. W. Barber 1981) and by the examination of peripheral buildings, including bake- and brew-houses, within the enclosure of Hoddom, Dumfriesshire (Lowe *et al*. 1991). Artefacts, such as cross-marked quern stones with a restricted distribution in the Atlantic West, testify both to the processing of cereals and to relations between monastic and secular centres (Campbell 1987). It is noteworthy that deer are strongly represented in the animal remains from Iona (Chapter 6): they may have been consumed primarily during Lent.

Plate 12.6 Aerial view of cropmarks near Boysack in the Lunan Valley, Angus. In the foreground, on the left of Invergighty Cottage, is a cemetery of square and round barrows; on the opposite bank of the stream, land boundaries and a possible roundhouse are visible against other cropmarks of palaeohydrological origin. Crown Copyright: Aberdeen Archaeological Surveys

EXTERNAL CONTACTS

Aristocratic tastes, and the requirement for wine for Christian liturgical practice, contributed to the maintenance of contact along the Atlantic seaways from Iberia and western France. Imports, most visibly pottery vessels, are found more commonly in western Scotland than further east but are nowhere abundant. In the fifth century, there were links with the eastern Mediterranean and perhaps Constantinople (Fulford 1989), represented by small quantities of material that may have been redistributed from south-western England (Lane 1994), but thereafter kitchen- and table-wares from western France dominated (A.C. Thomas 1981b, 1990). According to his biographer, St Columba encountered sailors from Gaul at 'caput regionis', an important, but unidentified, west Scottish centre, in the sixth century. This may have been Dunadd.

The west Scottish material shows that the principal point of contact seems to have been at secular elite centres (Campbell 1987; Lane 1994), such as Dunadd and

Dumbarton Rock, rather than either monastic settlements or beach trading stations. The latter have been increasingly recognized in Scandinavia and may be considered likely in Scotland. Current excavations at Whithorn in Galloway should help to clarify the role of ecclesiastical centres in relation to long-distance exchange, but the evidence from Iona suggests that they were relatively insignificant (Lane 1994). Much more work needs to be done to establish patterns of external contacts with eastern Scotland, from which such material is generally rarer.

CONCLUSIONS

The emergence of states

Probably the major theme to emerge from the bringing together of the archaeological and historical records for Scotland in the first millennium AD is the distancing of ultimate power, whether in secular terms in the development of larger-scale political units, or in the adoption of the Roman form of Christianity, with its continental-scale networks. Even the grandest of the settlements, both secular and ecclesiastical, associated with these major changes remain relatively small scale however, with none exceeding 10 ha in extent.

Driscoll (1992, 12) has remarked that the second half of the first millennium AD witnessed 'the disparate kingdoms of northern Britain develop into the embryonic Scottish nation'. Like the rest of the British Isles, Scotland remained outside the pattern of large, if unstable, political units (such as the Ottonian empire) that emerged on the Continent. The Scotto-Pictish kingdom may be seen as a northern British equivalent of territorial units, not dissimilar in size, such as Flanders and Burgundy.

On very few sites – and here the Brough of Birsay, and perhaps in some of its phases, Whithorn, may be exceptions – is it possible to argue from currently available evidence that industry was organized on more than a workshop level. In general, settlement and associated crafts seem to have remained essentially small scale. Equally, the formal division of large tracts of landscape seems to have remained largely absent, The Deil's Dyke in Nithsdale for example transpiring not to be of first millennium AD date, as had been surmised (J. W. Barber et al. 1982). Larger settlement sites, conceivably of the first millennium AD, certainly exist. The 23 ha fort on Tap o' Noth, overlooking the cluster of Pictish Class I sculpture from Rhynie, Aberdeenshire, or the sizeable promontory fort on the Mull of Galloway, with its wide outlook south to the Isle of Man, may stand as examples, but their dating is unknown. The principal archaeological concomitants of early statehood remain (along with the historical records, the sculptured stones and other rich material culture) several of the classes of settlement site mentioned above. Defining the subsistence strategies practised at these sites, and the characteristics of their contemporary landscapes, remain substantially tasks for the future.

13   The Early Norse Period

JOHN R. HUNTER

INTRODUCTION

This chapter covers a period of intense activity brought about by the movement of Scandinavian peoples in the ninth and tenth centuries AD. During those years Viking exploits and influences stretched from Russia to the New World: Atlantic islands were colonized; trade routes were opened up; raiding took place throughout north-west Europe; and the affairs of Britain became dominated by Scandinavian matters for several centuries. These events, which have an immense popular appeal, have been recounted, interpreted and discussed at length from a variety of historical, archaeological and art-historical standpoints. Part of this literature has taken a broad geographical view (e.g. Jones 1968; Graham-Campbell 1980; Roesdahl 1992) but much emphasis has also been placed on problems peculiar to the British Isles, in particular to England (e.g. Sawyer 1971; Wilson 1976; Loyn 1977; Smyth 1977). Scotland has received less individual attention, but no less detail of analysis (e.g. Duncan 1975; Crawford 1987; Ritchie 1993), and has additionally been the subject of much antiquarian and regional study (e.g. Grieg 1940; Fell *et al.* 1983; Morris 1985; Bigelow 1992; Batey *et al.* 1993).

GEOGRAPHY AND CHRONOLOGY

On the basis of Scandinavian influence interpreted from the amalgamation of evidence from graves, hoards, place-names and documentary references, the relevant area for discussion appears to lie mostly to the north and west of the Great Glen – itself navigable by portages – with particular emphasis on the Northern and Western Isles (Figure 13.1). For the purposes of this short contribution, the Isle of Man is excluded despite its strong Scandinavian character and its influential position in west Scottish affairs. Excluded too are parts of the Scottish southern lowlands where a less well evidenced Scandinavian presence is also attested. Perhaps inevitably in a study of this type, the perceived extent of this geographical province is to some extent an irrelevance. Discussion tends to be based on evidence. This is sharpest where research has been directed most, and here the bias lies firmly towards the north.

Scotland: Environment and Archaeology, 8000 BC – AD 1000. Edited by Kevin J. Edwards and Ian B. M. Ralston.

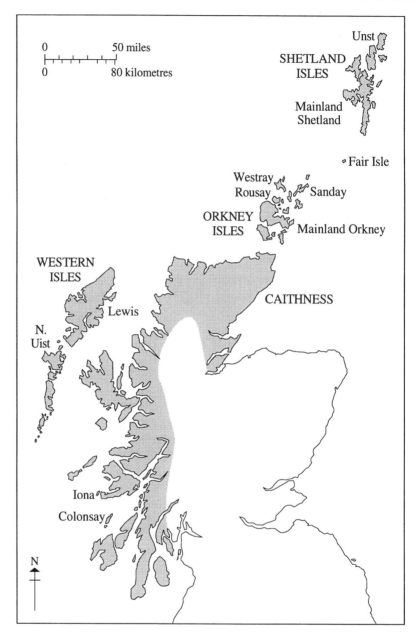

Figure 13.1 General area of Norse influence in north and west Scotland based on settlement sites, burials, hoards, place-names and documentary references collated from Crawford (1987) and Ritchie (1993)

The early Scandinavian phase of settlement sits awkwardly within the continuities of those records which form the substance of this volume. Its beginnings are vague; they reflect, within the broader context of maritime movement in the Atlantic and North Sea, a growing influence of Scandinavian culture reconstructed from material remains, place-names and limited documentary support. This influence appears to have occurred most indelibly along those coasts adjacent to the developing sea routes, notably from Scandinavia to Ireland, and within island groups used as navigation points along the northern and western approaches.

Inference rather than evidence points to an initial process of contact leading ultimately to the selective taking of land for settlement, although archaeology is hard-pressed to distinguish one from the other. Documentation, such as it is, records an event here or a raid there, little of which is archaeologically tangible. The duration and internal chronology of this primary process was almost certainly influenced by local factors, not least of which was the reaction of the extant population, but there are no absolute dating markers and the chronology of localized landnam (primary taking of land) is untestable by current scientific dating means. For working purposes, students of the period have tended to use AD 800 as a convenient starting point for a Scandinavian presence in north Britain, although the arbitrariness of this has given the period a degree of artificiality which has been difficult to shake off. On documentary grounds Scandinavian influence began with recorded raids at a slightly earlier date, and settlement occurred somewhat after the conventional start date. These two elements require clear distinction.

This is not an historical period as such, given that reliable documentation is rare in Scotland and Scandinavia before the late twelfth century. Opinions differ as to whether the eighth, ninth and tenth centuries are more properly a part of prehistory or some no-man's land of textual uncertainty. Sources are few, and those contemporary with the events they record even fewer, and almost all are biased or require interpretation. Their relative reliability has been much discussed by historians as part of a more common focusing of scholarly thought among those specializing in the different components of the period. Wainwright's classic 1962 study of the relevant disciplines (history, linguistics and archaeology) promoted the view that the period could be properly understood only by complementary means. 'Viking studies' is the term traditionally used to describe the investigation of this primary phase of raiding and movement of Scandinavian peoples. The field has since witnessed a slow move away from structural morphology and art history – favourite topics of invasion theorists – to embrace additional disciplines including a greater awareness of ethnography and antiquarian accounts (e.g. Baldwin 1978; Fenton 1978; Hunter 1991), the use of sociological and theoretical modelling (e.g. Samson 1992) and, most relevant here, the introduction of systematic palaeoenvironmental research (e.g. McGovern *et al.* 1988). These developments are to some extent a natural evolution of the subject area and in some measure directions imposed by a desire for compatibility with other archaeological period or geographical interests. As their basis, however, all utilize a broad historical framework dangerously built from disparate records: those of afflicted Christians (the contemporary literate majority); external chronicle entries; and memorabilia embedded within the later dynastic literature of Iceland.

Nevertheless, the historical framework is reasonably clear. Scandinavians, mostly but not exclusively Norwegian (the Norse of the title of this chapter), began to spread across the North Sea, fuelled initially by piracy, and facilitated by unprecedended skills of seamanship on which much has been written. It is not always possible to separate the various cultural components involved; later groups may have included first-generation mixed-race Scandinavians of Norwegian, Danish and Celtic stock, but all were unanswerable to any single authority. They were nomadic and appeared intent solely on the quest for portable wealth. Their impact was not exclusively northern. During the last decade of the eighth century, sporadic Scandinavian raiding occurred along many parts of the British coastline, recorded vividly in the Anglo-Saxon Chronicle at Portland, Dorset, and at Lindisfarne off the Northumbrian coast. Irish sources, more specifically concerned with monastic matters, also document contemporary raids along the western sea routes, including attacks on Iona. The Northern Isles, suitably remote from the Scottish mainland, and pivotal for either eastern or western approach, may even have provided bases for these sorties (O'Còrrain 1972, 82); these archipelagos were almost certainly a primary sighting point for vessels sailing from Norway and were therefore arguably most susceptible to settlement.

Even the Norwegian coast may have been raided from bases in north Britain, a fact which implies a certain degree of Scandinavian settlement on these shores, if only for wintering purposes. According to some early sources (e.g. Anderson 1922, I, 331), such raids formed the underlying reason why Orkney and Shetland may have been hastily established as an Earldom under the sovereignty of Norway, although opinions differ on this (Crawford 1987, 55). Nevertheless, the creation of the Earldom towards the end of the ninth century might be seen as an *official* endorsement of settlement which paved the way for the imposition of Norse rule and administration in those areas. Slightly later, and in contrast to the settlement of the Northern Isles (and hence by definition *unofficial*), increasing Scandinavian influence can be recognized in the southern parts of Pictland. This more likely reflects the eastward movement of mixed race groups spreading from the Irish Sea Basin, but the movement here and in many other parts of the western seaboard is obscured by linguistic and cultural difficulties.

Under the influence of the Earldom in the later ninth century, settlement was subsequently extended into the fertile areas of Caithness and Easter Ross where the place-name 'Dingwall' almost certainly represents a central meeting place. Elsewhere in the north, only in Mainland Shetland, at Tingwall, has such a site been formally identified, lying on a narrow tongue of land approached by a causeway leading out into a loch. The 'thing' (ON *þing*, assembly) was a communal meeting place and attests to the presence of an organized and active Scandinavian population in the vicinity. Towards the end of the tenth century when the southern part of the British Isles became the target of more pervasive Scandinavian impacts, the Earldom was still sufficiently important to be drawn into more widely based political affairs. During the time of Earl Sigurd the Stout, further geographical expansion of the Earldom took place and at the time of his death at the battle of Clontarf (in Ireland) in 1014 – an event of great historical import and a fitting end to the chronology of this chapter – the Earldom was arguably at the zenith of its power. It is inescapable that the historical, linguistic and archaeological bias of this period should reflect its impact.

RESEARCH DESIGNS

Although there has been sporadic investigation of Norse settlement sites in the Western Isles (e.g. Marshall 1964; MacLaren 1974; Crawford 1974, 1981) and in Caithness (Batey 1987), Orkney has continued to receive the lion's share of attention in view of its early political importance and the significance attached to it by modern researchers (e.g. Renfrew 1990). Shetland was relatively late in receiving systematic study, or for that matter even a resident archaeologist, despite a strong antiquarian tradition (e.g. Goudie 1904). Mainland Shetland also boasts the one Norse site, Jarlshof (Hamilton 1956), which has been used as a control (albeit to an unsatisfactory degree) for the interpretation of all other Norse sites. More recent excavations (e.g. Small 1966; Bigelow 1985; Crawford 1985, 1991) have demonstrated a potential no less than that of Orkney.

Orkney has by far the most complete sites and monuments record based on detailed fieldwork (e.g. Lamb 1980, 1984) with Shetland following at a considerable distance. The quality of the archaeological database for other parts of Norse-influenced Scotland is more variable: there are Royal Commission inventories for Argyll (e.g. RCAHMS 1984), but some other areas have received much less attention, academic and modern political priorities being directed elsewhere. The Western Isles also suffer from persistent natural sand movement which results in many stray finds, including those from burials, being discovered without the integrity of an archaeological context. There is the additional problem that although Norse place-names may be in the majority in some regions, Norse settlements are not the easiest to find, although a number of predictive models have been proposed (Marwick 1952; Small 1968; Alcock and Alcock 1980). On Orkney, where the greatest fieldwork emphasis has been placed, it is only recently that research designs have been devised to consider more than single sites (e.g. Morris 1989). Moreover, funding sources are such that investigation has tended to be biased towards coastal erosion sites of which the boat burial at Scar, Sanday, is the most dramatic recent example (Plate 13.1; Owen and Dalland forthcoming).

In the other Atlantic islands, by contrast, modern programmes targeted at specific themes of palaeoeconomy or human ecology have been implemented, often as part of more broadly based research designs (e.g. McGovern et al. 1988; Buckland et al. 1991; Christensen 1991; Keller 1991; Sveinbjarnardøttir 1991), but there the database, and hence the potential, differs. In some colonies the absence of both earlier and later land use has enabled the landscape and ecological effects of landnam settlement to be monitored, to the extent of identifying relict Norse field systems (Mahler 1991) as well as interpreting social infrastructure from the distribution of building remains (Durrenburger 1991). In north Britain, survival factors and the level of continued land use are such that comparative studies of this type are unrealistic, although there are a few exceptional areas where it has been possible, for example on Foula (Baldwin 1984).

SETTLEMENT

One of the reasons for the movement overseas was the shortage of good land in Norway, although other political and social factors were also pertinent. It was,

Plate 13.1 The eroded boat burial at Scar, Sanday, Orkney during excavation. Crown Copyright: reproduced by permission of Historic Scotland

therefore, inevitable that settlement would occur along the maritime approaches, particularly along the route to Ireland through the Western Isles, where the landscapes bore a close similarity to those of the homeland (Plate 13.2). This occurred without historical comment at a time when the affairs of the Orkney Earldom may have been more newsworthy, and is attested mostly from place names, although graves such as those on Colonsay (e.g. Ritchie 1981) or Lewis (Welander *et al.* 1987; CFA 1994, 6) and elsewhere in the Hebrides (Grieg 1940; for discussion see Ritchie 1993, 79–89) provide more tangible evidence. A process of 'ness-taking' may have preceded settlement proper, although this may be confused with the later enclosure of headlands for cattle stockading or by the naming of prominent features for navigation purposes (Fellows-Jensen 1984, 149). On more favourable soils, particularly in regions of machair where recent research has amplified the potential for identifying multi-period occupation (Parker Pearson and Webster 1994), the farm names point towards pockets of Scandinavian colonization,

Plate 13.2   Sorisdale on the island of Coll, featuring the fertile pocket of land and sheltered bay – characteristics favoured by early Norse settlers. Sorisdale bears the Scandinavian *-dalr* ('dale') suffix and lies along the sea route to Ireland. Copyright J. N. G. Ritchie

e.g. in the Inner Hebrides (Alcock and Alcock 1980, 61) and on Lewis (e.g. Cox 1989). Elsewhere in the Western Isles, on less productive land, the impact seems less pronounced. There, settlement viability may have depended on other factors. Such scattered populations were clearly sufficiently numerous and important in the mid tenth century to feature in Erik Bloodaxe's attempts to unify the various Norwegian elements located on both eastern and western coasts (Smyth 1979, 177). His efforts, ultimately futile, were to emphasize further the strategic and pivotal significance of the Northern Isles in contemporary politics.

A comparison between the grave goods (few though they are) recovered from the Northern and Western Isles respectively, has led to a hypothesis of regional settlement types based on cultural influence as much as natural resources. In the North, the Norse farmers capitalized on the rich soils and the administrative organization of Norwegian sovereignty, while their equivalents in the west, the 'Gaelicized petty chieftains', used trade (an activity supported by the occurrence of scales among burial goods) as well as less orthodox measures, to supplement the yield of poorer soils (Crawford 1987, 127). The occurrence of coin hoards in both the Western Isles and in Ireland dating to the later part of the tenth century (Graham-Campbell 1976a,b) might also support the increase of trading movement at that time. The decades of the later ninth century when such silver coin hoards were principally deposited are different from those in the Northern Isles (Graham-Campbell 1993), suggesting that dissimilar economic forces may have been at work in the two areas. One school of opinion, for example, has identified the comparatively rich burials of the Western Isles with the less stable elements of viking society for whom permanent settlement was a lower order of priority (Eldjárn 1984, 8). Such fundamental differences between the Western and Northern Isles, combined with problems of linguistic interpretation, clearly have implications for the nature of the archaeological record.

According to linguistic evidence, only in Orkney, Shetland and parts of Caithness are the names of natural features and habitations dominated by Scandinavian elements. In these places the colonization appears to have been wholesale, although its chronological sequences vary locally (Marwick 1952; Nicolaisen 1975, 1982). In Shetland, place-name evidence has been used to suggest that all the most useful land had already been settled and exploited by the end of the tenth century (Bigelow 1989, 185). It can be no accident that the fertile lowlands and coastal inlets of the Northern Isles could provide an environment of scattered homesteads following a regime of farming and fishing almost identical to that of coastal Norway. Elsewhere, in Sutherland and throughout the west coast and Western Isles, the Scandinavian elements may have become overwhelmed by Gaelic influences in the Middles Ages. The mixing of Gaelic elements may additionally reflect the general cultural confusion shared with Ireland in the ninth and tenth centuries. Oddly, the east coast seems never to have been settled; a few place-names of Scandinavian character occur around the mouth of the Forth, but neither the highlands to the north nor the lowlands to the south seem otherwise greatly affected. The reasons for this are unclear: hostile native populations; lack of easy contact with other Scandinavian groups in Britain; different soil types, or absence of suitable estuarine approaches may together have provided an adequate deterrent (Crawford 1987, 35). Only in the south-west lowlands, in Dumfries and Galloway and in the Isle of Man, are other place-name concentrations to be found. These might best be seen as associated with an Irish Sea influence or with the effects of a Danish-dominated England lying immediately adjacent, and from where the distribution of the distinctive hogback tombstone spread into southern Scotland during the tenth century (e.g. Lang 1994).

It is important to see all these events within the broader Atlantic context. By the time the Orkney Earldom had been established, Scandinavian settlement had occurred in the Faroes, Iceland had been discovered by Norwegian explorers and the eastern part of England lay under Danish rule. The history of studies of this period (McGovern 1990) makes clear the growing importance of an Atlantic context and, indirectly, highlights the ambiguity of Scotland's position. In a sense this is as much a period of Scandinavia's history as of Scotland's. The former's culture is imposed, its traditions are introduced and the relationships and parallels are undeniably external. Scotland's part-colonization, therefore, is to be seen as one element of a larger process of population movement and land-taking. But it is distinctive: the other Atlantic colonies were, apart from occasional references to papar (Irish monks), skraelings (natives) and the less plausible unipeds (one-legged creatures), uninhabited. They were otherwise virgin landscapes lying wholly exposed to the imposition of Norse land use and culture.

Such landscapes provide potential controls for assessment of anthropogenic influences through the palaeoenvironmental record; in the case of Iceland some 60% of the natural vegetation may have been destroyed in the clearance process (Zutter 1992, 139). The same contexts may also provide controls for behaviour and social

THE EARLY NORSE PERIOD

evolution, for example in terms of settlement type, interaction between local and distant groups, trade, and the establishment of specific crafts (e.g. Martens 1992). Around the Scottish coasts and islands there was already a long-established population. Clearance had taken place centuries previously, agricultural regimes had been devised, and Pictish and Gaelic culture developed. It was within this local context that the primary Scandinavian impact was made.

NATIVE AND INCOMER

Documentation gives little clue to the relationship between native and incomer. The late ninth-century Irish Annals report Norse contacts of a violent nature, but these deal with raiding. Settlement is a later and separate issue, the nature of which has become entwined in supposition and mythology. Archaeology has addressed the matter of native–incomer contacts in a range of ways (e.g. Crawford 1981; Bigelow 1992), but is not ideally suited to provide many answers. Interpretation of a Norse presence has traditionally been defined on the increasingly unreliable basis of the morphology of buildings and on diagnostic changes in artefact types and materials. Of the last, the most commonly used indicator is steatite, a hydrous magnesium silicate (talc) with the propensity for being worked into vessels. It occurs widely in western Norway but also outcrops in Shetland, particularly in Unst, and an impressive quarry exists at Cunningsburgh on Mainland. On examination, the change from clay to steatite as the preferred medium for the manufacture of containers, usually considered to coincide with the early Norse horizon, is not well defined; evidence for steatite trade from Shetland being known from the Bronze Age onwards (Buttler 1989, 194). In Scotland, Norse period sites are more validly recognized by the occurrence of steatite in some quantity, rather than by the occasional sherd. Grass-tempered pottery, native in origin, has also been noted to persist throughout the Norse occupation (MacSween 1991). Thus a gradual review of cultural indicators has occurred, such that new research strategies may now be proposed (Bigelow 1992). Part of this corrective process includes the assessment of palaeo-environmental data, initially to dispell the orthodoxy which ascribes much of the character of the modern farmscape, including the species of sheep, the size of horses, the varieties of sea-fowl, and even the presence of certain mice to the influences of Vikings (Hunter 1991, 192; cf. Chapter 6).

In only a handful of excavations has the nature of cultural change in the early Norse period been a specific focus, although the interpretations have not been entirely unanimous. The sites are too few, too geographically biased, and arguably unrepresentative. Only in one, however, at the Udal, North Uist, was the excavator convinced that colonization had been 'sudden and totally obliterative in terms of local native culture' (Crawford 1981, 267). In Orkney at Skaill (Plate 13.3), on the east coast of Mainland, the excavator sensed a 'clean break' in the cultural continuity of the site although integration between the two populations was still argued (Gelling 1984, 38; Buteux forthcoming), and at Westness, Rousay earlier native graves appeared to have been respected by subsequent Norse burials (Kaland 1993, 312). At two other Orkney sites, the 'farmstead' of Buckquoy (Ritchie 1977) and the adjacent Brough of Birsay (Hunter 1986) – two sites arguably related – the

Plate 13.3 The outline of House 1, the primary Norse farmstead on site 2 at Skaill, Deerness, Orkney. The building measured approximately 7.6 × 5.2 m and consisted of two rooms, a main chamber and a smaller D-shaped, cell-like room visible on the right. Copyright: Simon Buteux

Norse impact appears to indicate a degree of assimilation and architectural adaptation. The Brough of Birsay itself, a small tidal island off the north-west tip of Mainland, became a seat of the Earldom for which Buckquoy may have provided hinterland services. This islet, surrounded by cliffs, largely inaccessible and housing a string of Norse buildings without any adjacent farmland, contrasts sharply with the expected model of individual farmsteads located with respect to good land and a sheltered maritime approach.

Also on Orkney, at Pool, Sanday, a settlement, the interpretation of which can avoid the awkward problems of status associated with the Birsay sites, has been excavated. Its location better satisfies the traditional Atlantic model (e.g. Small 1968), and a similar degree of assimilation to that estimated for Birsay has been observed on the basis of structural continuity and the persistence of native pottery. Norse settlement at Pool was preceded by nucleated late Iron Age occupation which reached a peak around the seventh century (Hunter 1990). Subsequent contraction, but not depopulation of the type suggested in Egil's Saga (Pálsson and Edwards 1978, IV), left the site, its materials and its fertile lands (for which soils Sanday was renowned in the Middle Ages) invitingly exposed. What followed was not so much a cultural change as a regeneration in which new cultural influences made them-selves apparent (Hunter *et al.* 1993). Although one Iron Age building remained occupied, the physical focus of the site shifted: two new sub-rectangular structures were built, one reutilizing extant lengths of walling in a manner not dissimilar to that recognized at Skaill (Gelling 1984, 36), and the other containing an inner timber framework. Contemporary assemblages included steatite, pottery and objects typologically diagnostic of both cultures. This interface period within the sequence of occupation of the site also sees the introduction of more advanced iron-manu-facturing techniques. Knives were distinctive from those of the previous phase and showed evidence for composite iron and steel structures as well as pattern welding; both indicate craft specialization (Berg and McDonnell forthcoming). Comparison of slag inclusions between knives and items of a more domestic nature (e.g. nails and washers) suggested different manufacturing sources, implying that the knives seem to have been imported while the other material may have been produced locally. It is difficult to pin down the period during which all these changes occurred, but on the basis of radiocarbon dates they almost certainly took place well before the establishment of the Earldom (Hunter *et al.* 1993), and possibly even had their starting point within the late eighth century.

Reinvigoration of the site at Pool is also evident in the palaeoenvironmental record, but in a manner that indicates innovation rather than change; the level of data capture is unprecedented and merits comment. The macrobotanical remains (Bond forthcoming a) show a decrease in the ubiquity of cereals and associated weeds in the earliest part of this interface period, followed by a marked increase in its later part, during which *Hordeum vulgare* (hulled six-row barley) and *Avena sativa* (cultivated oats) reach their highest values. There is a parallel increase in weed species preferring light dry sandy soils and those of rich, well-cultivated soils, and a decrease of those favouring damper habitats; this might be interpreted as reflecting intensification in manuring and cultivation and of the absorbing into cultivation of lighter, sandier soils, possibly with implications for the development of an infield/outfield system. Manuring may have been carried out using seaweed which occurs as carbonized concentrations of the wracks *Fucus vesiculosus* and *F. serratus*, although these deposits may have had some other specialized use. The contemporary animal bone record (Bond forthcoming b) shows increasing evidence from the older cattle of pathologies (wear of joints and infections) which might be attributable to a greater emphasis on traction (Plate 13.4). By contrast the horse shows no such eburnation; this factor and possible evidence of breeding (identified from young animals less than 18 months old at death) points towards a pronounced increase in the importance attached to the horse in this resurgent occupation. Furthermore, the persistent, if low, levels of charcoal of alder (*Alnus glutinosa*), hazel (*Corylus avellana*) and willow (*Salix* sp.) suggest that some scrubland still remained to be utilized.

Flax (*Linum usitatissimum*) seed also appears in the interface period record in a quantity that argues for a substantial shift in the agricultural and economic base of the site. Flax cultivation, for which the local soils were particularly suited and which might be cropped for both oil and fibre (Bond and Hunter 1987), is known from late Iron Age contexts at Howe, near Stromness, on Mainland Orkney (Dickson 1994, 135). Its occurrence in this phase at Pool is innovative; it reinforces a discernible intensification of agriculture at this time. This interface horizon is also marked by only a small number of buildings. It may thus be that agricultural production was to a limited extent driven by market considerations.

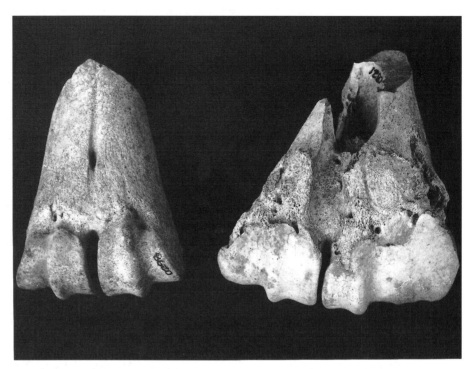

Plate 13.4   Increasing use of cattle for traction at Pool, Orkney, illustrated by metapodial (on right) showing extension of articular end and infection, probably resulting from arthritis. Normal metapodial shown on the left. Photograph by Jean Brown

There are infrequent but persistent remains of red deer of all ages, arguably indicative of local hunting (Plate 13.5). It is questionable whether this reflects communal hunting activity at a level commensurate with that required for deer, seals and caribou within the open landscapes of other Atlantic colonies (e.g. McGovern 1981; Martens 1992, 3), or whether the waters surrounding the island provided a natural limit sufficient to make communal effort less essential. A number of antlers recovered from Pool had been shed naturally rather than hacked from the skull, a fact which might indicate reasonable access to herds as much as trade. According to the faunal data, a decline in red deer numbers only occurs after the period under discussion here. This might be supported by a later reference to the earls travelling from Orkney to Caithness to hunt deer in the earlier part of the eleventh century (Pálsson and Edwards 1978, II).

Elsewhere in Orkney the environmental record has been less rewarding for this early interface period, although the discovery of waterlogged and organic remains at Tuquoy on Westray promises much (Owen 1993). The special characteristics of the Brough of Birsay, considered either as a pre-Norse monastery or as a secular Norse seat, biases the economic record. The needs of its population seem to have been served by, and probably selected from, the resources of a wider hinterland (Seller 1986, 215). As such it fails to provide representative data. Across the causeway from this islet lie the sites in Birsay Bay where the most research attention

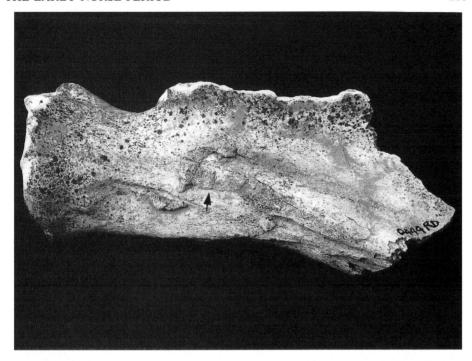

Plate 13.5 Scapula from a red deer at Pool, Orkney, where red deer were still available for hunting in the early Norse period. The scapula shows new bone formation (arrowed) resulting from wound to tissue which seems likely to have been caused by a projectile. Photograph by Jean Brown

has been directed (e.g. Ritchie 1977; Donaldson *et al.* 1981; Morris 1989). The value of botanical material from the sites is limited by sample size or dating range; other important comparable data, including those from Skaill, await publication. In Shetland, sampling at Jarlshof (Hamilton 1956) and Underhoull, Unst (Small 1966) was of a minor nature; in the Western Isles much potential rides on the unpublished sequence of material from the Udal on the north tip of North Uist (Crawford and Switsur 1977). A later trend towards increased reliance on marine resources seen at Buckquoy (Wheeler 1977, 214) may indicate a combination of land saturation and population increase, also evident at both Jarlshof and Underhoull where the records are artefactual rather than ecofactual. This trend is impressively demonstrated at Freswick on the east coast of Caithness, where the later exploitation of these marine resources is described as 'overwhelming' (Morris and Rackham 1992, 98).

CONSOLIDATION

If, as seems likely, Pool's subsistence economy can be considered representative, it suggests that many other fertile parts of north Britain may have experienced a period of confident agricultural stewardship during the ninth and tenth centuries. In essence, the economy adheres to a formula well attested both in the Norwegian homeland (e.g. Martens 1992) and in other parts of the Norse Atlantic colonies (e.g. Hansen 1991; Amorosi *et al*. 1992). The relative weightings of pastoralism and cultivation depended on latitude and were supplemented by the exploitation of appropriate local resources such as fowl and fish. It is the detail of all these which is now emerging. The marginal areas of north Britain, including the Northern and Western Isles, lay inside the climatic zone which allowed most latitude to this basic economy, and hence differed from the more northerly stations of Iceland and Greenland where climatic conditions were harsher and the seasonal cycles more critical. It is worth remarking that at Pool the longhouse, where cattle and family were housed together, appeared realtively late in the sequence.

It is also clear, however, that there were wider geographical shifts within the overall economic framework which may have had underlying features in common. In north Britain and in other parts of the Scandinavian world, including the Norwegian homeland, much interest has focused on farm mounds (Bertelsen 1979, 1984; Davidson *et al*. 1983, 1986) – characteristic landscape features composed of organic materials and argued to represent a by-product of exceptional economic factors. Their formation, normally attributed to the Viking period, has been consistently explained in terms of changes in land use or soil management. More detailed review and wider awareness of the phenomenon has shown a range of types and indeed dates: many belong to later periods of settlement consolidation or even to the Middle Ages (Bertelsen 1991; Bertelsen and Lamb 1993), although some may start contemporaneously with the initial Norse colonization. In north Britain, according to current research, farm mounds are unique to the Orkney islands of Sanday and North Ronaldsay where a small number still survive surmounted by nineteenth century farmsteads. Most appear to be related to prime land (Davidson *et al*. 1983); some may even predate the Viking period proper and, if this view can be substantiated, their continued use may attest further to the limitations of Norse cultural impact on established practices.

Even in north Britain neither climate nor resources were sufficiently flexible to permit fundamental change to subsistence methods (Hunter 1991, 193). What is apparent from the economic data recovered from Pool is the successful extension of the range of subsistence production at a time when favourable environmental conditions prevailed around the north Atlantic. This strength is evident with the advantage of hindsight: the end of the period of climatic warming which had facilitated settlement throughout the north Atlantic towards the end of the first millennium AD had serious implications for the more northerly colonies, notably Greenland, where even the most rudimentary subsistence routines were disrupted (McGovern 1981). North Britain it seems, because of its latitude, was largely unaffected, and here Norse traditions survived relatively unscathed.

ACKNOWLEDGEMENTS

I am grateful to Dr Barbara Crawford and Dr Julie Bond for their helpful advice and comments in the preparation of this chapter, and to Dr Graham Ritchie for kindly supplying the photograph of Coll. Simon Buteux generously made available his completed text of the Skaill excavations in advance of publication.

14 Environment and Archaeology in Scotland: Some Observations

KEVIN J. EDWARDS AND IAN B. M. RALSTON

The presentation of environment and archaeology in this volume has taken the form, inevitably, of a series of interlinked perspectives on themes currently judged to be of significance. The survey of some elements – vegetation, for instance – might seem relatively comprehensive owing to the existence of a widespread fossil record. Studies of vertebrate faunas, however, suffer from a dearth of suitable contexts for preservation in a territory where acid soils are frequent and where conditions for the survival of bones on archaeological sites are often unfavourable. Similar disparities afflict the survival opportunities of different forms of evidence within the archaeological record. For example, it seems likely from palaeoecological data that Mesolithic hunter–gatherers were present in the Western Isles, but rising sea levels, acid soils, peat and sand accretions have served to obscure or remove the artefactual evidence for a human presence. Contrastingly, archaeological evidence for human activity from the Neolithic period onwards can be visibly represented in quantity in some parts of Scotland; the stone-built monuments of the Northern Isles form a well-known instance. Even the best of environmental and archaeological records, though, have their lacunæ, and the grasp of past geographies is far too weak to enable a scaling-down of the research effort, not least at the level of the identification of further sites for study.

That several possible explanations can frequently be advanced for phenomena observable in the environmental and archaeological records is an indication of the task that will continue to face researchers. Features which are frequently seen, such as desolate moorland, eroded hillslopes, waterlogged valley floors and even abandoned meanders, may be the result of either human activities, or the products of natural instabilities in the landscape, or indeed both. The elm decline of *c.* 5100 BP may thus be a response to disease, soil deterioration, climate change, forest clearance or humanly-imposed stresses arising from woodland management; it may represent a collapse in elm populations arising from several of these factors. One combination of causes appropriate for the Borders, however, may be inapplicable in western Scotland. Equifinality (a common result arising from different causes) must therefore

Scotland: Environment and Archaeology, 8000 BC – AD 1000. Edited by Kevin J. Edwards and Ian B. M. Ralston.
© 1997 The editors and contributors. Published in 1997 by John Wiley & Sons Ltd.

be entertained in seeking explanations for such phenomena, and this is even more the case where human behaviour is involved. Competing explanations for the use of brochs may be advanced to illustrate this point: can the available evidence be used to discriminate whether brochs were the residences of an elite as opposed to those of more humble members of society; were they prestige monuments with non-utilitarian features, or a sensible response to defensive needs? What of polished stone 'axes'? Are they foresters' or woodworkers' tools, weapons, ard points, currency, symbols of power, all of these or none? These instances are proffered to emphasize the possible futility of expecting monocausal explanations for phenomena, especially where the features or structures concerned demonstrate considerable variability.

A propensity to extract as much meaning as possible from data is to be encouraged. This brings with it the danger of speculative forays portrayed as established schema which may be seen in retrospect as little better than slavish adherence to a current fashion or a particular ideological leaning. This, however, must not be interpreted as a signal to confine ourselves to description, but rather that the reader must critically examine any inferences made and explanations offered.

THE ON- AND OFF-SITE RECORDS

Much of the environmental research presented in these pages has been concerned with off-site rather than on-site situations (Edwards 1991). This partly reflects the interests and aims of many environmental specialists whose work has not been archaeological in primary orientation, but whose findings, perceived to be useful, have been drawn upon by archaeologists. As freely drained sites were as popular for habitation in the past as they are today, then obviously their on-site environmental (and archaeological) records would be expected to constitute desirable sources for information. Unfortunately, the factors which make such sites fitting for human occupation reduce opportunities for the survival of evidence. Thus, well-drained non-acid soils may be poor preservers of microfossils, and ecofacts and artefacts alike may be mixed by earthworms; conversely, soils of high acidity may dissolve bones or can destroy stratigraphic integrity by facilitating the transport of microfossils within the soil profile. Such problems are compounded by the shallow soil profiles frequently encountered on excavations in Scotland.

Furthermore, human activities are likely to disrupt sediment accretion over many parts of a site, and it may not be possible to provide answers to questions because of the impossibility of obtaining samples at a satisfactory resolution. Subsequent agricultural, domestic and industrial activities may lead at best to partial hiatuses in the depositional record; at worst, to the destruction of enormous numbers of archaeological sites, perhaps severely biasing site and distributional data (Stevenson 1975; Barclay 1992). The major deficiency in the off-site record – its spatial displacement from the precise focus of human activity – may be compensated for by its capability to provide a continuous, sensitive, high-resolution record of both general environmental and nearby anthropogenic events. The richness of the pollen record from lake and peat sites adjacent to former settlements, for example, may only be exceeded archaeologically by the material remains recovered from wetland contexts (cf. Coles 1992). Off-site palaeoecological records may also contain a strong signal of human occupation and exploitation for which the on-site con-comitant may no longer be apparent, at least on the surface, thus providing a challenge to the archaeologist (Whittington and Edwards 1994; Edwards 1996a). Quite clearly, all data sets have their strengths, weaknesses and place.

THE ON-SITE RECORD: A SURVEY

For many years, environmental information gleaned during the course of archaeo-logical projects, has featured in accounts of their results (e.g. Newstead [Curle 1911]; Chapter 6). In order to ascertain the degree to which environmental information has been sought in archaeological investigations in more recent times, although not to be taken as a measure of its integration, it might be instructive to consider briefly the inclusion of environmental studies within the excavation product. This has been assessed by looking at the last 20 volumes of *Proceedings of the Society of Antiquaries of Scotland* (volumes 105–124, issued between 1975 and 1995), the premier journal for the publication of Scottish archaeology, and evaluating every paper concerned with the excavation of Scottish sites of Norse or greater age. Some 201 relevant sites were considered further, varying from cist burials to complex multi-period settlement sites. It is not here possible to enter at length into the methods employed, but sites that had been excavated long before publication were ignored as being inappropriate to an assessment of changing practice; cemeteries were regarded as single sites; series of spatially discrete sites reported in the same article were treated as separate items; human skeletal information was excluded (most reports are anatomical descriptions and involve neither environmental comment nor isotopically-based considerations of palaeodiet); and sites not simply attributable to conventional periods (e.g. late Neolithic/early Bronze Age) were scored fractionally.

The numbers of sites published for each period are not distributed regularly – values of 5 for Viking/Norse and 11 for the Mesolithic are unsatisfactorily low for even the simple exercise undertaken here, while the Bronze Age has a much healthier 72 (Figure 14.1). When environmental contributions are expressed in percentage terms, then plausible patterns do emerge, and Figure 14.1 also indicates the proportion of papers containing some specialist environmental information within the designated archaeological periods. The value of about 73% of papers for the Mesolithic at least, where there is a long tradition of environmental study (cf. Clarke 1954), is no surprise, while the Norse figure of 90% may be a fair reflection of recent excavation practice for areas in Scotland and elsewhere in the North Atlantic region (cf. Morris and Rackham 1992; Batey *et al.* 1993). The value of 39% of papers with an environmental component for the Bronze Age requires further comment in that it disguises a major disparity: individual cist excavations account for exactly half of the Bronze age component, yet only 17% of their number involved an environmental dimension, while conversely, 62% of multiple-burial or non-funerary sites featured environmental study. The lowest score derives from Roman period excavations, where environmental analyses were published for only 23% of sites which appeared in the *Proceedings.* Many of these excavations, however, were small-scale interventions on the Antonine frontier.

258

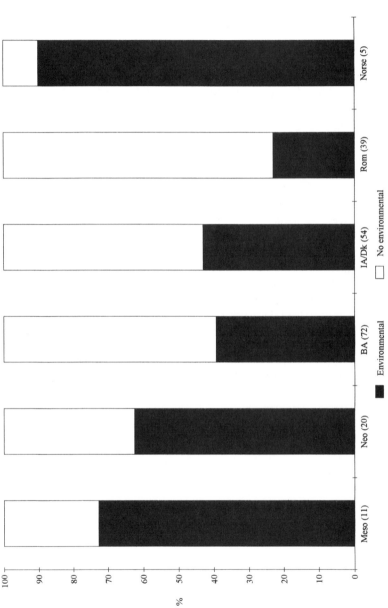

Figure 14.1 The number of excavations (in parentheses) assigned to archaeological periods, showing the proportions of Scottish excavations featuring environmental and non-environmental investigations. Meso = Mesolithic; Neo = Neolithic; BA = Bronze Age; IA/Dk = Iron Age/Dark Age (including sites described as Romano-British, early Christian and early Historic; Rom = Roman; Norse = Viking/Norse. Data extracted from *Proceedings of the Society of Antiquaries of Scotland*, volumes 105–124, published between 1975 and 1995

The major types of environmental analysis are displayed in Figure 14.2. The numerical superiority of plant macrofossils (around 37% of all environmental analyses) and animal bones (21%, including fish and bird bones), partly reflects their visibility within excavation deposits. The plant macrofossil component is also undoubtedly greatly over-represented in that on-site charcoal was extracted with the primary aim of obtaining material for radiocarbon dating; taxonomic identification seems to have been a by-product, given the often scant attention paid to the potential ecological implications of the charcoal, and sieving for seeds or other plant macrofosils is infrequently reported. Analyses of microfossils such as pollen (18% of environmental analyses), insects (2%) and to an extent mollusca (8%), require a positive decision to exploit the environmental potential of a site. It is perhaps surprising that soil, the medium in which most excavation takes place, has attracted so little analytical attention. In the present analysis, simple descriptions of soil components in and around the sites are excluded; an attempt at some pedological analysis (e.g. pH, particle-size, phosphate and even micromorphology), took place on only 15% of sites.

The available data set is probably insufficiently large for the establishment of meaningful patterns of particular environmental analyses through time. What can be illustrated, however inadequately, is the changing application of all environmental analyses over the review period. The unsmoothed data display an annual variability which is barely reduced by a five-term running average (Figure 14.3). The relatively low mean value of 48.5% (range 11.5–83.3%) is disconcerting when it is considered that virtually *all* excavations will be susceptible to and should profit from some form of environmental analysis. Figures such as those given here understandably evoke once again reactions akin to those voiced in a *Scottish Archaeological Forum* publication of almost two decades ago, *viz.*

> The archaeologist, ignoring [the] potential of his site, might be considered to be potentially as destructive of palaeogeographical information as he in turn views the developer of being of archaeology. (Edwards and Ralston 1979, 79)

and

> There is a need for many archaeologists to understand that the sites they excavate are often the sole repositories of evidence for the construction of past environments. (Whittington 1979, 84)

It cannot be denied, of course, that those archaeologists willing to seek funds for environmental research may find it difficult to get hard-pressed specialists to work on their sites (added to which is the fact that there is little incentive in producing a specialist environmental appendix to a publication, especially if such a report is relegated to microfiche); and archaeologists may find it even harder to understand the significance of some of the reports that are eventually produced! However, the discrepancies shown above between say the take-up of environmental analyses on prehistoric sites and their lowly adoption in Roman studies (or at least those published in the *Proceedings*), can only reflect a difference in ethos. With an awareness of the environmental potential of archaeological sites now a fundamental part of good archaeological training, the state of affairs outlined above – even

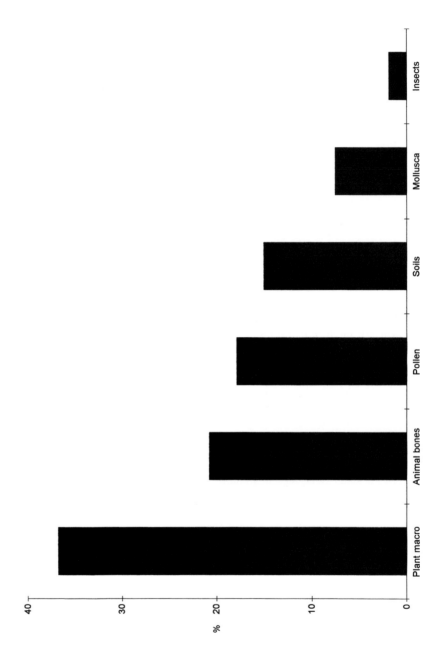

Figure 14.2 The major types of environmental data obtained from Scottish excavations expressed as percentages of all environmental investigations (data sources as for Figure 14.1)

261

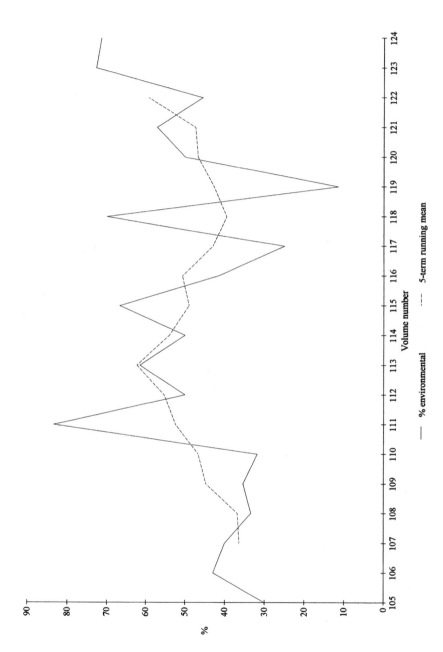

Figure 14.3 The percentages of Scottish excavation reports featuring environmental investigations over the review period (data sources as for Figure 14.1)

Plate 14.1   Deflation of machair sands by wind is destroying this Bronze Age settlement at Cladh Hallan, South Uist, here undergoing rescue excavation. Copyright: K. J. Edwards

allowing for the lag between research design and published outcome – indicates that a serious deficiency remains to be addressed.

LOOKING TO THE FUTURE

Mineral exploitation, changing farming patterns, urban sprawl, coastal erosion (either aeolian [Plate 14.1] or from rabbit and rat infestation [Plate 14.2]), and perhaps climate change are leading to a diminution of all the resources which can be used to reconstruct the past. While some of these threats result in rescue-oriented work, there is a need to formulate major research questions to enhance the study of the past. It is also necessary to contribute an environmental archaeological dimension to wider justifications for landscape conservation. The utility of environmental data is likely to be increased if the specialist providers and those engaged in other forms of archaeological work can focus on common problems which are amenable to solution by appropriate environmental methods. This is not to deny the fact that not all archaeological questions need environmental insights for their resolution, any more than environmental problems must have an archaeological dimension. If, however, a useful integration of the two types of information is seen as a worthwhile aim, then it might be asked what themes could be pursued to advantage.

Studies of climate change now point overwhelmingly to a rapid rise in temperatures in the early Holocene, but problems surround the reality and rapidity of climatic downturns from Neolithic times onwards, as different evidence produces different conclusions. It seems vital that studies of climate change founded on

Plate 14.2   These rat holes have pierced the sandy soils associated with a possible Iron Age settlement on Sandray. Copyright: K. J. Edwards

environmentally-sensitive insect species in Scotland, such as the Coleoptera and perhaps the Diptera, should be extended throughout the Holocene and correlated with sensitive ice-core data (Mayewski *et al.* 1996). In turn, this would enable valuable comparisons to be made with palynological data (cf. Huntley 1990, 1994) where the recognition of subtle climatic changes may be less readily appreciable after Lateglacial times because of vegetational lag effects and complications arising from human impacts on flora. Inferred human responses to supposed changes in climate could then be placed on a much surer footing than hitherto (cf. A. Harding 1982), with cultural data assessed against more objective climatic criteria.

Volcanic tephra is likely to prove of more use in providing temporally-precise isochrones for archaeological and environmental contexts (e.g. Hall *et al.* 1994; Dugmore *et al.* 1995) than in the investigation of catastrophic environmental impacts on marginal settlement (Baillie 1989; Burgess 1989; Grattan and Gilbertson 1994). The exploitation of such an amazing resource as tephra isochrones has barely been tapped, and awareness of its ability to furnish precise chronological horizons has been overshadowed by tabloid tales of nuclear winter and the demise of Bronze Age society (Keys 1988; Ezard 1996). The extension of tephrochronology would benefit from technological improvements aimed at speeding the process of tephra detection and identification.

The Mesolithic–Neolithic period of transition continues to fascinate and perplex. Further support needs to be sought for arguments for and against the precocious, indigenous development of practices similar to those employed in farming. If wholesale in-migrations of agriculturalists have fallen from favour, the Continental contribution to some fourth millennium cal BC types of evidence needs fuller evaluation; for example, the extent to which cereal cultivation practices followed Continental practice, as much as the indebtedness of certain architectural forms, such as the timber hall at Balbridie, to Continental models (Fairweather and Ralston 1993). Various scenarios for the longevity and viability of hunter–gatherer life-styles subsequent to the local adoption of agriculture can be posited, but for the moment seem incapable of resolution in the absence of fuller investigation of rare, stratified occupation deposits and improved and detailed palaeoecological informa-tion. The latter may be able to show continuity or variations in economic practice, including woodland management, cereal cultivation, animal exploitation and fishing.

From the Neolithic onwards, many of the key questions have to do with matters of scale rather than simply presence or absence. Such concerns are highlighted by the size of some individual lowland sites, such as the Cleaven Dyke (Barclay *et al.* 1995), and by the recognition of 'archaeological landscapes', including fields and houses, in both upland and lowland areas. The tasks ahead include the fuller determination of types and durations of farming strategies, and the reasons for expansion, contraction or abandonment of settlements and the land that supported them. Whilst it is reasonable to envisage the patterning as the outcome of complex interplays of human decision-making and environmental constraints, the latter are likely to be more amenable to field investigation.

The period of Roman influence may have been short-lived, but the impact of Rome went far beyond its frontiers. The precise nature of the relationship between the occupying forces and the indigenous populations remains debatable, although recent perspectives tend to downplay the scale of the impact of Roman upon native (Hanson and Breeze 1991; Chapter 11). Pollen data show agriculture and reduced woodland in central and southern areas (Boyd 1984; Dickson 1989; Dumayne and Barber 1994), as indeed they do for the pre-Roman Iron Age. This may indicate continuity of land use with either minimal interference, or the modification of native agricultural systems in order to supply at least some local Roman military require-ments. In north Fife and inland Aberdeenshire, however, evidence suggests that major woodland regeneration was underway in the early centuries AD. This may signify social and economic collapse in response to punitive military actions (Whittington and Edwards 1993), or may be a reflection of land use change towards intensification, either externally or internally induced, and either outwith, or par-tially within, the pollen catchments. The case for extensive integration of environ-mental and archaeological information is obvious and the requirement for far more pollen, plant macrofossil and animal bone data has been made (Hanson and Breeze 1991); the archaeological evidence for settlement during the first half of the first millennium AD north of the Forth also requires definition and investigation in order for it to be considered more fully in relation to perceived environmental change.

For much of the first millennium AD, consideration of the exploitation of the landscape forms a subset of a larger problem: to what extent should different practices be anticipated on the part of the groups recorded historically – the Britons, Picts, Scots, Angles and Norse? For example, palaeoeconomic evidence for the Pictish–Norse transition seems to indicate no significant differences in the style of agriculture by native and immigrant farmers. Yet the saga evidence, principally *Orkneyinga saga* (Pálsson and Edwards 1981), would imply major changes in the structure of society as the Earldom was consolidated. Environmental and palaeo-economic interpretations may provide indications of agricultural stability or innovations, at times and places where the historical evidence indicates frequent disruption as early states expanded and contracted in the conflicts recorded in annalistic and other sources. Contrasts between secular and ecclesiatical agricultural practices may also become apparent. More integrated studies are required if such questions are to be explored.

The study of landscapes at the mesoscale is far more likely to provide areally meaningful insights into continuity, dislocation and change than is the investigation of individual sites. Although there is nothing in Scotland on the scale of collabora-tion involved in the multidisciplinary Ystad Project for southern Sweden (Berglund 1991; and cf. the Småland Uplands Project [Lagerås 1996]), some researchers have been willing to explore landscape at a not dissimilar spatial scale. The Sheffield Environmental and Archaeological Research Campaign in the Hebrides (SEARCH), covering substantial areas of South Uist and Barra, as well as parts of Vatersay, Sandray, Pabbay and Mingulay, offers one model featuring intensive survey, excavation and environmental investigation, leading *inter alia* to extensive monograph publication (e.g. Branigan and Foster 1995; Gilbertson *et al.* 1996; Parker Pearson forthcoming). A number of smaller-scale exercises also exist, including the Lairg (McCullagh 1992a) and Bowmont Valley (Mercer and Tipping 1994) projects, while the Southern Hebrides Mesolithic Project (Mithen and Lake 1996) is very wide-ranging but is centred on a single cultural period. Given appropriate resources, there is a need for bolder initiatives in the study of environment and archaeology at the landscape scale. Such undertakings could apply in particular to transects across the divide between upland and lowland zones, thus profiting from their differing strengths in terms of preservation and visibility.

The paucity of excavated wetland sites in Scotland is disappointing, although underwater survey in particular has demonstrated how widespread the potential can be (Morrison 1985). Both threat-oriented and research projects are underway, but the scale of the response may still be considered insufficient given the wealth of the Scottish wetland resource. Its most celebrated components are the crannogs, in some instances threatened by drainage (Barber and Crone 1993). Survey in Loch Tay in particular has identified an extensive series of sites, some now radiocarbon-dated, and that at Oakbank under excavation (Dixon 1984). The important post-Roman example at Buiston, Ayrshire, known since last century, has also been re-examined (Crone 1991, 1993b). Other waterlogged settlement sites of a variety of dates have been excavated, albeit partially: Armit's work at Loch Olabhat, North Uist, has demonstrated the existence of superimposed Neolithic building horizons (Armit 1992a), and the examination of the complex Atlantic roundhouse at Loch na Berie, Lewis, has revealed internal waterlogged deposits (Harding and Armit 1990). Peatland sites are also threatened by both extraction and afforestation before their palaeoecological and archaeological potential can always be assessed, and there has yet to be any investigation in Scotland at the scale of those in the Somerset Levels (Coles and Coles 1986) or Flag Fen (Pryor 1991). In assessing areas of peatland where little structural evidence is readily visible, multiple-core predictive palynology may be useful, but assays with ground-penetrating radar would be a faster means of detection. Wetland sites provide both archaeological and environmental data in abundance, even if the organic artefacts are not always as common, or as ornate, as might have been anticipated (cf. Evans 1989). Excavation might routinely be widened to encompass more wetland sites, including those only intermittently submerged (for which practical limitations on work would be less severe) such as riverine and inter-tidal zones. The sentiment of Doran (1992, 132) is apposite: that archaeologists should look at the waterlogged deposits in a site and exclaim 'Fantastic, we've got wet materials!', instead of 'Stop digging, we've hit water'.

Whatever the nature of funding and excavation, however, the disappearance of sites may ultimately thwart all our research plans. Such concerns are not new, of course, and Scottish commentators have noted the loss of archaeological monuments since these latter were first described, as in the first quarter of the last century:

> In the course, however, of a short time, many of these have already disappeared; and, in a little while longer, from the rapid improvements of the country, many more of these monuments of antiquity will be totally obliterated. The vestiges of encampments, tumuli, barrows, and fields of battle, will be levelled by the plough; the pillars and circles of stones, already much diminished in number, will be built into walls and houses; and every memorial of past ages buried in oblivion. (Stuart 1822, 55)

This finds modern-day echoes, as for the Western Isles (Armit 1996, 235):

> Year by year the wind and weather tear at the most fragile remains in the islands. The processes that formed the Hebridean machair and preserved its archaeology also ultimately destroy it.

A case for the integrated conservation of environment and archaeology should be seen as a priority (cf. Bell and Walker 1992, 203–212; Wickham-Jones and Macinnes 1992), and this presents a challenge of co-ordination now being addressed by public bodies charged with managing our heritage, such as Scottish Natural Heritage and Historic Scotland. The funding for these comes from government, and ultimately from tax-payers. The lack of a unifying protective law to cover ancient monuments, wildlife and landscape, as in Denmark (Kristiansen 1989), means that we may lose our heritage by default (cf. Hunter and Ralston 1993). Knowledge can beget awareness, vigilance and urgency, and it is hoped that this book may go some way to fostering the bases for such an outlook.

References

Aaby, B. 1978 Cyclic changes in climate during 5,500 yrs, reflected in Danish raised bogs. *Danske Meteorolgisken Institutet Klimatologiske Meddelelser* **4**, 18–26.

Aaby, B. 1986 Trees as anthropogenic indicators in regional pollen diagrams from eastern Denmark. In Behre, K.-E. (ed.) *Anthropogenic Indicators in Pollen Diagrams.* Rotterdam: A. A. Balkema, 73–93.

Affleck, T. L. 1986 Excavation at Starr, Loch Doon 1985. *Glasgow Archaeological Society Bulletin* **22**, 10–21.

Affleck, T. L., Edwards, K. J. and Clarke, A. 1988 Archaeological and palynological studies at the Mesolithic pitchstone and flint site of Auchareoch, Isle of Arran. *Proceedings of the Society of Antiquaries of Scotland* **118**, 37–59.

Albrethsen, S. E. and Brinch Petersen, E. 1976 Excavation of a Mesolithic cemetry at Vedbæk, Denmark. *Acta Archaeologica* **47**, 1–28.

Alcock, L. 1981 Early historic fortifications in Scotland. In Guilbert, G. (ed.) *Hill-Fort Studies. Essays for A. H. A. Hogg.* Leicester: Leicester University Press, 150–180.

Alcock, L. 1987 Pictish studies: present and future. In Small, A. (ed.) *The Picts: A New Look at Old Problems.* Dundee: University of Dundee, 80–92.

Alcock, L. 1988a *Bede, Eddius and the Forts of the North Britons.* Jarrow: Jarrow lecture for 1988.

Alcock, L. 1988b The Rhind Lectures 1988–89 ('An heroic age: war and society in northern Britain AD 450–850'): a synopsis. *Proceedings of the Society of Antiquaries of Scotland* **118**, 327–334.

Alcock, L. 1988c Activities of potentates in Celtic Britain, A.D. 500–800: a positivist view. In Driscoll, S. T. and Nieke, M. R. (eds) *Power and Politics in Early Medieval Britain and Ireland.* Edinburgh: Edinburgh University Press, 22–46.

Alcock, L. 1993a *The Neighbours of the Picts: Angles, Britons and Scots at War and at Home.* Rosemarkie: Groam House Museum Trust.

Alcock, L. 1993b Image and icon in Pictish sculpture. In Spearman, R. M. and Higgitt, J. (eds) *The Age of Migrating Ideas.* Edinburgh: National Museums of Scotland, 230–236.

Alcock, L. and Alcock, E. A. 1980 Scandinavian settlement in the Inner Hebrides: recent research on placenames and in the field. *Scottish Archaeological Forum* **10**, 61–73.

Alcock, L. and Alcock, E. A. 1987 Reconnaissance excavations on Early Historic fortifications and other royal sites in Scotland 1974–84: 2, Excavations at Dunollie Castle, Oban, Argyll, 1978. *Proceedings of the Society of Antiquaries of Scotland* **117**, 119–147.

Alcock, L. and Alcock, E. A. 1990 Reconnaissance excavations on Early Historic fortifications and other royal sites in Scotland 1974–84: 4, Excavations at Alt Clut, Clyde Rock, Strathclyde 1974–75. *Proceedings of the Society of Antiquaries of Scotland* **120**, 95–149.

Alcock, L. and Alcock, E. A. 1992 Reconnaissance excavations on Early Historic fortifications and other royal sites in Scotland 1974–84: 5; A, Excavations and other fieldwork at Forteviot, Perthshire, 1981; B, Excavations at Urquhart Castle, Inverness-shire, 1983; C, Excavations at Dunnottar, Kincardineshire, 1984. *Proceedings of the Society of Antiquaries of Scotland* **122**, 215–289.

Alcock, L., Alcock, E. A. and Driscoll, S. T. 1989 Reconnaissance excavations on Early Historic sites in Scotland 1974–84: 3, Excavations at Dundurn, St Fillans, Perthshire, 1976 and 1977. *Proceedings of the Society of Antiquaries of Scotland* **119**, 189–226.

Allan, C. and Edwards, K. J. 1987 The distribution of lithic materials of possible Mesolithic age on the Isle of Arran. *Glasgow Archaeological Journal* **14**, 19–24.

Allen, J. R. and Anderson, J. 1993 *The Early Christian Monuments of Scotland*, **1**. Balnavies, Angus: Pinkfoot Press (reprint of 1903 edition).

Alley, R. B., Meese, D. A., Shuman, C. A., Gow, A. J., Taylor, K. C., Grootes, P. M., White, J. W. C., Ram, M., Waddington, E. D., Mayewski, P. A. and Zielinski, G. A. 1993 Abrupt increase in Greenland snow accumulation at the end of the Younger Dryas Event. *Nature* **362**, 527–529.

Amorosi, T., Buckland, P. C., Olafsson, G., Sadler, J. P. and Skidmore, P. 1992 Site status and the palaeoecological record: a discussion of the results from Bessastaðir, Iceland. In Morris, C. D. and Rackham, D. J. (eds) *Norse and Later Settlement and Subsistence in the North Atlantic*. Glasgow: University of Glasgow Department of Archaeology Occasional Paper Series **1**, 169–191.

Andersen, S. T. 1979 Identification of wild grass and cereal pollen. *Danmarks Geologiske Undersøgelse Årbog 1978*, 69–92.

Andersen, S. T. *et al.* 1990 Making cultural ecology relevant to Mesolithic research: I. A data base of 413 Mesolithic fauna assemblages. In Vermeersch, M. and van Peer, P. (eds) *Contributions to the Mesolithic in Europe*. Leuven: Leuven University Press, 23–51.

Anderson, A. O. 1922 *Early Sources of Scottish History AD 500–1286*, volumes 1 and 2. Edinburgh: Oliver and Boyd.

Anderson, J. 1895 Notice of a cave recently discovered at Oban, containing human remains and a refuse heap of shells and bones of animals, and stone and bone implements. *Proceedings of the Society of Antiquaries of Scotland* **29**, 211–230.

Anderson, J. 1898 Notes on the contents of a small cave or rock shelter at Druimvargie, Oban; and of three shell-mounds in Oronsay. *Proceedings of the Society of Antiquaries of Scotland* **32**, 298–313.

Andrews, M. V., Gilbertson, D. D., Kent, M. and Mellars, P. A. 1985 Biometric studies of morphological variation in the intertidal gastropod *Nucella lapillus* (L.): environmental and palaeoeconomic significance. *Journal of Biogeography* **12**, 71–87.

Armit, I. 1988 *Excavations at Loch Olabhat, North Uist 1988*. Edinburgh: Department of Archaeology University of Edinburgh Project Paper **10**.

Armit, I. 1990a Broch-building in northern Scotland: the context of innovation. *World Archaeology* **21**, 435–445.

Armit, I. 1990b Epilogue. In Armit, I (ed.) *Beyond the brochs*. Edinburgh: Edinburgh University Press, 194–210.

Armit, I. 1990c *The Loch Olabhat Project 1989*. Edinburgh: Department of Archaeology University of Edinburgh Project Paper **12**.

Armit, I. (ed.) 1990d *Beyond the brochs*. Edinburgh: Edinburgh University Press.

Armit, I. 1991 The Atlantic Scottish Iron Age: five levels of chronology. *Proceeding of the Society of Antiquaries of Scotland* **121**, 181–214.

Armit, I. 1992a The Hebridean Neolithic. In Sharples, N. M. and Sheridan, A. (eds) *Vessels for the Ancestors*. Edinburgh: Edinburgh University Press, 307–321.

Armit, I. 1992b *The Later Prehistory of the Western Isles of Scotland*. Oxford: British Archaeological Reports British Series **221**.

Armit, I. 1996 *The Archaeology of Skye and the Western Isles*. Edinburgh: Edinburgh University Press.

Armit, I. and Finlayson, B. 1992 Hunter–gatherers transformed: the transition to agriculture in northern and western Europe, *Antiquity* **66**, 664–676.

Ashmore, P. J. 1996. *Neolithic and Bronze Age Scotland*. London: Batsford/Historic Scotland.

Ashworth, A. C. 1972 A Late-glacial insect fauna from Red Moss, Lancashire, England. *Entomologica Scandinavica* **3**, 211–224.

Askew G. P., Payton, R. W. and Shiel, R. S. 1985 Upland soils and land clearance in Britain during the second millennium BC. In Spratt, D. and Burgess, C. B. (eds) *Upland Settlement in Britain. The Second Millennium B.C. and After*. Oxford: British Archaeological Reports British Series **143**, 5–33.

Atkinson, R. J. C. 1962 Fishermen and farmers. In Piggott, S. (ed.) *The Prehistoric Peoples of Scotland.* London: Routledge and Kegan Paul, 1–38.

Atkinson, R. J. C. 1968 Old mortality: some aspects of burial and population in Neolithic England. In Simpson, D. D. A. and Coles, J. M. (eds) *Studies in Ancient Europe.* Leicester: Leicester University Press, 83–93.

Atkinson, T. C., Briffa, K. R. and Coope, G. R. 1987 Seasonal temperatures in Britain during the past 22,000 years, reconstructed using beetle remains. *Nature* **325**, 587–592.

Audouze, F. and Büchsenschütz, O. E. 1992 *Towns, Villages and Countryside of Celtic Europe.* London: Batsford.

Bacchus, M. 1980 Beetles. In Berry, R. J. and Johnston, J. L. (eds) *The Natural History of Shetland.* London: Collins, 306–311.

Baillie, M. 1989 Do Irish bog oaks date the Shang Dynasty? *Current Archaeology* **117**, 310–313.

Baldwin, J. R. 1978 *Scandinavian Shetland. An Ongoing Tradition?* Edinburgh: Scottish Society for Northern Studies.

Baldwin, J. R. 1984 Hogin and hametoun: thoughts on the stratification of a Foula tun. In Crawford, B. E. (ed.) *Essays in Shetland History.* Lerwick: Shetland Times, 33–64.

Balfour-Browne, F. 1953 The aquatic Coleoptera of the western Scottish Islands, with a discussion on their origin and means of arrival. *Entomologist's Gazette* **4**, 79–127.

Ball, D. F. 1975 Processes of soil degradation: a pedological point of view. In Evans, J. G., Limbrey, S. and Cleere, H. (eds) *The Effect of Man on the Landscape: The Highland Zone.* London: Council for British Archaeology Research Report **11**, 20–27.

Ballantyne, C. K. 1986 Landslides and slope failures in Scotland: a review. *Scottish Geographical Magazine* **102**, 134–150.

Ballantyne, C. K. 1990 The Late Quaternary glacial history of the Trotternish Escarpment, Isle of Skye, Scotland, and its implications for ice-sheet reconstruction. *Proceedings of the Geologists' Association* **101**, 171–186.

Ballantyne, C. K. 1991a Holocene geomorphic activity in the Scottish Highlands. *Scottish Geographical Magazine* **107**, 84–98.

Ballantyne, C. K. 1991b Late Holocene erosion in upland Britain: climatic deterioration or human influence? *The Holocene* **1**, 81–85.

Ballantyne, C. K. and Eckford, J. D. 1984 Characteristics and evolution of two relict talus slopes in Scotland. *Scottish Geographical Magazine* **100**, 20–33.

Bang-Andersen, S. 1989 Mesolithic adaptations in the southern Norwegian highlands. In Bonsall, J. C. (ed.) *The Mesolithic in Europe.* Edinburgh: John Donald, 338–350.

Bannerman, J. 1974 *Studies in the History of Dalriada.* Edinburgh and London: Scottish Academic Press.

Barber, J. W. 1981 Excavations on Iona. *Proceedings of the Society of Antiquaries of Scotland* **111**, 282–380.

Barber, J. W. 1982a The investigation of some plough-truncated features at Kinloch Farm, Collessie in Fife. *Proceedings of the Society of Antiquaries of Scotland* **112**, 524–533.

Barber, J. W. 1982b Arran. *Current Archaeology* **83**, 358–363.

Barber, J. W. 1988 Isbister, Quanterness and the Point of Cott: the formulation and testing of some middle range theories. In Barrett, J. C. and Kinnes, I. L. A. (eds) *The Archaeology of Context in the Neolithic and Bronze Age: Recent Trends.* Sheffield: Department of Archaeology and Prehistory University of Sheffield, 57–62.

Barber, J. W. forthcoming The excavation of the tomb at Point of Cott. *Proceedings of the Prehistoric Society.*

Barber, J. W. and Brown, M. M. 1984 An Sithean, Islay. *Proceedings of the Society of Antiquaires of Scotland* **114**, 161–188.

Barber, J. W. and Crone, B. A. 1993 Crannogs – a diminishing resource? A survey of the crannogs of South West Scotland and excavations at Buiston crannog. *Antiquity* **67**, 520–533.

Barber, J. W., Maté, I. D. and Tabraham, C. J. 1982 The Deil's Dyke, Nithsdale. *Transactions of the Dumfries and Galloway Natural History and Antiquarian Society* **54**, 29–50.

Barber, K. E. 1981 *Peat stratigraphy and climatic change: a palaeoecological test of the theory of cyclic peat bog regeneration.* Rotterdam: A. A. Balkema.

Barber, K. E. 1982 Peat-bog stratigraphy as a proxy climate record. In Harding, A. F. (ed.) *Climatic Change in Later Prehistory.* Edinburgh: Edinburgh University Press, 103–133.

Barber, K. E. 1985 Peat stratigraphy and climatic changes: some speculations. In Tooley, M. J. and Sheail, G. M. (eds) *The Climatic Scene: Essays in Honour of Gordon Manley.* London: Allen and Unwin, 175–185.

Barber, K. E. 1994 Deriving Holocene palaeoclimates from peat stratigraphy: some misconceptions regarding the sensitivity and continuity of the record. *Quaternary Newsletter* **72**, 1–10.

Barber, K. E., Chambers, F. M., Dumayne, L., Haslam, D. J., Maddy, D. and Stoneman, R. E. 1994a Climatic change and human impact in north Cumbria: peat stratigraphic and pollen evidence from Bolton Fell Moss and Walton Moss. In Boardman, J. and Walden, J. (eds) *The Quaternary of Cumbria: Field Guide.* Oxford: Quaternary Research Association, 20–49.

Barber, K. E., Chambers, F. M., Maddy, D., Stoneman, R. E. and Brew, J. S. 1994b A sensitive high-resolution record of Late Holocene climatic change from a raised bog in northern England. *The Holocene* **4**, 198–205.

Barclay, G. J. 1983a Sites of the third millennium bc to the first millennium ad at North Mains, Strathallan, Perthshire. *Proceedings of the Society of Antiquaries of Scotland* **113**, 122–281.

Barclay, G. J. 1983b The excavation of a settlement of the Later Bronze Age and Iron Age at Myrehead, Falkirk District. *Glasgow Archaeological Journal* **10**, 41–71.

Barclay, G. J. 1985 Excavations at Upper Suisgill, Sutherland. *Proceedings of the Society of Antiquaries of Scotland* **115**, 159–198.

Barclay, G. J. 1989 The cultivation remains beneath the North Mains, Strathallan barrow. *Proceedings of the Society of Antiquaries of Scotland* **119**, 59–61.

Barclay, G. J. 1992 The Scottish gravels: a neglected resource. In Fulford, M. and Nichols, E. (eds) *Developing Landscapes of Lowland Britain. The Archaeology of the British Gravels: A Review.* London: Society of Antiquaries of London Occasional Paper **14**, 106–124.

Barclay, G. J. 1996 Neolithic buildings in Scotland. In Darvill, T. and Thomas, J. (eds) *Neolithic Houses in North-West Europe and Beyond.* Oxford: Oxbow Monograph **57**, 61–75.

Barclay, G. J. and Fairweather, A. D. 1984 Rye and ergot in the the Scottish later Bronze Age. *Antiquity*, **58**, 126.

Barclay, G. J., Maxwell, G. S., Simpson, I. A. and Davidson, D. A. 1995 The Cleaven Dyke: a Neolithic cursus monument/bank barrow in Tayside Region, Scotland. *Antiquity* **69**, 317–326.

Barclay, G. J. and Russell-White C. J. (eds) 1993 Excavations in the ceremonial complex at Balfarg/Balbirnie, Glenrothes, Fife, *Proceedings of the Society of Antiquaries of Scotland* **123**, 43–210.

Barker, G. 1983 The animal bones. In Hedges, J. W. *Isbister. A Chambered Tomb in Orkney.* Oxford: British Archaeological Reports British Series **113**, 133–150.

Barnetson, L. P. D. 1982 Animal husbandry – clues from Broxmouth. In Harding, D. W. (ed.) *Later Prehistoric Settlement in South-East Scotland.* Edinburgh: Department of Archaeology University of Edinburgh Occasional Paper **8**, 101–105.

Barnetson, L. P. D. 1988 Animal bones. In Thomas, G. D., Excavations at the Roman civil settlement at Inveresk, 1976–77. *Proceedings of the Society of Antiquaries of Scotland* **118**, 172–173.

Barratt, J. 1995 *'Few Know an Earl in Fishing Clothes'. Fish Middens and the Economy of the Viking Age and Late Norse Earldoms of Orkney and Caithness, Northern Scotland.* Unpublished PhD thesis, University of Glasgow.

Barrett, J. C. 1989 Food, gender and metal: questions of social reproduction. In Stig Sorensen, M. L. and Thomas, R. (eds) 1989 *The Bronze Age–Iron Age Transition in Europe. Aspects of Continuity and Change in European Societies c. 1200 to 500 BC.* Oxford: British Archaeological Reports British Series, **S483**, vol. 2, 304–320.

Barrett, J. C. 1994 *Fragments from Antiquity*. Oxford: Blackwell.

Barrett, J. C. and Downes, J. 1994 Tayside: North Pitcarmick (Kirkmichael Parish). *Discovery and Excavation in Scotland 1994*, 87.

Barrett, J. C. and Downes, J. M. in preparation *North Pitcarmick, North East Perthshire: The Early Medieval Inhabitation of a Prehistoric Landscape*.

Barrett, J. C. and Foster, S. M. 1991 Passing the time in Iron Age Scotland. In Hanson, W. S. and Slater, E. A. (eds) *Scottish Archaeology: New Perceptions*. Aberdeen: Aberdeen University Press, 44–56.

Barton, R. N. E. 1989 Long blade technology in southern Britain. In Bonsall, J. C. (ed.) *The Mesolithic in Europe*. Edinburgh: John Donald, 264–271.

Batey, C. E. 1987 *Freswick Links, Caithness, A Re-appraisal of the Late Norse Site in its Context*. Oxford: British Archaeological Reports British Series, **179**.

Batey, C. E., Jesch, J. and Morris, C. D. (eds) 1993 *The Viking Age in Caithness, Orkney and the North Atlantic*. Edinburgh: Edinburgh University Press.

Behre, K.-E. 1981 The interpretation of anthropogenic indicators in pollen diagrams. *Pollen et Spores* **23**, 225–245.

Bell, M. 1983 Valley sediments as evidence of prehistoric land-use on the South Downs. *Proceedings of the Prehistoric Society* **49**, 119–150.

Bell, M. 1990 *Brean Down excavations 1983–1987*. London: English Heritage Archaeological Report **15**.

Bell, M. and Walker, M. J. C. 1992 *Late Quaternary Environmental Change: Physical and Human Perspectives*. Harlow: Longman.

Benn, D. I. 1992 The genesis and significance of 'hummocky moraine': evidence from the Isle of Skye, Scotland. *Quaternary Science Reviews* **11**, 781–799.

Benn, D. I., Lowe, J. J. and Walker, M. J. C. 1992 Glacier response to climatic change during the Loch Lomond Stadial and early Flandrian: geomorphological and palynological evidence from the Isle of Skye, Scotland. *Journal of Quaternary Science* **7**, 125–144.

Bennett, K. D. 1984 The Post-Glacial history of *Pinus sylvestris* in the British Isles. *Quaternary Science Reviews* **3**, 133–156.

Bennett, K. D. 1989 A provisional map of forest types for the British Isles 5000 years ago. *Journal of Quaternary Science* **4**, 141–144.

Bennett, K. D. and Birks, H. J. B. 1990 Postglacial history of alder (*Alnus glutinosa* (L.) Gaertn.) in the British Isles. *Journal of Quaternary Science* **5**, 123–133.

Bennett, K. D., Boreham, S., Hill, K., Packman, S., Sharp, M. J. and Switsur, V. R. 1993 Holocene environmental history at Gunnister, north Mainland, Shetland. In Birnie, J. F., Gordon, J. E., Bennett, K. D. and Hall, A. (eds) *The Quaternary of Shetland: Field Guide*. Cambridge: Quaternary Research Association, 83–98.

Bennett, K. D., Boreham, S., Sharp, M. J. and Switsur, V. R. 1992 Holocene history of environment, vegetation and human settlement on Catta Ness, Lunnasting, Shetland. *Journal of Ecology* **80**, 241–273.

Bennett, K. D., Fossitt, J. A., Sharp, M. J. and Switsur, V. R. 1990 Holocene vegetational and environmental history at Loch Lang, South Uist, Western Isles, Scotland. *New Phytologist* **114**, 281–298.

Bennett, K. D. and Humphrey, R. W. 1995 Analysis of late-glacial and Holocene rates of vegetational change at two sites in the British Isles. *Review of Palaeobotany and Palynology* **85**, 263–287.

Bennett, M. R. and Boulton, G. S. 1993 A reinterpretation of Scottish 'hummocky moraine' and its significance for the deglaciation of the Scottish Highlands during the Younger Dryas or Loch Lomond Stadial. *Geological Magazine* **130**, 301–318.

Berg, S. and McDonnell, G. forthcoming The ironwork. In Hunter, J. R., Dockrill, S. J., Bond, J. M. and Smith, A. N. (eds) *Archaeological Investigations on Sanday, Orkney*. Edinburgh: Society of Antiquaries of Scotland Monograph Series.

Berglund, B. E. 1986 The cultural landscape in a long-term perspective. Methods and theories behind the research on land-use and landscape dynamics. *Striae* **24**, 79–87.

Berglund, B. E. (ed.) 1991 *The Cultural Landscape during 6000 Years in Southern Sweden – The Ystad Project*. Copenhagen: Ecological Bulletins **41**.

Berry, R. J. 1969 History in the evolution of *Apodemus sylvaticus* at one edge of its range. *Journal of the Zoological Society of London* **159**, 311–328.

Berry, R. J. 1979 The Outer Hebrides: where genes and geography meet. *Proceedings of the Royal Society of Edinburgh* **77B**, 21–46.

Berry, R. J. and Johnston, J. L. 1980 *The Natural History of Shetland*. London: Collins.

Berry, R. J. and Rose, F. E. N. 1975 Islands and the evolution of *Microtus arvalis* (Microtinae). *Journal of the Zoological Society of London* **177**, 395–405.

Bertelsen, R. 1979 Farm mounds in North Norway, a review of recent research. *Norwegian Archaeological Review* **12**, 48–56.

Bertelsen, R. 1984 Farm mounds of the Harstad area. *Acta Borealia* **1**, 7–25.

Bertelsen, R. 1991 A North-East Atlantic perspective. *Acta Archaeologica* **61**, 22–28.

Bertelsen, R. and Lamb, R. G. 1993 Settlement mounds in the north Atlantic. In Batey, C. E., Jesch, J. and Morris, C. D. (eds) *The Viking Age in Caithness, Orkney, and the North Atlantic*. Edinburgh: Edinburgh University Press, 544–554.

Beveridge, E. 1911 *North Uist: Its Archaeology and Topography*. Edinburgh: Brown.

Bibby, J. S., Douglas, H. A., Thomasson, A. J. and Robertson, J. S. 1982 *Land Capability Classification for Agriculture*. Aberdeen: Monograph of the Soil Survey of Scotland, Macaulay Institute for Soil Research.

Bibby, J. S., Heslop, R. E. F. and Hartnup, R. 1988 *Land Capability Classification for Forestry in Britain*. Aberdeen: Monograph of the Soil Survey of Scotland, Macaulay Institute for Soil Research.

Bidwell, P. and Speak, S. 1989 South Shields. *Current Archaeology* **116**, 283–287.

Bigelow, G. F. 1985 Sandwick, Unst and the Late Norse Shetland economy. In Smith, B. (ed.) *Shetland Archaeology*. Lerwick: Shetland Times, 95–127.

Bigelow, G. F. 1989 Life in medieval Shetland: an archaeological perspective. *Hikuin* **15**, 183–192.

Bigelow, G. F. 1992 Issues and prospects in Shetland Norse archaeology. In Morris, C. D. and Rackham, D. J. (eds) *Norse and Later Settlement and Subsistence in the North Atlantic*. Glasgow: Department of Archaeology University of Glasgow Occasional Paper Series **1**, 19–32.

Binford, L. R. 1978 *Nunamiut Ethnoarchaeology*. New York: Academic Press.

Birks, H. H. 1975 Studies in the vegetational history of Scotland. IV. Pine stumps in Scottish blanket peats. *Philosophical Transactions of the Royal Society of London* **B270**, 181–226.

Birks, H. J. B. 1980 *Quaternary Vegetational History of West Scotland*. Cambridge: 5th International Palynological Conference Excursion C8 Guidebook.

Birks, H. J. B. 1982 Holocene (Flandrian) chronostratigraphy of the British Isles: a review. *Striae* **16**, 99–105.

Birks, H. J. B. 1989 Holocene isochrone maps and patterns of tree-spreading in the British Isles. *Journal of Biogeography* **16**, 503–540.

Birks, H. J. B. 1990 Changes in vegetation and climate during the Holocene of Europe. In Boer, W. M. and De Groot, R. S. (eds) *Landscape – Ecological Impact of Climatic Change*. Amsterdam: IOS Press, 133–158.

Birks, H. J. B. and Birks, H. H. 1980 *Quaternary Palaeoecology*. London: Edward Arnold.

Birks, H. J. B. and Line, J. M. 1992 The use of rarefaction analysis for estimating palynological richness from Quaternary pollen-analytical data. *The Holocene* **2**, 1–10.

Birks, H. J. B. and Madsen, B. J. 1979 Flandrian vegetational history of Little Loch Roag, Isle of Lewis, Scotland. *Journal of Ecology* **67**, 825–842.

Birks, H. J. B. and Williams, W. 1983 Late-Quaternary vegetational history of the Inner Hebrides. *Proceedings of the Royal Society of Edinburgh* **83B**, 269–292.

Bishop, A. H. 1914 An Oronsay shell-mound – a Scottish pre-Neolithic site. *Proceedings of the Society of Antiquaries of Scotland* **48**, 52–108.

Bishop, W. W. and Coope, G. R. 1977 Stratigraphical and faunal evidence for Lateglacial and Flandrian environments in south-west Scotland. In Gray, J. M. and Lowe, J. J. (eds) *Studies in the Scottish Lateglacial environment*. Oxford: Pergamon Press, 61–88.

Blackford, J. J. 1993 Peat bogs as sources of proxy climatic data: past approaches and future research. In Chambers, F. M. (ed.) *Climate Change and Human Impact on the Landscape*. London: Chapman and Hall, 47–56.

Blackford, J. J. and Chambers, F. M. 1991 Proxy records of climate change from blanket mires: evidence for a Dark Age (1400 BP) climatic deterioration in the British Isles. *The Holocene* **1**, 63–67.

Blackford, J. J., Edwards, K. J., Dugmore, A. J., Cook, G. T. and Buckland, P. C. 1992 Icelandic volcanic ash and the mid-Holocene Scots pine (*Pinus sylvestris*) pollen decline in northern Scotland. *The Holocene* **2**, 260–265.

Bohncke, S. J. P. 1988 Vegetation and habitation history of the Callinish area, Isle of Lewis, Scotland. In Birks, H. H., Birks, H. J. B., Kaland, P. E. and Moe, D. (eds) *The Cultural Landscape – Past, Present and Future*. Cambridge: Cambridge University Press, 445–461.

Bond, J. M. forthcoming a The botanical remains. In Hunter, J. R., Dockrill, S. J., Bond, J. M. and Smith, A. N. (eds) *Archaeological Investigations on Sanday, Orkney*. Society of Antiquaries of Scotland Monograph Series.

Bond, J. M. forthcoming b The faunal remains. In Hunter, J. R., Dockrill, S. J., Bond, J. M. and Smith, A. N. (eds) *Archaeological Investigations on Sanday, Orkney*. Society of Antiquaries of Scotland Monograph Series.

Bond, J. M. and Hunter, J. R. 1987 Flax-growing in Orkney from the Norse period to the 18th century. *Proceedings of the Society of Antiquaries of Scotland* **117**, 175–181.

Bonsall, J. C. 1981 The coastal factor in the Mesolithic settlement of north-west England. In Gramsch, B. (ed.) *Mesolithikum in Europa*. Berlin: Deutscher Verlag der Wissenschaften, 451–472 (Veröffentlichungen des Museums für Ur- und Frühgeschichte Potsdam, **14–15**).

Bonsall, J. C. 1988 Morton and Lussa Wood, the case for early Flandrian settlement of Scotland: comment on Myers. *Scottish Archaeological Review* **5**, 30–33.

Bonsall, J. C. 1992 Archaeology. In Walker, M. J. C., Gray, J. M. and Lowe, J. J. (eds) *The South-West Scottish Highlands: Field Guide*. Cambridge: Quaternary Research Association, Cambridge, 27–34.

Bonsall, J. C. 1996 The 'Obanian problem': coastal adaptation in the Mesolithic of western Scotland. In Pollard, T. and Morrison, A. (eds) *The Early Prehistory of Scotland*. Edinburgh: Edinburgh University Press, 183–197.

Bonsall, J. C. and Smith, C. 1989 Late Palaeolithic and Mesolithic bone and antler artifacts from Britain: first reactions to accelerator dates. *Mesolithic Miscellany* **10**, 33–38.

Bonsall, J. C. and Smith, C. 1990 Bone and antler technology in the British Late Upper Palaeolithic and Mesolithic: the impact of accelerator dating. In Vermeersch, P. M. and Van Peer, P. (eds) *Contributions to the Mesolithic in Europe*. Leuven: Leuven University Press, 359–368.

Bonsall, J. C. and Sutherland, D. G. 1992 The Oban caves. In Walker, M. J. C., Gray, J. M. and Lowe, J. J. (eds) *The South-West Scottish Highlands: Field Guide*. Cambridge: Quaternary Research Association, 115–121.

Boulton, G. S., Peacock, J. D. and Sutherland, D. G. 1991 Quaternary. In Craig, G. Y. (ed.) *Geology of Scotland*. London: The Geological Society, 503–543.

Bowen, D. Q., Rose, J., McCabe, A. M. and Sutherland, D. G. 1986 Correlation of Quaternary glaciations in England, Ireland, Scotland and Wales. *Quaternary Science Reviews* **5**, 299–340.

Bown, C. J. and Shipley, B. M. 1982 *Soil and Land Capability for Agriculture: South-East Scotland*. Aberdeen: The Macaulay Institute for Soil Research.

Boyd, W. E. 1984 Environmental change and Iron Age land management in the area of the Antonine Wall. *Glasgow Archaeological Journal* **11**, 75–81.

Boyd, W. E. 1985a Palaeobotanical evidence from Mollins. *Britannia* **16**, 37–48.

Boyd, W. E. 1985b Palaeobotanical report. In Keppie, L. J. F. (ed.) Excavations at the Roman fort of Bar Hill, 1978–82. *Glasgow Archaeological Journal* **12**, 79–81.

Boyd, W. E. 1988 Cereals in Scottish antiquity. *Circaea* **5**, 101–110.

Bradley, R. J. 1978 *The Prehistoric Settlement of Britain*. London: Routledge and Kegan Paul.

Bradley, R. J. 1984 *The Social Foundations of Prehistoric Britain: Themes and Variations in the Archaeology of Power*. London: Longman.

Bradley, R. J. 1985 *Consumption, Change and the Archaeological Record.* Edinburgh: Department of Archaeology University of Edinburgh Occasional Paper **13**.

Bradley, R. J. 1993 *Altering the Earth.* Edinburgh: Society of Antiquaries of Scotland Monograph Series **8**.

Bradley, R. J. and Edmonds, M. 1988 Fieldwork at Great Langdale, Cumbria, 1985–7: preliminary report. *Antiquaries Journal* **68**, 181–209.

Bradley, R. J. and Edmonds, M. 1993 *Interpreting the Axe Trade.* Cambridge: Cambridge University Press.

Bradley, R. J., Harding, J. and Mathews, M. 1993 The siting of prehistoric rock art in Galloway, south-west Scotland. *Proceedings of the Prehistoric Society* **59**, 269–283.

Bramwell, D. 1977 Bird and vole bones from Buckquoy, Orkney. In Ritchie, A. (ed.) Excavation of Pictish and Viking-Age farmsteads at Buckquoy, Orkney. *Proceedings of the Society of Antiquaries of Scotland,* **108**, 209–211.

Bramwell, D. 1979 The bird bones. In Renfrew, A. C. (ed.) *Investigations in Orkney.* London: Society of Antiquaries of London Research Report **38**, 138–143.

Bramwell, D. 1983a Appendix 5: Bird bones from Knap of Howar, Orkney. In Ritchie, A. (ed.) Excavation of a Neolithic farmstead at Knap of Howar, Papa Westray, Orkney. *Proceedings of the Society of Antiquaries of Scotland,* **113**, 100–103.

Bramwell, D. 1983b The bird remains. In Hedges, J. W. (ed.) *Isbister. A Chambered Tomb in Orkney.* Oxford: British Archaeological Reports British Series **113**, 159–170.

Bramwell, D. (with B. B. Smith) 1994 The bird remains. In Smith, B. B. (ed.) *Howe, Four Millennia of Orkney Prehistory.* Edinburgh: Society of Antiquaries of Scotland Monograph Series **9**, 153–157.

Branigan, K. and Foster, P. 1995 *Barra. Archaeological Research on Ben Tangaval.* Sheffield: Sheffield Academic Press.

Braund, D. C. 1991 Roman and native in Transcaucasia from Pompey to Successianus. In Maxfield, V. A. and Dobson, M. J. (eds) *Roman Frontier Studies 1989.* Exeter: University of Exeter Press, 419–423.

Brayshay, B. 1992 *Pollen Analysis and the Vegetational History of Barra and South Uist in the Outer Hebrides, Scotland.* Unpublished PhD thesis, University of Sheffield.

Brayshay, B. A. and Edwards, K. J. 1996 Lateglacial and Holocene vegetational history of South Uist and Barra. In Gilbertson, D. D., Kent, M. and Grattan, J. P. (eds) *The Environment of the Outer Hebrides: the Last 14,000 Years.* Sheffield: Sheffield Academic Press, pp. 13–26.

Brazier, V. and Ballantyne, C. K. 1989 Late Holocene debris cone evolution in Glen Feshie, western Cairngorm Mountains, Scotland. *Transactions of the Royal Society of Edinburgh: Earth Sciences* **80**, 17–24.

Brazier, V., Whittington, G. and Ballantyne, C. K. 1988 Holocene debris cone evolution in Glen Etive, Western Grampian Highlands, Scotland. *Earth Surface Processes and Landforms* **13**, 525–531.

Breeze, D. J. 1982 *The Northern Frontiers of Roman Britain.* London: Batsford.

Breeze, D. J. 1984 Demand and supply on the northern frontier. In Miket, R. and Burgess, C. B. (eds) *Between and Beyond the Walls: Essays on the Prehistory and History of North Britain in Honour of George Jobey.* Edinburgh: John Donald, 264–286.

Breeze, D. J. 1985 Roman forces and native populations. *Proceedings of Society of Antiquaries of Scotland* **115**, 223–228.

Breeze, D. J. 1986 The manufacture of pottery in Roman Scotland. *Proceedings of Society of Antiquaries of Scotland* **116**, 185–189.

Breeze, D. J. 1989 The impact of the Roman army on north Britain. In Barrett, J. C., Fitzpatrick, A. P., and Macinnes, L. (eds) *Barbarians and Romans in North-West Europe from the Later Republic to Late Antiquity.* Oxford: British Archaeological Reports International Series **S471**, 227–234.

Breeze, D. J. 1994 The imperial legacy – Rome and her neighbours. In Crawford, B. E. (ed.) *Scotland in Dark Age Europe.* St Andrews: St John's House Papers **5**, 13–19.

Breeze, D. J. and Dobson, B. 1987 *Hadrian's Wall,* 3rd edition. Harmondsworth: Penguin Books.

Bridge, M. C., Haggart, B. A. and Lowe, J. J. 1990 The history and palaeoclimatic significance of subfossil remains of *Pinus sylvestris* in blanket peats from Scotland. *Journal of Ecology* **78**, 77–99.

Britnell, W. J. 1982 The excavation of two barrows at Trelystan, Powys. *Proceedings of the Prehistoric Society* **48**, 133–202.

Brown, M. 1983 New evidence for Anglian settlement in East Lothian. *Scottish Archaeological Review* **2.2**, 156–163.

Buckland, P. C. 1981 The early dispersal of insect pests of stored products as indicated by archaeological research. *Journal of Stored Product Research* **17**, 1–12.

Buckland, P. C. 1988 North Atlantic faunal connections – introduction or endemics? *Entomologica scandinavica Supplement* **32**, 7–29.

Buckland, P. C. and Dinnin, M. H. 1993 Holocene woodlands: the fossil insect evidence. In Kirby, M. and Drake, C. M. (eds) *Dead Wood Matters: The Ecology and Conservation of Saproxylic Invertebrates in Britain.* Peterborough: English Nature Science **7**, 6–20.

Buckland, P. C., Dugmore, A. J., Perry, D. W., Savory, D. and Sveinbjarnadòttir, G. 1991 Holt in Eyjafjallasveit, Iceland. A palaeoecological study of the impact of landnam. *Acta Archaeologica*, **61**, 252–271.

Buckland, P. C. and Edwards, K. J. 1984 The longevity of pastoral episodes of clearance activity in pollen diagrams – the rôle of post-occupation grazing. *Journal of Biogeography* **11**, 243–249.

Buckland, P. C. and Sadler, J. P. 1989 A biogeography of the human flea, *Pulex irritans* L. (Siphonaptera: Pulicidae). *Journal of Biogeography* **16**, 115–120.

Buckley, T. E. and Harvey-Brown, J. A. 1891 *A Vertebrate Fauna of the Orkney Islands.* Edinburgh: David Douglas.

Buckley, V. (ed.) 1990 *Burnt Offerings. International Contributions to Burnt Mound Archaeology.* Dublin: Wordwell.

Bunting, M. J. 1994 Vegetation history of Orkney, Scotland: pollen records from two small basins in west Mainland. *New Phytologist* **128**, 771–792.

Burgess, C. B. 1976 Meldon Bridge: a Neolithic defended promontory complex near Peebles. In Burgess, C. B. and Miket, R. (eds) *Settlement and Economy in the Third and Second Millennia BC*, Oxford: British Archaeological Reports British Series **33**, 151–179.

Burgess, C. B. 1980 *The Age of Stonehenge.* London: Dent.

Burgess, C. B. 1989 Volcanoes, catastrophe and the global crisis of the late second millennium BC. *Current Archaeology* **117**, 325–329.

Burgess, C. B. 1995 Bronze Age settlements and domestic pottery in north Britain: some suggestions. In Kinnes, I. and Varndell, G. (eds) *'Unbaked Urns of Rudely Shape': Essays on British and Irish Pottery for Ian Longworth.* Oxford: Oxbow Monograph Series **55**, 145–158.

Burl, H. A. W. 1976 *The Stone Circles of the British Isles.* London: Yale University Press.

Burl, H. A. W. 1980 Science or symbolism: problems of archaeoastronomy. *Antiquity* **54**, 191–200.

Burl, H. A. W. 1984 Report on the excavation of a Neolithic mound at Boghead, Speymouth Forest, Fochabers, Moray, 1972 and 1974. *Proceedings of the Society of Antiquaries of Scotland* **114**, 35–73.

Burl, H. A. W. 1991 *Prehistoric Henges.* Princes Risborough: Shire Publications.

Buteux, S. (ed.) forthcoming *Settlements at Skaill, Deerness, Orkney.* Oxford: British Archaeological Reports British Series.

Butler, S. 1989 Pollen analysis from the west rampart. In Frere, S. S. and Wilkes, J. J. (eds) *Strageath: Excavations within the Roman Fort.* London: Society for the Promotion of Roman Studies, 272–274.

Buttler, S. 1989 Steatite in Norse Shetland. *Hikuin* **15**, 193–206.

Callander, J. G. 1929 Land movements in Scotland in prehistoric and recent times. *Proceedings of the Society of Antiquaries of Scotland* **63**, 314–322.

Callander, J. G., Cree, J. E. and Ritchie, J. 1927 Preliminary report on caves containing Palaeolithic relics near Inchnadamph, Sutherland. *Proceedings of the Society of Antiquaries of Scotland* **61**, 169–172.

Campbell, E. 1987 A cross-marked quern from Dunadd and other evidence for relations between Dunadd and Iona. *Proceedings of the Society of Antiquaries of Scotland* **117**, 105–117.

Campbell, E. and Lane, A. 1993 Celtic and Germanic interaction in Dalriada: the 7th century metalworking site at Dunadd. In Spearman, R. M. and Higgitt, J. (eds) *The Age of Migrating Ideas*. Edinburgh: National Museums of Scotland, 52–63.

Carter, P. L., Phillipson, D. and Higgs, E. S. 1965 The animal bones. In Hastings, D. and Cunliffe, B. W. (eds) The excavation of an Iron Age farmstead at Hawks Hill, Leatherhead. *Surrey Archaeological Collections* **62**, 40–42.

Carter, S. P. 1993 Tulloch Wood, Forres, Moray: the survey and dating of a fragment of prehistoric landscape. *Proceedings of the Society of Antiquaries of Scotland* **123**, 215–233.

Carter, S. P. 1994 Radiocarbon dating evidence for the age of narrow cultivation ridges in Scotland. *Tools and Tillage* **7**, 83–91.

Caseldine, C. J. 1979 Early land clearance in south-east Perthshire. *Scottish Archaeological Forum* **9**, 1–15.

Caseldine, C. J. and Hatton, J. 1993 The development of high moorland on Dartmoor: fire and the influence of Mesolithic activity on vegetation change. In Chambers, F. M. (ed.) *Climate Change and Human Impact on the Landscape*. London: Chapman and Hall, 119–131.

Caulfield, S. 1978 Neolithic fields: the Irish evidence. In Bowen, C. and Fowler, P. J. (eds) *Early Land Allotment in the British Isles: A Survey of Recent Work*. Oxford: British Archaeological Reports British Series **48**, 137–143.

CFA 1994 *Centre for Field Archaeology Report no. 2*. Edinburgh: University of Edinburgh.

Chambers, F. M. 1988 Archaeology and the flora of the British Isles: the moorland experience. In Jones, M. (ed.) *Archaeology and the Flora of the British Isles*. Oxford: Oxford University Committee for Archaeology Monograph **14**, 107–115.

Chambers, F. M. and Elliott, L. 1989 Spread and expansion of *Alnus* Mill, in the British Isles: timing, agencies and possible vectors. *Journal of Biogeography* **16**, 541–550.

Chaplin, R. E. 1971 *The Study of Animal Bones from Archaeological Sites*. London: Seminar Press.

Charman, D. J. 1992 Blanket mire formation at the Cross Lochs, northern Scotland. *Boreas* **21**, 53–72.

Childe, V. G. 1925 *The Dawn of European Civilization*. London: Kegan Paul, Trench and Trubner.

Childe, V. G. 1935 *The Prehistory of Scotland*. London: Kegan Paul, Trench and Trubner.

Childe, V. G. 1946 *Scotland before the Scots*. London: Methuen.

Christensen, K. M. B. 1991 Aspects of the Norse economy in the western settlement in Greenland. *Acta Archaeologica* **61**, 158–165.

Clark, J. G. D. 1947 Whales as an economic factor in Prehistoric Europe. *Antiquity* **21**, 84–104.

Clark, J. G. D. 1952 *Prehistoric Europe: The Economic Basis*. London: Methuen.

Clark, J. G. D. 1954 *Excavations at Star Carr*. Cambridge: Cambridge University Press.

Clark, J. G. D. 1956 Notes on the Obanian with special reference to antler- and bone-work. *Proceedings of the Society of Antiquaries of Scotland* **89**, 91–106.

Clark, J. G. D. 1980 *Mesolithic Prelude: the Palaeolithic–Neolithic Transition in Old World Prehistory*. Edinburgh: Edinburgh University Press.

Clarke, A. 1990 Bone. In Wickham-Jones, C. R. *Rhum: Mesolithic and Later Sites at Kinloch, Excavations 1984–86*. Edinburgh: Society of Antiquaries of Scotland Monograph Series **7**, 126.

Clarke, A. and Griffiths, D. 1990 The use of bloodstone as a raw material for flaked stone tools in the west of Scotland. In Wickham-Jones, C. R. (ed.) *Rhum: Mesolithic and Later Sites at Kinloch: Excavations 1984–86*. Edinburgh: Society of Antiquaries of Scotland Monograph Series **7**, 149–156.

Clarke, A. and Wickham-Jones, C. R. 1988 The ghost of Morton revisited: comment on Myers. *Scottish Archaeological Review* **5**, 35–37.

Clarke, D. L. 1976 Mesolithic Europe: the economic basis. In Sieveking, G. de G.,

Longworth, I. H. and Wilson, K. E. (eds) *Problems in Economic and Social Archaeology*. London: Duckworth, 449–481.

Clarke, D. V. 1976a *The Neolithic Village at Skara Brae, Orkney, Excavations 1972–3: An Interim Report*. Edinburgh: HMSO.

Clarke, D. V. 1976b Excavations at Skara Brae: a summary account. In Burgess, C. B. and Miket, R. (eds) *Settlement and Economy in the Third and Second Millennia BC*. Oxford: British Archaeological Reports British Series **33**, 233–250.

Clarke, D. V. 1983 Rinyo and the Orcadian Neolithic. In O'Connor, A. and Clarke, D. V. (eds) *From the Stone Age to the 'Forty-five*. Edinburgh: John Donald, 45–56.

Clarke, D. V., Cowie, T. G. and Foxon, A. 1985 *Symbols of Power at the time of Stonehenge*. Edinburgh: HMSO.

Clarke, D. V., Hope, R. and Wickham-Jones, C. 1978 The Links of Noltland. *Current Archaeology* **61**, 44–46.

Clarke, D. V. and Sharples, N. M. 1985 Settlement and subsistence in the third millennium BC. In Renfrew, A. C. (ed.) *The Prehistory of Orkney BC 4000–1000 AD*. Edinburgh: Edinburgh University Press, 54–82.

Close-Brooks, J. 1984 Pictish and other burials. In Friell, J. G. P. and Watson, W. G. (eds) *Pictish Studies. Settlement, Burial and Art in Dark Age Northern Britain*. Oxford: British Archaeological Reports British Series **125**, 87–114.

Close-Brooks, J. 1986 Excavations at Clatchard Craig, Fife 1953–4 and 1959–60. *Proceedings of the Society of Antiquaries of Scotland* **116**, 117–184.

Clutton-Brock, J. 1979 Report on the mammalian remains other than rodents from Quanterness. In Renfrew, A. C. *Investigations in Orkney*. London: Society of Antiquaries of London Research Report **38**, 112–134.

Clutton-Brock, J. and MacGregor, A. 1988 An end to medieval reindeer in Scotland. *Proceedings of the Society of Antiquaries of Scotland* **118**, 23–35.

COHMAP 1988 Climatic changes of the last 18 000 years: observations and model simulations. *Science* **241**, 1043–1052.

Coles, B. (ed.) 1992 *The Wetland Revolution in Prehistory*. Exeter: The Prehistoric Society/ WARP.

Coles, B. and Coles, J. M. 1986 *Sweet Track to Glastonbury: The Somerset Levels in Prehistory*. London: Thames and Hudson.

Coles, J. and Simpson, D. D. A. 1965 The excavation of a Neolithic round barrow at Pitnacree, Perthshire, Scotland. *Proceedings of the Prehistoric Society* **31**, 34–57.

Coles, J. M. 1971 The early settlement of Scotland: excavations at Morton, Fife. *Proceedings of the Prehistoric Society* **37**, 284–366.

Coles, J. M. 1983 Morton revisited. In O'Connor, A. and Clarke, D. V. (eds) *From the Stone Age to the 'Forty-five*. Edinburgh: John Donald, 9–18.

Coles, J. M., Heal, S. V. E. and Orme, B. J. 1978 The use and character of wood in prehistoric Britain and Ireland. *Proceedings of the Prehistoric Society* **44**, 1–45.

Colley, S. M. 1983 The marine faunal remains. In Hedges, J. W. (ed.) *Isbister. A Chambered Tomb in Orkney*. Oxford: British Archaeological Reports British Series **113**, 151–158.

Colley, S. M. 1989 The fish remains. In Morris, C. D. *The Birsay Bay Project* **1**. Durham: Department of Archaeology University of Durham Monograph **1**, 248–259.

Colley, S. M. 1994 The fish remains. In Smith, B. B. (ed.) *Howe: Four Millennia of Orkney Prehistory*. Edinburgh: Society of Antiquaries of Scotland Monograph Series 9, 157–160.

Connock, K. D., Finlayson, B. and Mills, A. C. M. 1993 The excavation of a shell midden site at Carding Mill Bay, near Oban, Scotland. *Glasgow Archaeological Journal* **17**, 25–38.

Coppock, J. T. 1976 *An Agricultural Atlas of Scotland*. Edinburgh: John Donald.

Coope, G. R. 1962 Coleoptera from a peat interbedded between two boulder clays at Burnhead near Airdrie. *Transactions of the Geological Society of Glasgow* **24**, 279–286.

Coope, G. R. 1968 Fossil beetles collected by James Bennie from Late Glacial silts at Corstorphine, Edinburgh. *Scottish Journal of Geology* **4**, 339–348.

Coope, G. R. 1975 Climatic fluctuations in northwest Europe since the Last Interglacial indicated by fossil assemblages of Coleoptera. In Wright, A. E. and Moseley, F. (eds) *Ice*

Ages: Ancient and Modern. Liverpool: Seel House Press, 153–168. (Geological Journal Special Issue **6**.)

Coope, G. R. 1981 Report on the Coleoptera from an eleventh-century house at Christ Church Place, Dublin. In Bekker-Nielsen, H., Foote, P. and Olsen, O. (eds) *Proceedings of the Eighth Viking Congress.* Odense: Odense University Press, 51–56.

Corbet, G. B. 1979 Report on the rodent remains. In Renfrew, A. C. *Investigations in Orkney.* London: Society of Antiquaries of London Research Report **38**, 135–137.

Cormack, W. F. 1970 A Mesolithic site at Barsalloch, Wigtownshire. *Transactions of the Dumfriesshire and Galloway Natural History and Antiquarian Society* **47**, 63–80.

Cormack, W. F. and Coles, J. M. 1964 A Mesolithic site at Low Clone, Wigtownshire. *Transactions of the Dumfriesshire and Galloway Natural History and Antiquarian Society* **41**, 67–98.

Courty, M. A., Goldberg, P. and Macphail, R. I. 1989 *Soils and Micromorphology in Archaeology.* Cambridge: Cambridge University Press.

Cowan, E. J. 1984 Myth and identity in Early Medieval Scotland. *Scottish Historical Review* **63**, 111–135.

Cowie, T. G. 1978 Excavations at the Catstane, Midlothian, 1977. *Proceedings of the Society of Antiquaries of Scotland* **109**, 166–201.

Cowie, T. G. 1979 Carwinning Hill. *Discovery and Excavation in Scotland 1978*, 28.

Cowie, T. G. 1988 *Magic Metal: Early Metalworkers in the North-East.* Aberdeen: Anthropological Museum, University of Aberdeen.

Cox, R. A. V. 1989 Place-name evidence in the west of Lewis: approaches and problems in establishing a Norse settlement. *Scottish Archaeological Review* **6**, 107–115.

Craig, G. Y. (ed.) 1991 *Geology of Scotland.* London: The Geological Society.

Cramp, S. (ed.) 1983 *The Birds of the Western Palaearctic*, **3**. Oxford: Oxford University Press.

Crawford, B. E. 1985 The Biggins, Papa Stour: a multi-disciplinary investigation. In Smith, B. (ed.) *Shetland Archaeology.* Lerwick: Shetland Times, 125–158.

Crawford, B. E. 1987 *Scandinavian Scotland.* Leicester: Leicester University Press.

Crawford, B. E. 1991 Excavations at the Biggins, Papa Stour. *Acta Archaeologica* **61**, 36–43.

Crawford, I. A. 1974 Scot? Norseman and Gael. *Scottish Archaeological Forum* **6**, 1–16.

Crawford, I. A. 1981 War or peace – Viking colonisation in the Northern and Western Isles of Scotland. In Bekker-Nielsen, H., Foote, P. and Olsen, O. (eds) *Proceedings of the Eighth Viking Congress 1977.* Odense: Odense University Press, 259–270.

Crawford, I. A. and Switsur, R. 1977 Sandscaping and C14: the Udal, North Uist. *Antiquity* **51**, 124–136.

Crone, B. A. 1991 Buiston crannog. *Current Archaeology* **127**, 295–297.

Crone, B. A. 1993a Excavation and survey of sub-peat features of Neolithic, Bronze and Iron Age date at Bharpa Carinish, North Uist, Scotland. *Proceedings of the Prehistoric Society* **59**, 361–382.

Crone, B. A. 1993b Crannogs and chronologies. *Proceedings of the Society of Antiquaries of Scotland* **123**, 245–254.

Crowson, R. A. 1971 Some records of Curculionidae (Coleoptera) from southern Scotland. *Entomologist's monthly Magazine* **107**, 47–52.

Crowson, R. A. 1981 *The Biology of the Coleoptera.* London: Academic Press.

Cullingford, R. A., Caseldine, C. J. and Gotts, P. E. 1980 Early Flandrian land and sea level changes in lower Strathearn. *Nature* **284**, 159–161.

Cunliffe, B. W. 1983 The Iron Age of northern Britain: a view from the South. In Chapman, J. C. and Mytum, H. C. (eds) *Settlement in North Britain 1000 BC–AD 1000.* Oxford: British Archaeological Report British Series **118**, 83–102.

Curle, C. L. 1982 *Pictish and Norse finds from the Brough of Birsay 1934–1974.* Edinburgh: Society of Antiquaries of Scotland Monograph Series **1**.

Curle, J. 1911 *A Roman Frontier Post and its People: The Fort of Newstead in the Parish of Melrose.* Glasgow: Maclehose.

Dalland, M. 1992 Long cist burials at Four Winds, Longniddry, East Lothian. *Proceedings of the Society of Antiquaries of Scotland* **122**, 197–206.

Darvill, T. 1987 *Prehistoric Britain*. London: Batsford.

Davenport, C. A., Ringrose, P. S., Becker, A., Hancock, P. and Fenton, C. 1989 Geological investigations of late and post glacial earthquake activity in Scotland. In Gregersen, S. and Basham, P. (eds) *Earthquakes at North Atlantic Passive Margins: Neotectonics and Postglacial Rebound*. Dordrecht: Kluwer Scientific Publishers, 127–141.

Davidson, D. A., Harkness, D. D. and Simpson, I. A. 1986 The formation of farm mounds on the island of Sanday. *Geoarchaeology* **1**, 45–60.

Davidson, D. A., Jones, R. L. and Renfrew, C. 1976 Palaeoenvironmental reconstruction and evaluation: a case study from Orkney. *Transactions of the Institute of British Geographers*, New Series **1**, 346–361.

Davidson, D. A., Lamb, R. G. and Simpson, I. A. 1983 Farm mounds in north Orkney: a preliminary report. *Norwegian Archaeological Review* **16**, 39–44.

Davidson, D. A. and Simpson, I. A. 1984 The formation of deep topsoils in Orkney. *Earth Surface Processes and Landforms* **9**, 75–81.

Davidson, D. A. and Smout, T. C. in press Soil change in Scotland: the legacy of past land improvement processes. In Usher, M. and Taylor, A. (eds) *Soil Sustainability in Scotland*. Edinburgh: HMSO.

Davies, G. and Turner, J. 1979 Pollen diagrams from Northumberland. *New Phytologist* **82**, 783–804.

Davies, R. W. 1971 The Roman military diet. *Britannia* **2**, 122–142.

Davis, M. and Sheridan, A. 1993 Scottish prehistoric 'jet' jewellery: some new work. *Proceedings of the Society of Antiquaries of Scotland* **123**, 455–456.

Davis, M. B. and Botkin, D. B. 1985 Sensitivity of cool-temperate forests and their fossil pollen record to rapid temperature change. *Quaternary Research* **23**, 327–340.

Dawson, A. G. 1979 *Raised Shorelines of Jura, Scarba and NE Islay*. Unpublished PhD thesis, University of Edinburgh.

Dawson, A. G. 1980 Shore erosion by frost: an example from the Scottish Lateglacial. In Lowe, J. J., Gray, J. M. and Robinson, J. E. (eds) *Studies in the Lateglacial of North-West Europe*. Oxford: Pergamon Press, 45–53.

Dawson, A. G. 1984 Quaternary sea-level changes in western Scotland. *Quaternary Science Reviews* **3**, 345–368.

Dawson, A. G., Long, D. and Smith, D. E. 1988 The Storegga Slides: evidence from eastern Scotland for a possible tsunami. *Marine Geology* **82**, 271–276.

Dawson, A. G., Smith, D. E. and Long, D. 1990 Evidence for a tsunami from a Mesolithic site in Inverness, Scotland. *Journal of Archaeological Science* **17**, 509–512.

Deith, M. R. 1983 Molluscan calendars: the use of growth-line analysis to establish seasonality of shellfish collection at the Mesolithic site of Morton, Fife. *Journal of Archaeological Science* **10**, 423–440.

Deith, M. R. 1986 Subsistence strategies at a Mesolithic camp site: evidence from stable isotope analysis of shells. *Journal of Archaeological Science* **13**, 61–78.

Delair, J. B. 1969 North of the hippopotamus belt: a brief review of Scottish fossil mammals. *Bulletin of the Mammal Society of the British Isles* **31**, 16–21.

Dennell, R. W. 1983 *European Economic Prehistory – A New Approach*. London: Academic Press.

Dennell, R. W. 1985 The hunter–gatherer/agricultural frontier in prehistoric temperate Europe. In Green, S. W. and Perlman, S. M. (eds) *The Archaeology of Frontiers and Boundaries*. London: Academic Press, 113–139.

Dent, A. 1977 Orkney pigs. *Ark, The Journal of the Rare Breeds Preservation Trust* **4**, 304.

Dickson, C. A. 1988 Distinguishing cereal from wild grass pollen: some limitations. *Circaea* **5**, 67–72.

Dickson, C. A. 1989 The Roman army diet in Britain and Germany. *Archäobotanik. Dissertationes Botanicae* **133**, 135–154.

Dickson, C. A. 1994 Plant remains. In Smith, B. B. (ed.) *Howe. Four millennia of Orkney Prehistory, Excavations 1978–1982*. Edinburgh: Society of Antiquaries of Scotland Monograph Series **9**, 125–139.

Dickson, C. A. forthcoming Pollen analysis. In Hanson, W. S. *Elginhaugh: A Flavian Fort and its Annexe*. London: Roman Society.

Dickson, J. H. 1992 Scotland's woodlands: their ancient past and precarious present. *Botanical Journal of Scotland* **46**, 155–165.

Dickson, J. H., Dickson, C. A. and Breeze, D. J. 1979 Flour or bread in a Roman military ditch at Bearsden, Scotland. *Antiquity* **53**, 47–51.

Dickson, J. H., Stewart, J. H., Thompson, R., Turner, G., Baxter, M. S., Drndarsky, N. D. and Rose, J. 1978 Palynology, palaeomagnetism and radiometric dating of Flandrian marine and freshwater sediments of Loch Lomond. *Nature* **274**, 548–553.

Dimbleby, G. W. 1962 *The Development of British Heathlands and Their Soils*. Oxford: Oxford Forestry Memoir **23**.

Dimbleby, G. W. 1985 *The Palynology of Archaeological Sites*. London: Academic Press.

Dinnin, M. H. 1993 *Islands within islands*. Unpublished PhD thesis, University of Sheffield.

Dixon, K. and Southern, P. 1992 *The Roman Cavalry*. London: Batsford.

Dixon, T. N. 1984 Oakbank crannog. *Current Archaeology* **90**, 217–220.

Dore, J. N. and Gillam, J. P. 1979 *The Roman Fort at South Shields*. Newcastle: Society of Antiquaries of Newcastle upon Tyne.

Dockrill, S. J. 1987 *Excavations at Tofts Ness, Sanday. Interim 1987*. Bradford: University of Bradford.

Dockrill, S. J. and Simpson, I. A. in press The identification and interpretation of prehistoric anthropogenic soils in the Northern Isles using an integrated sampling framework. *Archaeological Prospection*.

Donaldson, A. M., Morris, C. D. and Rackham, D. J. 1981 The Birsay Bay project. In Brothwell, D. and Dimbleby, G. (eds) *Environmental Aspects of Coasts and Islands*. Oxford: British Archaeological Reports International Series **S94**, 65–85.

Donner, J. J. 1957 The geology and vegetation of Late-glacial retreat stages in Scotland. *Transactions of the Royal Society of Edinburgh* **63**, 221–264.

Doran, G. 1992 Problems and potential of wet sites in North America: the example of Windover. In Coles, B. (ed.) *The Wetland Revolution in Prehistory*. Exeter: The Prehistoric Society/WARP, 125–134.

Driscoll, S. T. 1988 Power and authority in Early Historic Scotland: Pictish symbol stones and other documents. In Gledhill, J., Bender, B. and Larsen, M. T. (eds) *State and Society. The Emergence and Development of Social Hierarchy and Political Centralization*. London: Unwin Hyman, 215–236. (One World Archaeology **4**.)

Driscoll, S. T. 1991 The archaeology of state formation in Scotland. In Hanson, W. S. and Slater, E. A. (eds) *Scottish Archaeology: New Perceptions*. Aberdeen: Aberdeen University Press, 81–111.

Driscoll, S. T. 1992 Discourse on the frontiers of history: material culture and social reproduction in early Scotland. *Historical Archaeology* **26.3**, 12–24.

Dubois, A. D. and Ferguson, D. K. 1985 The climatic history of pine in the Cairngorms based on radiocarbon dates and stable isotope analysis, with an account of the events leading up to its colonization. *Review of Palaeobotany and Palynology* **46**, 55–80.

Dugmore, A. J., Larsen, G. and Newton, A. J. 1995 Seven tephra isochrones in Scotland. *The Holocene* **5**, 257–266.

Dumayne, L. 1993a Invader or native? Vegetation clearance in northern Britain during Romano-British time. *Vegetation History and Archaeobotany* **2**, 29–36.

Dumayne, L. 1993b Iron Age and Roman vegetation clearance in northern Britain: further evidence. *Botanical Journal of Scotland* **46**, 385–392.

Dumayne, L. 1994 The effect of the Roman occupation on the environment of Hadrian's Wall: a pollen diagram from Fozy Moss, Northumbria. *Britannia* **25**, 217–224.

Dumayne, L. and Barber, K. E. 1994 The impact of the Romans on the environment of northern England: pollen data from three sites close to Hadrian's Wall. *The Holocene* **4**, 165–173.

Duncan, A. A. M. 1975 *Scotland: The Making of the Kingdom*. Edinburgh: Oliver and Boyd. (The Edinburgh History of Scotland **1**.)

Durno, S. E. 1965 Pollen analytical evidence of 'landnam' from two Scottish sites. *Transactions of the Botanical Society of Edinburgh* **40**, 347–351.

Durrenberger, I. C. 1991 Production in medieval Iceland. *Acta Archaeologica* **61**, 14–21.

Earwood, C. 1990 The wooden artefacts from Loch Glashan crannog, Mid-Argyll. *Proceedings of the Society of Antiquaries of Scotland* **120**, 79–94.

Earwood, C. 1991 Two early historic bog butter containers. *Proceedings of the Society of Antiquaries of Scotland* **121**, 231–240.

Edmonds, M. R. 1992 Their use is wholly unknown. In Sharples, N. M. and Sheridan, A. (eds) *Vessels for the Ancestors*. Edinburgh: Edinburgh University Press, 179–193.

Edwards, K. J. 1974 A half-century of pollen-analytical research in Scotland. *Transactions of the Botanical Society of Edinburgh* **42**, 211–222.

Edwards, K. J. 1978 *Palaeoenvironmental and archaeological investigations in the Howe of Cromar, Grampian Region, Scotland*. Unpublished PhD thesis, University of Aberdeen.

Edwards, K. J. 1979a Palynological and temporal inference in the context of prehistory. *Journal of Archaeological Science* **6**, 255–270.

Edwards, K. J. 1979b Environmental impact in the prehistoric period. *Scottish Archaeological Forum* **9**, 27–42.

Edwards, K. J. 1985 Radiocarbon dating. In Edwards, K. J. and Warren, W. P. (eds) *The Quaternary History of Ireland*. London: Academic Press, 280–293.

Edwards, K. J. 1988 The hunter–gatherer/agricultural transition and the pollen record in the British Isles. In Birks, H. H., Birks, H. J. B., Kaland, P. E. and Moe, D. (eds) *The Cultural Landscape – Past, Present and Future*. Cambridge: Cambridge University Press, 255–266.

Edwards, K. J. 1989a Meso-Neolithic vegetational impacts in Scotland and beyond: palynological considerations. In Bonsall, J. C. (ed.) *The Mesolithic in Europe*. Edinburgh: John Donald, 143–155.

Edwards, K. J. 1989b The cereal pollen record and early agriculture. In Milles, A., Williams, D. and Gardner, N., (eds) *The Beginnings of Agriculture*. Oxford: British Archaeological Reports International Series, **S496**, 113–135.

Edwards, K. J. 1990 Fire and the Scottish Mesolithic: evidence from microscopic charcoal. In Vermeersch, P. M. and van Peer, P. (eds) *Contributions to the Mesolithic in Europe*. Leuven: Leuven University Press, 71–79.

Edwards, K. J. 1991 Using space in cultural palynology: the value of the off-site pollen record. In Harris, D. R. and Thomas, K. D. (eds) *Modelling Ecological Change*. London: Institute of Archaeology University College London, 61–73.

Edwards, K. J. 1993a Models of mid-Holocene forest farming for north-west Europe. In Chambers, F. M. (ed.) *Climate Change and Human Impact on the Landscape*. London: Chapman and Hall, 133–145.

Edwards, K. J. 1993b Human impact on the prehistoric environment. In Smout T. C. (ed.) *Scotland Since Prehistory. Natural Change and Human Impact*. Aberdeen: Scottish Cultural Press, 17–27.

Edwards, K. J. 1996a A Mesolithic of the Western and Northern Isles of Scotland? Evidence from pollen and charcoal. In Pollard, T. and Morrison, A. (eds) *The Early Prehistory of Scotland*. Edinburgh: Edinburgh University Press, 23–38.

Edwards, K. J. 1996b Tom Affleck and his contribution to the study of the Mesolithic of southwest Scotland. In Pollard, T. and Morrison, A. (eds) *The Early Prehistory of Scotland*. Edinburgh: Edinburgh University Press, 108–122.

Edwards, K. J., Ansell, M. and Carter, B. A. 1983 New Mesolithic sites in south-west Scotland and their significance as indicators of inland penetration. *Transactions of the Dumfriesshire and Galloway Natural History and Antiquarian Society* **58**, 9–15.

Edwards, K. J. and Berridge, J. M. A. 1994 The Late-Quaternary vegetational history of Loch a'Bhogaidh, Rinns of Islay S.S.S.I., Scotland. *New Phytologist* **128**, 749–769.

Edwards, K. J., Hirons, K. R. and Newell, P. J. 1991 The palaeoecological and prehistoric context of minerogenic layers in blanket peat: a study from Loch Dee, southwest Scotland. *The Holocene* **1**, 29–39.

Edwards, K. J. and Hirons, K. R. 1982 Date of blanket peat initiation and rates of spread –
a problem of research design. *Quaternary Newsletter* **36**, 32–37.

Edwards, K. J. and Hirons, K. R. 1984 Cereal pollen grains in pre-Elm Decline deposits:
implications for the earliest agriculture in Britain and Ireland. *Journal of Archaeological
Science* **11**, 71–80.

Edwards, K. J., Dugmore, A. J., Buckland, P. C., Blackford, J. J. and Cook, G. T. in press
Hekla-4 ash, the pine decline in Northern Ireland and the effective use of tephra
isochrones: a comment on Hall, Pilcher and McCormac. *The Holocene.*

Edwards, K. J. and McIntosh, C. J. 1988 Improving the detection rate of cereal-type pollen
grains from *Ulmus* decline and earlier deposits from Scotland. *Pollen et Spores* **30**, 179–
188.

Edwards, K. J. and Mithen, S. 1995 The colonization of the Hebridean islands of western
Scotland: evidence from the palynological and archaeological records. *World Archaeology*
26, 348–365.

Edwards, K. J. and Moss, A. G. 1993 Pollen data from the Loch of Brunatwatt, west
Mainland. In Birnie, J. F., Gordon, J. E., Bennett, K. D. and Hall, A. M. (eds) *The
Quaternary of Shetland: Field Guide.* Cambridge: Quaternary Research Association, 126–
129.

Edwards, K. J. and Ralston, I. B. M. 1978 New dating and environmental evidence from
Burghead Fort, Moray. *Proceedings of the Society of Antiquaries of Scotland* **109**, 202–210.

Edwards, K. J. and Ralston, I. B. M. 1979 Archaeology and environment in Scotland: at the
cross-roads? *Scottish Archaeological Forum* **9**, 78–81.

Edwards, K. J. and Ralston, I. B. M. 1984 Postglacial hunter-gatherers and vegetational
history in Scotland. *Proceedings of the Society of Antiquaries of Scotland* **114**, 15–34.

Edwards, K. J. and Rowntree, K. M. 1980 Radiocarbon and palaeoenvironmental evidence
for changing rates of erosion at a Flandrian stage site in Scotland. In Cullingford, R. A.,
Davidson, D. A. and Lewin, J. (eds) *Timescales in Geomorphology.* Chichester and New
York: J Wiley and Sons, 207–233.

Edwards, K. J. and Whittington, G. W. 1993 Aspects of the environmental and depositional
history of a rock basin lake in eastern Scotland, UK. In McManus, J. and Duck, R. W.
(eds) *Geomorphology and Sedimentology of Lakes and Reservoirs.* Chichester: John Wiley
and Sons, 155–180.

Edwards, K. J., Whittington, G. and Hirons, K. R. 1995 The relationship between fire and
long-term wet heath development in South Uist, Outer Hebrides, Scotland. In Thompson,
D. B. A., Hester, A. J. and Usher, M. B. (eds) *Heaths and Moorland: Cultural Landscapes.*
Edinburgh: HMSO, 240–248.

Eldjárn, K., 1984 Graves and grave goods: survey and evaluation. In Fenton, A. and
Pálsson, H. (eds) *The Northern and Western Isles in the Viking World.* Edinburgh: John
Donald, 2–11.

Elliot, W. 1991 Animal footprints on Roman bricks from Newstead. *Proceedings of the
Society of Antiquaries of Scotland* **121**, 223–226.

Engelstad, E. 1989 Mesolithic house sites in Arctic Norway. In Bonsall, J. C. (ed.) *The
Mesolithic in Europe.* Edinburgh: John Donald, 331–337.

Erdtman, G. 1923 Iakttagelser från en mikropaleontologisk undersökning av nord-skotska,
hebridiska, orkadiska och shetländska torvmarker. *Geologiska Föreningens i Stockholm
Förhandlingar* **45**, 538–545.

Erdtman, G. 1924 Studies in the micropalaeontology of postglacial deposits in northern
Scotland and the Scottish Isles, with especial reference to the history of woodlands. *Journal
of the Linnean Society* **46**, 449–504.

Evans, C. 1989 Perishables and worldly goods – artifact decoration and classification in the
light of wetlands research. *Oxford Journal of Archaeology* **8**, 179–201.

Evans, J. G. 1979 The palaeoenvironment of coastal blown-sand deposits in western and
northern Britain. *Scottish Archaeological Forum* **9**, 16–26.

Ewart, J. C. 1911 Animal remains. In Curle, J. *A Roman Frontier Post and its People: The
Fort at Newstead,* Glasgow: Maclehose.

Ezard, J. 1996 Modern peril seen in ashes of lost clan. *The Guardian,* 22 March 1996.

Fairhurst, H. and Taylor, D. B. 1971 A hut-circle settlement at Kilphedir, Sutherland. *Proceedings of the Society of Antiquaries of Scotland* **103**, 65–99.

Fairweather, A. D. and Ralston, I. B. M. 1993 The Neolithic timber hall at Balbridie, Grampian Region, Scotland: a preliminary note on dating and plant macrofossils. *Antiquity* **67**, 313–323.

Feachem, R. W. 1966 The hill-forts of northern Britain. In Rivet, A. L. F. (ed.) *The Iron Age in Northern Britain*, Edinburgh: Edinburgh University Press, 59–87.

Fell, C., Foote, P., Graham-Cambell, J. and Thomson, R. (eds) 1983 *The Viking Age in the Isle of Man*. London: Viking Society for Northern Research.

Fellows-Jensen, G. 1984 Viking settlement in the Northern and Western Isles – the placename evidence as seen from Denmark and the Danelaw. In Fenton, A. and Pálsson, H. (eds) *The Northern and Western Isles in the Viking World*. Edinburgh: John Donald, 148–168.

Fenton, A., 1970 Paring and burning and the cutting of turf and peat in Scotland. In Gailey, A. and Fenton, A. (eds) *The Spade in Northern and Atlantic Europe*. Belfast: Institute of Irish Studies, Queen's University, 155–193.

Fenton, A. 1974 The *cas-chrom*: a review of the Scottish evidence. *Tools and Tillage* **2**, 131–148.

Fenton, A. 1978 *The Northern Isles: Orkney and Shetland*. Edinburgh: John Donald.

Ferguson, R. I. 1981 Channel form and channel changes. In Lewin, J. (ed.) *British Rivers*. London: Allen and Unwin, 90–125.

Fernie, E. C. 1986 Early church architecture in Scotland. *Proceedings of the Society of Antiquaries of Scotland* **116**, 393–412.

Finlay, J. 1984 *Faunal Evidence for Prehistoric Economy and Settlement in the Outer Hebrides to c.1000 AD*. Unpublished PhD thesis, University of Edinburgh.

Finlay, J. 1991 Animal bone. In Campbell, E., Excavations of a wheelhouse and other Iron Age structures at Sollas, North Uist, by R. J. C. Atkinson in 1957. *Proceedings of the Society of Antiquaries of Scotland* **121**, 147–148.

Finlayson, B. 1990a *A Pragmatic Approach to the Functional Analysis of Chipped Stone Tools*. Unpublished PhD thesis, University of Edinburgh.

Finlayson, B. 1990b Lithic exploitation during the Mesolithic in Scotland. *Scottish Archaeological Review* **7**, 41–57.

Finlayson, B. 1995 Complexity in the Mesolithic of the western Scottish seaboard. In Fischer, A. (ed.) *Proceedings of the Man, Sea and the Mesolithic Conference, Horsholm*. Oxford: Oxbow Monograph **53**, 261–264.

Finlayson, B., Finlay, N. and Mithen, S. 1996 Mesolithic chipped stone assemblages: descriptive and analytical procedures used by the Southern Hebrides Mesolithic Project. In Pollard, T. and Morrison, A. (eds) *The Early Prehistory of Scotland*. Edinburgh: Edinburgh University Press, 252–266.

Firbas, F. 1949 *Spät- und nacheiszeitliche Waldgeschichte Mitteleuropas nördlich der Alpen*. Jena: Fischer.

Firth, C. R. 1992 Postglacial uplift in Scotland: evidence from shorelines. In Fenton, C. H. (ed.) *Neotectonics in North West Scotland: A Field Guide*. Glasgow: University of Glasgow, 16–20.

Fisher, J. and Waterston, G. 1941 The breeding distribution, history and population of the fulmar (*Fulmarus glacialis*) in the British Isles. *Journal of Animal Ecology* **10**, 204–272.

Fitzpatrick, A. P. 1989 The submission of the Orkney Islands to Claudius: new evidence? *Scottish Archaeological Review* **6**, 24–33.

Fitzpatrick, E. A. 1956 An indurated soil horizon formed by permafrost. *Journal of Soil Science* **7**, 248–254.

Fleming, A. 1988 *The Dartmoor Reaves. Investigating Prehistoric Land Divisions*. London: Batsford.

Flenley, J. and Pearson, M. C. 1967 Pollen analysis of a peat from the island of Canna (Inner Hebrides). *New Phytologist* **66**, 299–306.

Fojut, N. 1982 Towards a geography of Shetland brochs. *Glasgow Archaeological Journal* **9**, 38–59.

Forsyth, K. 1995 Language in Pictland, spoken and written. In Nicoll, E. H. (ed.) *A Pictish Panorama*. Balgavies, Angus: Pinkfoot Press, 7–10.

Fossitt, J. A. 1990 *Holocene Vegetation History of the Western Isles, Scotland*. Unpublished PhD thesis, University of Cambridge.

Fossitt, J. A. 1996 Late Quaternary vegetation history of the Western Isles of Scotland. *New Phytologist* **132**, 171–196.

Foster. S. M. 1989 Transformations in social space: Iron Age Orkney and Caithness. *Scottish Archaeological Review* **6**, 34–55.

Foster, S. M. 1992 The state of Pictland in the age of Sutton Hoo. In Carver, M. O. H. (ed.), *The Age of Sutton Hoo*. Woodbridge: Boydell, 217–234.

Foster, S. M. 1996 *Picts, Gaels and Scots*. London: Batsford/Historic Scotland.

Fowler, P. J. 1981 Wildscape to landscape: 'enclosure' in prehistoric Britain. In Mercer, R. J. (ed.) *Farming Practice in British Prehistory*. Edinburgh: Edinburgh University Press, 9–54.

Foxon, A. 1991 *Bone, Antler, Tooth and Horn Technology and Utilisation in Prehistoric Scotland*. Unpublished PhD thesis, University of Glasgow.

Fraser, D. 1983 *Land and Society in Neolithic Orkney*. Oxford: British Archaeological Reports British Series **117**.

Fredskild, B. and Humle, L. 1991 Plant remains from the Norse farm Sandnes in the Western Settlement, Greenland. *Acta Borealia* **1**, 69–81.

Fulford, M. G. 1984 Demonstrating Britannia's economic dependence in the first and second centuries. In Blagg, T. F .C. and King, A. C. (eds) 1984 *Military and Civilian in Roman Britain: Cultural Relationships in a Frontier Province*. Oxford: British Archaeological Reports British Series **136**, 129–142.

Fulford, M. G. 1989 Byzantium and Britain: a Mediterranean perspective on post-Roman Mediterranean imports in western Britain and Ireland. *Medieval Archaeology* **33**, 1–6.

Funnel, B. M. 1995 Global sea-level and the (pen-)insularity of later Cenozoic Britain. In Preece, R. C. (ed.) *Island Britain: A Quaternary Perspective*. London: The Geological Society Special Publication **96**, 3–13.

Gailey, A. and Fenton, A. (eds) 1970 *The Spade in Northern and Atlantic Europe*. Belfast: Institute of Irish Studies, Queen's University.

Garton, D. 1987 *Lismore Fields, Buxton: 1987 Summary Report*, Derbyshire: Trent and Peak Archaeological Trust.

Gates, T. 1982 Farming on the frontier: Romano-British fields in Northumberland. In Clack, P. and Haselgrove, S. (eds) *Rural Settlement in the Roman North*. Durham: Council for British Archaeology Group 3, 21–42.

Gates, T. and O'Brien, C. 1988 Cropmarks at Milfield and New Bewick and the recognition of Grübenhäuser in Northumberland. *Archaeologia aeliana*, **(5 ser) 16**, 1–9.

Gear, A. J. and Huntley, B. 1991 Rapid changes in the range limits of Scots pine 4000 years ago. *Science* **251**, 544–547.

Geikie, J. 1881 *Prehistoric Europe. A Geological Sketch*. London: Edward Stanford.

Geikie, J. 1894 *The Great Ice Age*, 3rd edition. London: Daldy, Isbister and Co.

Gelling, P. S. 1984 The Norse buildings at Skaill, Deerness, Orkney and their immediate predecessors. In Fenton, A. and Pálsson, H. (eds) *The Northern and Western Isles in the Viking World*. Edinburgh: John Donald, 12–38.

Gentles, D. 1993 Vitrified forts. *Current Archaeology* **133**, 18–20.

Gibson, A. 1992 Approaches to the later Neolithic and Bronze Age settlement of Britain. *Colloque International de Lons-le-Saunier, 16–19 mai 1990*. Lons-le-Saunier: Cercle jurassien, 41–48.

Gilbertson, D. D., Kent, M. and Grattan, J. P. (eds) 1996 *The Environment of the Outer Hebrides: The Last 14,000 Years*. Sheffield: Sheffield Academic Press.

Gilbertson, D. D., Kent, M., Schwenninger, J.-L., Wathern, P. A., Weaver, R. and Brayshay, B. A. 1995 The machair vegetation of South Uist and Barra in the Outer Hebrides of Scotland: its interacting ecological, geomorphic and historical dimensions. In Butlin, R. A. and Roberts, N. (eds) *Ecological Relations in Historical Times: Human Impact and Adaptation*. London: Blackwell, 17–44.

Girling, M. A. 1977 Fossil insect assemblages from Rowland's track. *Somerset Levels Papers* **3**, 51–60.

Girling, M. A. and Greig, J. R. A. 1977 Palaeoecological investigations of a site at Hampstead Heath, London. *Nature* **268**, 45–47.

Godwin, H. 1975 *History of the British Flora: A Factual Basis for Phytogeography*, 2nd edition. Cambridge: Cambridge University Press.

Goodburn, R. 1978 Roman Britain in 1977. *Britannia* **9**, 404–472.

Göransson, H. 1986 Man and the forests of nemoral broad-leaved trees during the Stone Age. *Striae* **24**, 145–152.

Göransson, H. 1987 *Neolithic Man and the Forest Environment around Alvastra Pile Dwelling*. Stockholm: Theses and Papers in North-European Archaeology, **20**.

Gore, A. J. P. (ed.) 1993 *Mires, Swamp, Bog, Fen and Moor*. Amsterdam: Elsevier.

Gowen, M. 1988 *Three Irish Gas Pipelines: New Archaeological Evidence in Munster*. Dublin: Wordwell.

Goudie, G. 1904 *The Celtic and Scandinavian Antiquities of Shetland*. London and Edinburgh: Blackwood.

Gourlay, R. and Barrett, J. 1984 Dail na Caraidh. *Current Archaeology* **94**, 347–349.

Graham, A. 1939 Cultivation terraces in south-eastern Scotland. *Proceedings of the Society of Antiquaries of Scotland* **73**, 289–315.

Graham-Campbell, J. 1976a The Viking-age silver and gold hoards of Scandinavian character from Scotland. *Proceedings of the Society of Antiquaries of Scotland* **107**, 114–135.

Graham-Campbell, J. 1976b The Viking-age silver hoards of Ireland. In Almqvist, B. and Green, D. (eds), *Proceedings of the Seventh Viking Congress 1973*. Dublin: Royal Irish Academy, 39–74.

Graham-Campbell, J. (ed.) 1980 *The Viking World*. New Haven: Yale University Press.

Graham-Campbell, J. 1993 The northern hoards of Viking-Age Scotland. In Batey, C. E., Jesch, J. and Morris, C. D. (eds) *The Viking Age in Caithness, Orkney and the North Atlantic*. Edinburgh: Edinburgh University Press, 173–186.

Grieg, S. 1940 *Viking Antiquities in Scotland*. Oslo: H Aschehoug, The Scientific Research Fund of 1919. (Shetelig, H. (ed.) Viking antiquities in Great Britain and Ireland **II**.)

Grieve, S. 1882 Notice on the discovery of remains of the great auk or garefowl (*Alca impennis*) on the island of Oronsay, Argyllshire. *Journal of the Linnean Society (Zoology)* **16**, 479–487.

Grigson, C. 1969 The uses and limitations of differences in absolute size in the distinction between the bones of aurochs (*Bos primigenius*) and domestic cattle (*Bos taurus*). In Ucko, P. J. and Dimbleby, G. W. (eds) *The Domestication and Exploitation of Plants and Animals*. London: Duckworth, 277–294.

Grigson, C. and Mellars, P. A. 1987 The mammalian remains from the middens. In Mellars, P. A. *Excavations on Oronsay: Prehistoric Human Ecology on a Small Island*. Edinburgh: Edinburgh University Press, 243–289.

Groenman-van Waateringe, W. 1983 The early agricultural utilization of the Irish landscape: the last word on the elm decline? In Reeves-Smyth T. and Hamond, F. (eds) *Landscape Archaeology in Ireland*. Oxford: British Archaeological Reports British Series **116**, 217–232.

Groenman-van Waateringe, W. 1986 Grazing possibilities in the Neolithic of the Netherlands based on palynological data. In Behre, K.-E., (ed.) *Anthropogenic Indicators in Pollen Diagrams*. Rotterdam: A. A. Balkema, 187–202.

Groenman-van Waateringe, W. 1989 Food for soldiers, food for thought. In Barrett, J. C., Fitzpatrick, A. P., and Macinnes, L. (eds) *Barbarians and Romans in North-West Europe from the Later Republic to Late Antiquity*. Oxford: British Archaeological Reports International Series **S471**, 96–107.

Groenman-van Waateringe, W. 1993 The effects of grazing on the pollen production of grasses. *Vegetation History and Archaeobotany* **2**, 157–162.

Groenman-van Waateringe, W. forthcoming Classical authors and the diet of Roman soldiers: true or false? In *Proceedings of the Sixteenth International Congress of Roman Frontier Studies.*

Grove, J. 1988 *The Little Ice Age.* London: Methuen.

Guiot, J., Pons, A., de Beaulieu, J. L. and Reille, M. 1989 A 140 000 year continental climate reconstruction from two European pollen records. *Nature* **338**, 309–313.

Grattan, J. P. and Gilbertson, D. D. 1994 Acid-loading from Icelandic tephra falling on acidified ecosystems as a key to understanding archaeological and environmental stress in northern and western Britain. *Journal of Archaeological Science* **21**, 851–859.

Haggart, B. A. 1987 Relative sea-level changes in the Moray Firth area, Scotland. In Tooley, M. J. and Shennan, I. (eds) *Sea-Level Changes.* London: Blackwell, 67–108.

Haggarty, A. M. 1988 Iona: some results from recent work. *Proceedings of the Society of Antiquaries of Scotland* **118**, 203–213.

Haggarty, A. M. 1991 Machrie Moor, Arran: recent excavations at two stone circles. *Proceedings of the Society of Antiquaries of Scotland* **121**, 51–94.

Haggarty, A. M. and Haggarty, G. 1983 Excavations at Rispain Camp, Whithorn, 1978–81. *Transactions of the Dumfriesshire and Galloway Natural History and Antiquarian Society* **58**, 21–51.

Hall, A. M. 1991 Pre-Quaternary landscape evolution in the Scottish Highlands. *Transactions of the Royal Society of Edinburgh: Earth Sciences* **82**, 1–26.

Hall, A. M. and Bent, A. J. A. 1990 The limits of the last British ice sheet in northern Scotland and the adjacent shelf. *Quaternary Newsletter* **61**, 2–12.

Hall, A. M. and Sugden, D. E. 1987 Limited modification of mid-latitude landscapes by ice sheets: the case of northeast Scotland. *Earth Surface Processes and Landforms* **12**, 531–542.

Hall, A. R., Kenward, H. K., Williams, D. and Grieg, J. R. A. 1983 *Environment and Living Conditions at Two Anglo-Scandinavian Sites.* London: Council for British Archaeology for the York Archaeological Trust. (The Archaeology of York **14/4**.)

Hall, V. A., Pilcher, J. R. and McVicker, S. J. 1994 Tephra-linked studies and environmental archaeology, with special reference to Ireland. *Circaea* **11**, 17–22.

Hallén, Y. 1994 The use of bone and antler at Foshigarry and Bac Mhic Connain, two Iron Age sites on North Uist, Western Isles. *Proceedings of the Society of Antiquaries of Scotland* **124**, 189–231.

Halliday, S. P. 1982 Later prehistoric farming in south-east Scotland. In Harding, D. W. (ed.) *Later Prehistoric Settlement in South-East Scotland.* Edinburgh: Department of Archaeology University of Edinburgh Occasional Paper **8**, 75–91.

Halliday, S. P. 1985 Unenclosed upland settlement in the east and south-east of Scotland. In Spratt, D. and Burgess, C. (eds) *Upland settlement in Britain: The Second Millennium BC and After.* Oxford: British Archaeological Reports British Series **143**, 231–252.

Halliday, S. P. 1986 Cord rig and early cultivation in the Borders. *Proceedings of the Society of Antiquaries of Scotland* **116**, 584–585.

Halliday, S. P. 1990 Patterns of fieldwork and the distribution of burnt mounds in Scotland. In Buckley, V. (ed.) *Burnt Offerings. International Contributions to Burnt Mound Archaeology.* Dublin: Wordwell, 60–61.

Halliday, S. P. 1993 Marginal agriculture in Scotland. In Smout T. C. (ed.) *Scotland Since Prehistory. Natural Change and Human Impact.* Aberdeen: Scottish Cultural Press, 64–78.

Halliday, S. P., Hill, P. J. and Stevenson, J. B. 1981 Early agriculture in Scotland. In Mercer, R. J. (ed.) *Farming Practice in British Prehistory.* Edinburgh: Edinburgh University Press, 55–65.

Halliday, S. P. and Stevenson, J. B. 1991 Surveying for the future: RCAHMS archaeological survey. In Hanson, W. S. and Slater, E. A. (eds) *Scottish Archaeology: New Perceptions.* Aberdeen: Aberdeen University Press, 129–139.

Halstead, P. forthcoming Remains of mammalian fauna from Baleshare and Hornish Point. In Barber, J. (ed.) *Excavations in the Outer Hebrides.* Edinburgh.

Hamilton, J. R. C. 1956 *Excavations at Jarlshof, Shetland.* Edinburgh: HMSO. (Ministry of Works Archaeological Report **1**.)

Hamilton-Dyer, S. and McCormick, F. 1993 The animal bones. In Connock, K. D.,

Finlayson, B. and Mills, A. C. M. (eds) The excavation of a shell midden site at Carding Mill Bay, near Oban, Scotland. *Glasgow Archaeological Journal* **17**, 34.

Hansen, S. S. 1991 Toftanes: A Faroese viking age farmstead from the 9–10th centuries AD. *Acta Archaeologica* **61**, 44–53.

Hanson, W. S. 1978a Roman campaigns north of the Forth-Clyde isthmus: the evidence of the temporary camps. *Proceedings of Society of Antiquaries of Scotland* **109**, 140–150.

Hanson, W. S. 1978b The organisation of the Roman military timber supply. *Britannia* **9**, 293–305.

Hanson, W. S. 1982 Roman military timber buildings: construction and reconstruction. In McGrail, S. (ed.) *Woodworking Techniques before AD 1500*. Oxford: British Archaeological Reports International Series **S129**, 169–186.

Hanson, W. S. 1991a *Agricola and the Conquest of the North*, 2nd edition. London: Batsford.

Hanson, W. S. 1991b Tacitus' *Agricola*: an archaeological and historical study. In Haase, W. (ed.) *Aufstieg und Niedergang der römischen Welt* **II.33.3.** Berlin: de Gruyter, 1741–1784.

Hanson, W. S. in press Forest clearance and the Roman army. *Britannia* **27**.

Hanson, W. S. forthcoming a *Elginhaugh: A Flavian Fort and its Annexe*. London: Roman Society.

Hanson, W. S. forthcoming b Building the forts and frontiers. In Breeze, D. J. (ed.) *The Frontiers of the Roman Empire*. London: Batsford.

Hanson, W. S. and Breeze, D. J. 1991 The future of Roman Scotland. In Hanson, W. S. and Slater, E. A. (eds) *Scottish Archaeology: New Perceptions*. Aberdeen: Aberdeen University Press, 57–80.

Hanson, W. S. and Macinnes, L. 1980 Forests, forts and fields, a discussion. *Scottish Archaeological Forum* **12**, 98–113.

Hanson, W. S. and Macinnes, L. 1991 Soldiers and settlement in Wales and Scotland. In Jones, R. F. J. (ed.) *Roman Britain: Recent Trends*. Sheffield: J. R. Collis Publications, 85–92.

Hanson, W. S. and Maxwell, G. S. 1986 *Rome's North-West Frontier: The Antonine Wall*, 2nd edition. Edinburgh: Edinburgh University Press.

Harding, A. (with Lee, G. E.) 1987 *Henge Monuments and Related Sites of Great Britain*. Oxford: British Archaeological Reports British Series **175**.

Harding, A. (ed.) 1982 *Climatic Change in Later Prehistory*. Edinburgh: Edinburgh University Press.

Harding, D. W. (ed.) 1982 *Later Prehistoric Settlement in South-East Scotland*. Edinburgh: Department of Archaeology University of Edinburgh Occasional Paper **8**.

Harding, D. W. and Armit, I. 1990 Survey and excavation in west Lewis. In Armit, I (ed.) *Beyond the Brochs*. Edinburgh: Edinburgh University Press, 71–107.

Harman, M. 1983 Animal remains from Ardnave, Islay. In Ritchie, J. N. G. and Welfare, H. Excavations at Ardnave, Islay. *Proceedings of the Society of Antiquaries of Scotland*, **113**, 343–350.

Harrington, P. and Pierpont, S. 1980 Port Charlotte chambered cairn, Islay: an interim note. *Glasgow Archaeological Journal* **7**, 113–115.

Harris, A. L. 1991 The growth and structure of Scotland. In Craig, G. Y. (ed.) *Geology of Scotland*. London: Geological Society, 1–24.

Harris, J. 1984 A preliminary survey of hut circles and field systems in SE Perthshire. *Proceedings of the Society of Antiquaries of Scotland* **114**, 199–216.

Harrison, C. J. O. and Cowles, G. S. 1977 The extinct large cranes of the North-West Palaearctic. *Journal of Archaeological Science* **4**, 25–28.

Harvey, A. M., Oldfield, F. and Baron, A. F. 1981 Dating of post-glacial landforms in the central Howgills. *Earth Surface Processes and Landforms* **6**, 401–412.

Harvey, A. M. and Renwick, W. H. 1987 Holocene alluvial fan and terrace formation in the Bowland Fells, northwest England. *Earth Surface Processes and Landforms* **12**, 249–257.

Haslam, C. J. 1987 *Late Holocene Peat Stratigraphy and Climatic Change – a Macrofossil Investigation from the Raised Mires of North Western Europe*. Unpublished PhD thesis, University of Southampton.

Hawksworth, D. L. 1960 Studies on the peat deposits of the island of Foula, Shetland. *Transactions of the Botanical Society of Edinburgh* **40**, 576–591.

Hawksworth, D. L. 1974 *The Changing Flora and Fauna of Britain*. London: Academic Press.

Hedeager, L. 1992 *Iron-Age Societies. From Tribe to State in Northern Europe 500 BC–AD 700*. Oxford: Blackwell.

Hedges, J. W. 1975 Excavation of two Orcadian burnt mounds at Liddle and Beaquoy. *Proceedings of the Society of Antiquaries of Scotland* **106**, 39–98.

Hedges, J. W. 1982 An archaeodemographical perspective on Isbister. *Scottish Archaeological Review* **1**, 5–20.

Hedges, J. W. 1987 *Bu, Gurness and the Brochs of Orkney*. Oxford: British Archaeological Reports British Series **163–165** (3 vols).

Hedges, R. E. M., Housley, R. A., Bronk-Ramsey, C. and van Klinken, G. J. 1993 Radiocarbon dates from the Oxford AMS system: Archaeometry datelist 16. *Archaeometry* **35**, 147–167.

Henderson, I. 1979 The silver chain from Whitecleugh, Shieldholm, Crawfordjohn, Lanarkshire. *Transactions of the Dumfriesshire and Galloway Natural History and Antiquarian Society* **54**, 20–28.

Henderson, I. 1987 Early Christian monuments of Scotland displaying crosses but no other ornament. In Small, A (ed.) *The Picts: A New Look at Old Problems*. Dundee: University of Dundee, 45–58.

Henshall, A. S. 1963 *The Chambered Tombs of Scotland*, vol. 1. Edinburgh: Edinburgh University Press.

Henshall, A. S. 1972 *The Chambered Tombs of Scotland*, vol. 2. Edinburgh: Edinburgh University Press.

Higham, N. J. 1989 Roman and native in England north of the Tees: acculturation and its limitations. In Barrett, J. C., Fitzpatrick, A. P., and Macinnes, L. (eds) *Barbarians and Romans in North-West Europe from the Later Republic to Late Antiquity*. Oxford: British Archaeological Reports International Series **S471**, 153–174.

Higham, N. J. 1991 Soldiers and settlement in northern England. In Jones, R. F. J. (ed.) *Roman Britain: Recent Trends*. Sheffield: J. R. Collis Publications, 93–101.

Hill, P. H. 1982a Towards a new classification of early houses. *Scottish Archaeological Review* **1**, 24–31.

Hill, P. H. 1982b Settlement and chronology. In Harding, D. W. (ed.) *Later Prehistoric Settlement in South-East Scotland*. Edinburgh: Department of Archaeology University of Edinburgh Occasional Paper **8**, 4–43.

Hill, P. H. 1982c Broxmouth hill-fort excavations, 1977–78: an interim report. In Harding, D. W. (ed.) *Later Prehistoric Settlement in South-East Scotland*. Edinburgh: Department of Archaeology University of Edinburgh Occasional Paper **8**, 141–188.

Hill, P. H. 1987 Traprain Law: the Votadini and the Romans. *Scottish Archaeological Review* **4**, 85–91.

Hill, P. H. 1991 *Whithorn 4: excavations 1990–1991*. Whithorn: The Whithorn Trust.

Hill, P. H. and Kucharski, K. 1990 Early medieval ploughing at Whithorn and the chronology of plough pebbles. *Transactions of the Dumfriesshire and Galloway Natural History and Antiquarian Society* **65**, 73–83.

Hingley, R. 1992 Society in Scotland from 700 BC to AD 200. *Proceedings of the Society of Antiquaries of Scotland* **122**, 7–53.

Hirons, K. R. and Edwards, K. J. 1986 Events at and around the first and second *Ulmus* declines: palaeoecological investigations in Co. Tyrone, Northern Ireland. *New Phytologist* **104**, 131–153.

Hirons, K. R. and Edwards, K. J. 1990 Pollen and related studies at Kinloch, Isle of Rhum, Scotland, with particular reference to possible early human impacts on vegetation. *New Phytologist* **116**, 715–727.

Hobley, A. S. 1989 The numismatic evidence for the post-Agricolan abandonment of the Roman frontier in northern Scotland. *Britannia* **20**, 69–74.

Hogg, A. H. A. 1979 *British Hill-Forts: An Index*. Oxford: British Archaeological Reports British Series **62**.

Holdsworth, P. 1991 Dunbar. *Current Archaeology* **127**, 315–317.

Holdsworth, P. 1993 Excavations at Castle Park, Dunbar: an interim report on the Anglian evidence. *Transactions of the East Lothian Antiquarian and Field Naturalists Society* **22**, 31–52.

Hooke, J. M., Harvey, A. M., Millar, S. Y. and Redmond, C. E. 1990 The chronology and stratigraphy of the alluvial terraces of the River Dane Valley, Cheshire, NW England. *Earth Surface Processes and Landforms* **15**, 717–737.

Hope-Taylor, B. K. 1977 *Yeavering: An Anglo-British centre of early Northumbria*. London: HMSO. (Department of the Environment Archaeological Reports **7**.)

Hope-Taylor, B. K. 1980 Balbridie . . . and Doon Hill. *Current Archaeology* **72**, 18–19.

Hoppe, G. 1965 Submarine peat in the Shetland Islands. *Geografiska Annaler* **47A**, 195–203.

Huff, D. 1954 *How to Lie with Statistics*. London: Victor Gollancz.

Hulme, P. D. and Shirriffs, J. 1985 Pollen analysis of a radiocarbon-dated core from North Mains, Strathallan, Perthshire. *Proceedings of the Society of Antiquaries of Scotland* **115**, 105–113.

Hunter, F. A. 1977 Ecology of pinewood beetles. In Bunce, R. G. H. and Jeffers, J. N. R. (eds) *Native pinewoods of Scotland*. Cambridge: Institute of Terrestial Ecology, 42–51.

Hunter, J. 1976 *The Making of the Crofting Community*. Edinburgh: John Donald.

Hunter, J. R. 1986 *Rescue Excavations at the Brough of Birsay 1974–82*. Edinburgh: Society of Antiquaries of Scotland Monograph Series **4**.

Hunter, J. R. 1990 Pool, Sanday, a case study for the Late Iron Age and Viking periods. In Armit, I. (ed.) *Beyond the Brochs*. Edinburgh: Edinburgh University Press, 175–193.

Hunter, J. R. 1991 The multi-period landscape. In Hanson, W. S. and Slater, E. A. (eds), *Scottish Archaeology: New Perceptions*. Aberdeen: Aberdeen University Press, 178–195.

Hunter, J. R., Bond, J. M. and Smith, A. N. 1993 Some aspects of early Viking settlement in Orkney. In Batey, C. E., Jesch, J. and Morris, C. D. (eds) *The Viking Age in Caithness, Orkney and the North Atlantic*. Edinburgh: Edinburgh University Press, 272–284.

Hunter, J. R. and Ralston, I. B. M. 1993 The structure of British archaeology. In Hunter, J. R. and Ralston, I. B. M. (eds) *Archaeological Resource Management in the UK: An Introduction*. Stroud: Alan Sutton/Institute of Field Archaeologists, 30–43.

Huntley, B. 1990 European post-glacial forests: compositional changes in response to climatic change. *Journal of Vegetation Science* **1**, 507–518.

Huntley, B. 1993 Rapid early-Holocene migration and high abundance of hazel (*Corylus avellana* L.): alternative hypotheses. In Chambers, F. M. (ed.) *Climate Change and Human Impact on the Landscape*. London: Chapman and Hall, 205–215.

Huntley, B. 1994 Late Quaternary and Holocene palaeoecology and palaeoenvironments of the Morrone Birkwoods, Scotland. *Journal of Quaternary Science* **9**, 311–336.

Huntley, B. and Birks, H. J. B. 1983 *An Atlas of Past and Present Pollen Maps for Europe: 0–13 000 Years Ago*. Cambridge: Cambridge University Press.

Ilett, M. J. 1980 *Aspects of Neolithic Settlement in North-West Europe and Britain*. Unpublished PhD thesis, University of Cambridge.

Inglis, J. 1987 Patterns in stone, patterns in population: symbol stones seen from beyond the Mounth. In Small, A. (ed.) *The Picts: A New Look at Old Problems*. Dundee: University of Dundee, 73–79.

Innes, J. L. 1983 Lichenometric dating of debris flow deposits in the Scottish Highlands. *Earth Surface Processes and Landforms* **8**, 579–588.

Isdo, S. B. 1989 A problem for palaeoclimatology. *Quaternary Research* **31**, 433–434.

Jackson, A. 1984 *The Symbol Stones of Scotland*. Stromness: Orkney Press.

Jackson, D. J. 1956 The capacity for flight of certain water beetles and its bearing on their origin in the Western Scottish Isles. *Proceedings of the Linnean Society of London* **167**, 76–96.

Jacobi, R. M. 1973 Aspects of the 'Mesolithic Age' in Great Britain. In Kozlowski, S. (ed.) *The Mesolithic in Europe*. Warsaw: Warsaw University Press, 237–266.

Jacobi, R. M. 1982 When did Man come to Scotland? *Mesolithic Miscellany* **3**, 8–9.

Jardine, W. G. 1977 Location and age of Mesolithic occupation sites on Oronsay, Inner Hebrides. *Nature* **267**, 138–140.

Jardine, W. G. 1987 The Mesolithic coastal setting. In Mellars, P. A. *Excavations on Oronsay: Prehistoric Human Ecology on a Small Island.* Edinburgh: Edinburgh University Press, 25–51.

Jardine, W. G. and Masters, L. J. 1977 A dug-out canoe from Catherinefield Farm, Locharbriggs, Dumfriesshire. *Transactions of the Dumfriesshire and Galloway Natural History and Antiquarian Society* **52**, 56–65.

Jessen, K. and Helbaek, H. 1944 Cereals in Great Britain and Ireland in prehistoric and early historic times. *Det Kongelige Danske Videnskabernes Selskab Biologiske Skrifter* **3**, 1–68.

Jobey, G. 1966 Homesteads and settlements in the frontier area. In Thomas, A. C. (ed.) *Rural settlement in Roman Britain.* London: Council for British Archaeology Research Report **7**, 1–14.

Jobey, G. 1974 Notes on some population problems in the area between the two Roman Walls, I. *Archaeologia Aeliana* **(5 ser) 2**, 17–26.

Jobey, G. 1976 Traprain Law: a summary. In Harding, D. W. (ed.) *Hillforts: Later Prehistoric Earthworks in Britain and Ireland.* London: Academic Press, 192–204.

Jobey, G. 1978 Burnswark Hill. *Transactions of the Dumfriesshire and Galloway Natural History and Antiquarian Society* **53**, 57–104.

Jobey, G. 1980 Green Knowe unenclosed platform settlement and Harehope Cairn, Peeblesshire. *Proceedings of the Society of Antiquaries of Scotland* **110**, 72–113.

Jobey, G. 1988 Gowanburn River Camp: an Iron Age, Romano-British and more recent settlement in north Tynedale. *Archaeologia Aeliana* **(5 ser) 16**, 11–28.

Jóhansen, J. 1975 Pollen diagrams from the Shetland and Faroe Islands. *New Phytologist* **75**, 369–387.

Jones, A. H. M. 1964 *The Later Roman Empire.* Oxford: Oxford University Press.

Jones, G. 1968 *A History of the Vikings.* Oxford: Oxford University Press.

Jones, G. D. B. 1986 Roman military site at Cawdor. *Popular Archaeology* **7.3**, 13–16.

Kaland, P. E. 1986 The origin and management of Norwegian coastal heaths as reflected by pollen analysis. In Behre, K.-E. (ed.), *Anthropogenic Indicators in Pollen Diagrams.* Rotterdam: A. A. Balkema, 19–36.

Kaland, S. 1993 The settlement of Westness, Rousay. In Batey, C. E., Jesch, J. and Morris, C. D. (eds) *The Viking Age in Caithness, Orkney and the North Atlantic.* Edinburgh: Edinburgh University Press, 308–317.

Keatinge, T. H. and Dickson, J. H. 1979 Mid-Flandrian changes in vegetation on Mainland, Orkney. *New Phytologist* **82**, 585–612.

Keith-Lucas, M. 1986 Vegetation development and human impact. In Whittle, A., Keith-Lucas, M., Milles, A., Noddle, B., Rees, S. and Romans, J. C. C. (eds) *Scord of Brouster: An Early Agricultural Settlement on Shetland. Excavations 1977–1979.* Oxford: Oxford University Committee for Archaeology Monograph **9**, 92–118.

Keller, C. 1991 Vikings in the West Atlantic: a model of Norse Greenlandic medieval society. *Acta Archaeologica* **61**, 126–141.

Kemp, R. A. 1985 *Soil Micromorphology and the Quaternary.* Cambridge: Quaternary Research Association Technical Guide **2**.

Kenward, H. K. 1975 The biological and archaeological implications of the beetle *Aglennus brunneus* (Gyllenhall) in ancient faunas. *Journal of Archaeological Science* **2**, 63–69.

Kenward, H. K. 1976 Further archaeological records of *Aglennus brunneus* (Gyll.) in Britain and Ireland, including confirmation of its presence in the Roman period. *Journal of Archaeological Science* **3**, 275–277.

Kenward, H. K., Allison, E. P., Morgan, L. M., Jones, A. K. G. and Hutchison, A. R. 1991 The insect and parasite remains. In McCarthy, M. R. (ed.) *The Structural Sequence and Environmental Remains from Castle Street, Carlisle.* Carlisle: Cumberland and Westmoreland Antiquarian and Archaeological Society Research Series **5**, 65–72.

Kenworthy, J. B. 1981 *Excavation of a Mesolithic Settlement Site at Nethermills Farm, Crathes, Near Banchory, Grampian, 1978–80: Interim Statement.* St Andrews: Department of Archaeology University of St Andrews.

Kerney, M. P. 1976 *Atlas of the Non-marine Mollusca of the British Isles.* Cambridge: Institute of Terrestrial Ecology.

Keys, D. 1988 Cloud of volcanic dust blighted north Britain 3000 years ago. *The Independent,* 16 August 1988.

King, A. C. 1984 Animal bones and the dietary identity of military and civilian groups in Roman Britain, Germany and Gaul. In Blagg, T. F. C. and King, A. C. (eds) *Military and Civilian in Roman Britain: Cultural Relationships in a Frontier Province.* Oxford: British Archaeological Reports British Series **136**, 187–217.

Kinnes, I. 1985 Circumstance not context: the Neolithic of Scotland as seen from outside. *Proceedings of the Society of Antiquaries of Scotland* **115**, 15–57.

Kinnes, I. 1988 The Cattleship Potemkin: reflections on the first Neolithic in Britain. In Barrett, J. C. and Kinnes, I. (eds) *The Archaeology of Context in the Neolithic and Bronze Age: Recent Trends.* Sheffield: Department of Archaeology and Prehistory University of Sheffield, 2–8.

Kinnes, I. 1994 The Neolithic. In Vyner, B. (ed.) *Building on the Past.* London: Royal Archaeological Institute, 90–102.

Knights, B. A., Dickson, C. A., Dickson, J. H. and Breeze, D. J. 1983 Evidence concerning the Roman military diet at Bearsden, Scotland, in the 2nd century A.D. *Journal of Archaeological Science* **10**, 139–152.

Knox, E. M. 1954 Pollen analysis of a peat at Kingsteps Quarry, Nairn. *Transactions of the Botanical Society of Edinburgh* **36**, 224–229.

Kristiansen, K. 1989 Perspectives on the archaeological heritage: history and future. In Cleere, H. F. (ed.) *Archaeological Heritage Management in the Modern World.* London: Unwin Hyman, 23–30. (One World Archaeology **9**.)

Kutzbach, J. E. and Guetter, P. J. 1986 The influence of changing orbital parameters and surface boundary conditions on climate simulations for the past 18000 years. *Journal of Atmospheric Sciences* **43**, 1726–1759.

Lacaille, A. D. 1954 *The Stone Age in Scotland.* Oxford: Oxford University Press. (Publications of the Welcome Historical Medical Museum **6**.)

Lagerås, P. 1996 A short presentation of the project: Man and environment in the Småland Uplands during the last 6000 years. In Lagerås, P. (ed.) *Vegetation and Land-Use in the Småland Uplands during the last 6000 Years.* Lund: Lundqua Thesis **36**, Appendix V.

Laing, D. 1976 *The Soils of the Country round Perth, Arbroath and Dundee.* Edinburgh: HMSO.

Lamb, H. H. 1966 Trees and climatic history in Scotland; a radiocarbon test and other evidence. In Lamb, H. H. (ed.) *The Changing Climate.* London: Methuen, 157–169.

Lamb, H. H. 1977 *Climate. Past, Present and Future: Vol. 2. Climatic History and the Future.* London: Methuen.

Lamb, R. G. 1980 *The Archaeological Sites and Monuments of Sanday and North Ronaldsay, Orkney.* Edinburgh: RCAHMS. (Archaeological Sites and Monuments Series **11**.)

Lamb, R. G. 1984 *The Archaeological Sites and Monuments of Eday and Stronsay.* Edinburgh: RCAHMS. (Archaeological Sites and Monuments Series **23**.)

Lambeck, K. 1995 Late Devensian and Holocene shorelines of the British Isles and North Sea from models of glacio-hydro-isostatic rebound. *Journal of the Geological Society* **152**, 437–448.

Lane, A. 1994 Fifth to seventh century trading systems in western Britain and Ireland. In Crawford, B. E. (ed.) *Scotland in Dark Age Europe.* St Andrews: St John's House Papers **5**, 104–115.

Lang, J. T. 1994 The Govan hogbacks: a reappraisal. In Ritchie, A. (ed.), *Govan and its Early Medieval Sculpture.* Gloucester: Sutton Publishing, 123–131.

Lanting, J. N. and van der Waals, J. D. 1972 British beakers as seen from the continent: a review article. *Helinium* **12**, 20–46.

Larsson, L. 1984 The Skateholm Project: a late Mesolithic settlement and cemetery complex at a southern Swedish bay. *Meddelanden från Lunds Universitets Historiska Museum* **5**, 5–38.

Larsson, L. 1989 Late Mesolithic settlements and cemeteries at Skateholm, Southern Sweden. In Bonsall, J. C. (ed.) *The Mesolithic in Europe.* Edinburgh: John Donald, 367–378.

Larsson, L., Meiklejohn, C. and Newell, R. R. 1981 Human skeletal material from the Mesolithic site of Agerod I: HC, Scania, southern Sweden. *Fornvännen* **76**, 161–168.

Lawson, T. J. 1981 The 1926–7 excavations of the Creag nan Uamh bone caves, near Inchnadamph, Sutherland, *Proceedings of the Society of Antiquaries of Scotland* **111**, 7–20.

Lawson, T. J. 1984 Reindeer in the Scottish Quaternary. *Quaternary Newsletter* **42**, 1–7.

Lawson, T. J. and Bonsall, J. C. 1986 The Palaeolithic of Scotland: a reconsideration of evidence from Reindeer Cave, Assynt. In Collcutt, S. N. (ed.) *The Palaeolithic of Britain and its Nearest Neighbours: Recent Trends.* Sheffield: Department of Archaeology and Prehistory University of Sheffield, 85–89.

Lee, R. B. and De Vore I. (eds) 1968 *Man the Hunter.* Chicago: Aldine.

Lekander, B., Bejer-Petersen, B., Kangas, E. and Bakke, A. 1977 The distribution of bark beetles in the Nordic countries. *Acta Entomologica Fennica* **32**, 1–115.

Leslie, A. F. 1990 Inveresk, East Lothian District. *Discovery and Excavation in Scotland 1990*, 29–30.

Lewis, F. J. 1905 The plant remains in Scottish peat mosses. Part I. The Scottish Southern Uplands. *Transactions of the Royal Society of Edinburgh* **41**, 699–723.

Lewis, F. J. 1911 The plant remains in the Scottish peat mosses. Part IV. The Scottish Highlands and Shetland, with an appendix on the Icelandic peat deposits. *Transactions of the Royal Society of Edinburgh* **47**, 793–833.

Lewis-Williams, D. and Dowson, T. A. 1993 On vision and power in the Neolithic: evidence from the decorated monuments. *Current Anthropology* **34**, 55–65.

Linton, D. L. 1951 Watershed breaching by ice in Scotland. *Transactions of the Institute of British Geographers* **15**, 1–15.

Linton, D. L. 1959 Morphological contrasts of eastern and western Scotland. In Miller, R. and Watson, J. W. (eds) *Geographical Essays in Memory of A. G. Ogilvie.* Edinburgh: Thomas Nelson and Sons, 16–45.

Livens, R. G. 1956 Three tanged flint points from Scotland. *Proceedings of the Society of Antiquaries of Scotland* **89**, 438–443.

Locker, A. 1994 The fish remains. In Smith, B. B. (ed.) *Howe. Four millennia of Orkney prehistory.* Society of Antiquaries of Scotland Monograph Series **9**, 157–159.

Long, D., Wickham-Jones, C. R. and Ruckley, N. A. 1986 A flint artifact from the northern North Sea. In Roe, D. (ed.) *Studies in the Upper Palaeolithic of Britain and Northwest Europe.* Oxford: British Archaeological Reports British Series **S296**, 55–62.

Longworth, I. H. 1967 Further discoveries at Brackmont Mill, Brackmont Farm and Tentsmuir, Fife. *Proceedings of the Society of Antiquaries of Scotland* **99**, 60–92.

Loveday, R. and Petchey, M. 1982 Oblong ditches: a discussion and some new evidence. *Aerial Archaeology* **8**, 17–24.

Lowe, C. E., Craig, D. and Dixon, D. 1991 New light on the Anglian 'Minster' at Hoddom. *Transactions of the Dumfriesshire and Galloway Natural History and Antiquarian Society* **56**, 11–35.

Lowe, J. J. 1984 A critical evaluation of pollen-stratigraphic investigations of pre-Late Devensian sites in Scotland. *Quaternary Science Reviews* **3**, 405–432.

Lowe, J. J. 1993 Isolating the factors in early- and mid-Holocene palaeobotanical records from Scotland. In Chambers, F. M. (ed.) *Climate Change and Human Impact on the Landscape.* London: Chapman and Hall, 67–82.

Loyn, H. R. 1977 *The Vikings in Britain.* London: Batsford.

Lynch, A. 1981 *Man and Environment in South-West Ireland 4000 BC–AD 800.* Oxford: British Archaeological Reports British Series **85**.

Macartney, E. 1984 Analysis of faunal remains. In Fairhurst, H. (ed.) *Excavations at Crosskirk Broch, Caithness.* Edinburgh: Society of Antiquaries of Scotland Monograph Series **3**, 133–147.

Macaulay Institute for Soil Research 1984 *Organization and Methods of the 1:250,000 Soil Survey of Scotland.* Aberdeen: Soil Survey of Scotland, Macaulay Institute for Soil Research.

Macaulay Institute for Soil Research undated *Land Capability Classification for Agriculture* (a pamphlet).

Macdonald, G. and Curle, A. O. 1929 The Roman fort at Mumrills, near Falkirk. *Proceedings of the Society of Antiquaries of Scotland,* **63**, 396–569.

Macdonald, G. and Park, A. 1906 The Roman fort at Bar Hill, Dunbartonshire excavated by Mr Alexander Whitelaw of Gartshore. *Proceedings of the Society of Antiquaries of Scotland* **40**, 403–456.

MacGregor, M. 1976 *Early Celtic Art in North Britain.* Leicester: Leicester University Press (2 vols).

Macinnes, L. 1982 Pattern and purpose: the settlement evidence. In Harding, D. W. (ed.) *Later Prehistoric Settlement in South-East Scotland.* Edinburgh: Department of Archaeology University of Edinburgh Occasional Paper **8**, 57–74.

Macinnes, L. 1984a Brochs and the Roman occupation of Lowland Scotland. *Proceedings of the Society of Antiquaries of Scotland* **114**, 235–250.

Macinnes, L. 1984b Settlement and economy: East Lothian and the Tyne-Forth province. In Miket, R. and Burgess, C. B. (eds) *Between and Beyond the Walls.* Edinburgh: John Donald, 176–198.

Macinnes, L. 1989 Baubles, bangles and beads: trade and exchange in Roman Scotland. In Barrett, J. C., Fitzpatrick, A. P., and Macinnes, L. (eds) *Barbarians and Romans in North-West Europe from the Later Republic to Late Antiquity.* Oxford: British Archaeological Reports International Series **S471**, 108–116.

MacKie, E. W. 1964 New Excavations on the Monamore Neolithic chambered cairn, Lamlash, Isle of Arran, in 1961. *Proceedings of the Society of Antiquaries of Scotland* **97**, 1–34.

MacKie, E. W. 1977 *Science and Society in Prehistoric Britain.* London: Paul Elek.

MacKie, E. W. 1993 Review of Renfrew, C. (ed.) 1990 *The prehistory of Orkney B.C. 4000–1000 A.D.*, 2 edition. *Glasgow Archaeological Journal* **16** (1989–90: 1993), 89–92.

Macklin, M. G. 1993 Holocene river alluviation in Britain. *Zeitschrift für Geomorphologie Supplement Band* **88**, 109–122.

Macklin, M. G., Rumsby, B. T. and Heap, T. 1992 Flood alluviation and entrenchment: Holocene valley-floor development and transformation in the British uplands. *Geological Society of America Bulletin* **104**, 631–643.

MacLaren, A. 1974 A Norse house on Drimore machair, S Uist. *Glasgow Archaeological Journal* **3**, 9–18.

Maclean, A. C. and Rowley-Conwy, P. A. 1984 The carbonised material from Boghead, Fochabers. In Burl, H. A. W. (ed.) Report on the excavation of a Neolithic mound at Boghead, Speymouth Forest, Fochabers, Moray, 1972 and 1974. *Proceedings of the Society of Antiquaries of Scotland* **114**, 69–71.

MacSween, A. 1991 *The Analysis of Neolithic and Iron Age Pottery Using XRF and Thin-section Petrology.* Unpublished PhD thesis, University of Bradford.

MacSween, A. 1992 Orcadian Grooved Ware. In Sharples, N. M. and Sheridan, A. (eds), *Vessels for the Ancestors.* Edinburgh: Edinburgh University Press, 259–271.

Magny, M. 1982 Atlantic and Sub-boreal: dampness and dryness? In Harding, A. F. (ed.) *Climatic Change in Later Prehistory.* Edinburgh: Edinburgh University Press, 33–43.

Mahler, D. L. D. 1991 Argisbrekka: new evidence of shielings in the Faroe Islands. *Acta Archaeologica* **61**, 60–72.

Mann, J. C. 1974 The northern frontier after AD 369. *Glasgow Archaeological Journal* **3**, 34–42.

Mann, J. C. 1985 Two *topoi* in the *Agricola*. *Britannia* **16**, 21–24.

Mann, J. C. 1988 The history of the Antonine Wall – a reappraisal. *Proceedings of Society of Antiquaries of Scotland* **118**, 131–137.

Mann, J. C. 1992 Loca. *Archaeologia aeliana* (**5 ser**) **20**, 53–55.

Manning, W. H. 1975 Economic influences on land use in the military areas of the Highland zone during the Roman period. In Evans, J. G., Limbrey, S. and Cleere, H. (eds) *The Effect of Man on the Landscape: The Highland Zone.* London: Council for British Archaeology Research Report **11**, 112–116.

Manning, W. H. 1985 The iron objects. In Pitts, L. and St. Joseph, J. K. (eds) *Inchtuthil: The*

Roman Legionary Fortress. London: Society for the Promotion of Roman Studies, 289–299.

Marshall, D. N. 1964 Report on the excavations at Little Dunagoil. *Transactions of the Buteshire Natural History Society* **16**, 30–69.

Martens, I. 1992 Some aspects of marginal settlement in Norway during the Viking Age and Middle Ages. In Morris, C. D. and Rackham, D. J. (eds) *Norse and Later Settlement and Subsistence in the North Atlantic*. Glasgow: Department of Archaeology University of Glasgow Occasional Paper Series **1**, 1–7.

Marwick, H. 1952 *Orkney Farm Names*. Kirkwall: W. M. Mackintosh.

Maxwell, G. S. 1983 Recent aerial survey in Scotland. In Maxwell, G. S. (ed.) *The Impact of Aerial Reconnaissance on Archaeology*. London: Council for British Archaeology Research Report **49**, 27–40.

Maxwell, G. S. 1987 Settlement in southern Pictland – a new overview. In Small, A. (ed.) *The Picts – A New Look at Old Problems*. Dundee: University of Dundee, 31–44.

Mayewski, P. A., Buckland, P. C., Edwards, K. J., Meeker, L. D. and O'Brien, S. 1996 Climate change events as seen in the Greenland ice core (GISP2). Implications for the Mesolithic of Scotland. In Pollard, T. and Morrison, A. (eds) *The Early Prehistory of Scotland*. Edinburgh: Edinburgh University Press, 74–84.

McConnell. P. 1968 *The Agricultural Notebook*, 15th edition. Butterworths, London.

McCormick, F. 1981 The animal bones from Ditch 1. In Barber, J. (ed.) Excavations on Iona, 1979. *Proceedings of the Society of Antiquaries of Scotland* **111**, 313–318.

McCormick, F. 1983 Dairying and beef production in early Christian Ireland: the faunal evidence. In Reeves-Smyth, T. and Hamond, F. (eds) *Landscape Archaeology in Ireland*. Oxford: British Archaeological Reports British Series **116**, 253–267.

McCormick, F. 1984 Large mammal bones. In Sharples, N. M. (ed.) Excavations at Pierowall Quarry, Westray, Orkney. *Proceedings of the Society of Antiquaries of Scotland* **114**, 108–111.

McCormick, F. 1987 The animal bones. In Manning, C. (ed.) Excavation at Moyne graveyard, Shrule, County Mayo. *Proceedings of the Royal Irish Academy* **87C**, 60–68.

McCormick, F. 1992 Early faunal evidence for dairying. *Oxford Journal of Archaeology* **11**, 201–209.

McCullagh, R. P. J. 1989 Excavations at Newton, Islay. *Glasgow Archaeological Journal* **15**, 23–51.

McCullagh, R. P. J. 1992a Lairg. *Current Archaeology* **131**, 457–459.

McCullagh, R. P. J. 1992b *Lairg. The Archaeology of a Changing Landscape*. Edinburgh: AOC (Scotland) Ltd.

McCullagh, R. P. J. 1996 An interim report on the results of the Lairg project 1988–1992. *Northern Studies*, in press.

McGovern, T. H. 1981 The economics of extinction in Norse Greenland. In Wigley, T. M. L. (ed.) *Climate and History*. Cambridge: Cambridge University Press, 404–434.

McGovern, T. H. 1985 The Arctic frontier of Norse Greenland. In Green, S. W. and Perlman, S. M. (eds) *The Archaeology of Frontiers and Boundaries*. New York: Academic Press, 275–323.

McGovern, T. H. 1990 The archaeology of the Norse North Atlantic. *Annual Reviews of Anthropology* **19**, 331–351.

McGovern, T. H., Bigelow, G. F., Amorosi, T. and Russell, D. 1988 Northern islands, human error, and environmental degradation: a view of social and ecological change in the medieval North Atlantic. *Human Ecology* **16**, 225–270.

McGrail, S. 1993 Prehistoric seafaring in the Channel. In Scarre, C. and Healy, F. (eds) *Trade and Exchange in Prehistoric Europe*. Oxford: Oxbow Monograph **33**, 199–210.

McVean, D. N. 1956a Ecology of *Alnus glutinosa* (L.) Gaertn. V. Notes on some British alder populations. *Journal of Ecology* **44**, 321–330.

McVean, D. N. 1956b Ecology of *Alnus glutinosa* (L.) Gaertn. VI. Post-glacial history. *Journal of Ecology* **44**, 331–333.

McVean, D. N. 1964 Pre-history and ecological history. In Burnett, J. H. (ed.) *The Vegetation of Scotland*. Edinburgh: Oliver and Boyd, 561–567.

McVean, D. N. and Ratcliffe, D. A. 1962 *Plant Communities of the Scottish Highlands.* London: HMSO.

Megaw, R. and Megaw, J. V. S. 1986 *Early Celtic Art in Britain and Ireland.* Aylesbury: Shire Publications.

Meiklejohn, C. and Denston, B. 1987 The human skeletal material. In Mellars, P. A. (ed.) *Excavations on Oronsay: Prehistoric Human Ecology on a Small Island.* Edinburgh: Edinburgh University Press, 290–300.

Mellars, P. A. 1974 The Palaeolithic and Mesolithic. In Renfrew, A. C. (ed.) *British Prehistory: A New Outline.* London: Duckworth, 41–99.

Mellars, P. A. 1976a Fire ecology, animal populations and man: a study of some ecological relationships in prehistory. *Proceedings of the Prehistoric Society* **42**, 15–45.

Mellars, P. A. 1976b Settlement patterns and industrial variability in the British Mesolithic. In Sieveking, G. de G., Longworth, I. H. and Wilson, K. E. (eds) *Problems in Economic and Social Archaeology.* London: Duckworth, 375–399.

Mellars, P. A. 1978 Excavation and economic analysis of Mesolithic shell middens on the island of Oronsay (Inner Hebrides). In Mellars, P. A. (ed.) *The Early Postglacial Settlement of Northern Europe: An Ecological Perspective.* London: Duckworth, 371–396.

Mellars, P. A. 1987 *Excavations on Oronsay: Prehistoric Human Ecology on a Small Island.* Edinburgh: Edinburgh University Press.

Mellars, P. A. and Wilkinson, M. R. 1980 Fish otoliths as indicators of seasonality in prehistoric shell middens: the evidence from Oronsay (Inner Hebrides). *Proceedings of the Prehistoric Society* **46**, 19–44.

Mercer, J. 1969 Stone tools from a washing-limit deposit of the highest Post Glacial transgression. *Proceedings of the Society of Antiquaries of Scotland* **100**, 1–46.

Mercer, J. 1970 Flint tools from the present tidal zone, Lussa Bay, Isle of Jura, Argyll. *Proceedings of the Society of Antiquaries of Scotland* **102**, 1–30.

Mercer, J. 1971 A regression-time stoneworkers' camp, 33 ft OD, Lussa River, Isle of Jura. *Proceedings of the Society of Antiquaries of Scotland* **103**, 1–32.

Mercer, J. 1972 Microlithic and Bronze Age camps 75–26ft O. D., N Carn, Isle of Jura. *Proceedings of the Society of Antiquaries of Scotland* **104**, 1–22.

Mercer, J. 1974 Glenbatrick Waterhole, a microlithic site on the Isle of Jura. *Proceedings of the Society of Antiquaries for Scotland* **105**, 9–32.

Mercer, J. 1980 Lussa Wood I: the Late-glacial and early Post-glacial occupation of Jura. *Proceedings of the Society of Antiquaries of Scotland* **110**, 1–32.

Mercer, J. and Searight, S. 1987. Glengarrisdale: confirmation of Jura's third microlithic phase. *Proceedings of the Society of Antiquaries of Scotland* **116**, 41–55.

Mercer, R. J. 1991 The survey of a hilltop enclosure on Ben Griam Beg, Caithness and Sutherland District, Highland Region. In Hanson, W. S. and Slater, E. A. (eds) *Scottish Archaeology: New Perceptions.* Aberdeen: Aberdeen University Press, 140–152.

Mercer, R. J. and Tipping, R. 1994 The prehistory of soil erosion in the northern and eastern Cheviot Hills, Anglo-Scottish Borders. In Foster, S. and Smout, T. C. (eds) *The History of Soils and Field Systems.* Aberdeen: Scottish Cultural Press, 1–25.

Merritt, J. W., Coope, G. R., Taylor, B. J. and Walker, M. J. C. 1990 Late Devensian organic deposits beneath till in the Teith Valley, Perthshire. *Scottish Journal of Geology,* **26**, 15–24.

Mills, C. M., Crone, A., Edwards, K. J. and Whittington, G. 1994 The excavation and environmental investigation of a sub-peat stone bank near Loch Portain, North Uist, Outer Hebrides. *Proceedings of the Society of Antiquaries of Scotland* **124**, 155–171.

Mithen, S. J. 1989 New evidence for Mesolithic settlement on Colonsay. *Proceedings of the Society of Antiquaries of Scotland* **119**, 33–41.

Mithen, S. J. 1990 Gleann Mor: a Mesolithic site on Islay. *Current Archaeology* **119**, 376–377.

Mithen, S. J. and Finlayson, B. 1991 Red deer hunters on Colonsay? The implications of Staosnaig for the interpretation of the Oronsay middens. *Proceedings of the Prehistoric Society* **57**, 1–8.

Mithen, S. J., Finlayson, B. and Finlay, N. 1994 A lower Palaeolithic handaxe from Scotland. *Lithics* **13**, 1–5.

Mithen, S. J., Finlayson, B., Finlay, N. and Lake, M. 1992 Excavation at Bolsay Farm, a Mesolithic settlement on Islay. *Cambridge Archaeological Journal* **2**, 242–253.

Mithen, S. J. and Lake, M. 1996 The Southern Hebrides Mesolithic Project: reconstructing Mesolithic settlement in western Scotland. In Pollard, T. and Morrison, A. (eds) *The Early Prehistory of Scotland*. Edinburgh: Edinburgh University Press, 123–151.

Moore, P. D. 1975 The origin of blanket mires. *Nature* **256**, 267–269.

Moore, P. D. 1985 Forests, man and water. *International Journal of Environmental Studies* **25**, 159–166.

Moore, P. D. 1988 The development of moorlands and upland mires. In Jones, M. (ed.) *Archaeology and the Flora of the British Isles*. Oxford: Oxford University Committee for Archaeology Monograph **14**, 116–122.

Moore, P. D. 1993 The origin of blanket mire, revisited. In Chambers, F. M. (ed.) *Climate Change and Human Impact on the Landscape*. London: Chapman and Hall, 217–224.

Moore, P. D., Webb, J. A. and Collinson, M. E. 1991 *Pollen Analysis*. Oxford: Blackwell Scientific Publications.

Moroney, M. J. 1965 *Facts from Figures*, 2nd edition. Harmondsworth: Allen Lane.

Morris, C. D. 1985 Viking Orkney: a survey. In Renfrew, A. C. (ed.) *The Prehistory of Orkney BC 4000–1000 AD*. Edinburgh: Edinburgh University Press, 210–242.

Morris, C. D. 1989 *The Birsay Bay Project*. Durham: Department of Archaeology University of Durham Monograph Series **1**.

Morris, C. D. and Rackham, D. J. (eds) 1992 *Norse and Later Settlement and Subsistence in the North Atlantic*. Glasgow: Department of Archaeology University of Glasgow Occasional Paper Series **1**.

Morris, C. D., Rackham, D. J., Batey, C. E., Huntley, J. P., Jones, A. K. G. and O'Connor, T. P. 1992 Excavations at Freswick Links, Caithness 1980–82: environmental column samples from the cliff-side. In Morris, C. D. and Rackham, D. J. (eds) *Norse and Later Settlement and Subsistence in the North Atlantic*. Glasgow: Department of Archaeology University of Glasgow Occasional Paper Series **1**, 43–102.

Morrison, A. 1980 *Early Man in Britain and Ireland*. London: Croom Helm.

Morrison, A. and Bonsall, J. C. 1989 The early-postglacial settlement of Scotland: a review. In Bonsall, J. C. (ed.) *The Mesolithic in Europe*. Edinburgh: John Donald, 134–142.

Morrison, I. A. 1983 Prehistoric Scotland. In Whittington, G. and Whyte, I.D. (eds) *An Historical Geography of Scotland*. London: Academic Press, 1–23.

Morrison, I. A. 1985 *Landscape with Lake Dwellings: The Crannogs of Scotland*. Edinburgh: Edinburgh University Press.

Movius, H. L. 1940a *The Irish Stone Age: Its Chronology, Development and Relationships*. Cambridge: Cambridge University Press.

Movius, H. L. 1940b An Early Post-Glacial archaeological site at Cushendun, County Antrim. *Proceedings of the Royal Irish Academy* **C46**, 1–84.

Muckelroy, K. 1981 Middle Bronze Age trade between Britain and Europe: a maritime perspective. *Proceedings of the Prehistoric Society* **47**, 275–297.

Mulholland, H. 1970 The Mesolithic industries of the Tweed Valley. *Transactions of the Dumfriesshire and Galloway Antiquarian and Natural History Society* **47**, 81–110.

Munro, R. 1899 *Prehistoric Scotland and its Place in European Civilisation*. Edinburgh and London: William Blackwood and Sons.

Murray, D. M. and Ralston, I. B. M. forthcoming The excavation of a square-ditched barrow and other cropmarks at Boysack Mills, Inverkeilor, Angus. *Proceedings of the Society of Antiquaries of Scotland*.

Murray, H. K., Murray, J. C., Shepherd, A. N. and Shepherd, I. A. G. 1992 Evidence of agricultural activity of the later second millennium BC at Rattray, Aberdeenshire. *Proceedings of the Society of Antiquaries of Scotland* **122**, 113–125.

Murray, N. A., Bonsall, C., Sutherland, D. G., Lawson, T. J. and Kitchener, A. C. 1993 Further radiocarbon determinations on reindeer remains of Middle and Late Devensian age from the Creag nan Uamh caves, Assynt, NW Scotland. *Quaternary Newsletter* **70**, 1–10.

Musson, C. R. 1971 A study of possible building forms at Durrington Walls, Woodhenge and the Sanctuary. In Wainwright, G. J. (ed.) *Durrington Walls: Excavations 1966–1968*. London: Society of Antiquaries of London Research Report **29**, 363–377.

Myers, A. M. 1987 All shot to pieces? Inter-assemblage variability, lithic analysis and Mesolithic assemblage 'types': some preliminary observations. In Brown, A. G. and Edmonds, M. R. (eds) *Lithic Analysis and Later British Prehistory*. Oxford: British Archaeological Reports British Series **162**, 137–154.

Myers, A. M. 1988 Scotland inside and outside of the British mainland Mesolithic. *Scottish Archaeological Review* **5**, 23–29.

Newell, P. J. 1988 A buried wall in peatland by Sheshader, Isle of Lewis. *Proceedings of the Society of Antiquaries of Scotland* **118**, 79–93.

Newell, P. J. 1990 *Aspects of the Flandrian Vegetational History of South-West Scotland, with Special Reference to Possible Mesolithic Impact*. Unpublished PhD thesis, University of Birmingham.

Nichols, H. 1967 Vegetational change, shoreline displacement and the human factor in the late Quaternary history of south-west Scotland. *Transactions of the Royal Society of Edinburgh* **67**, 145–187.

Nicolaisen, W. F. H. 1975 Scandinavian place-names. In MacNeill, P. and Nicholson, R. (eds) *An historical atlas of Scotland*. St Andrews: Conference of Scottish Medievalists, 6–7.

Nicolaisen, W. F. H. 1982 The Viking settlement of Scotland: the evidence of placenames. In Farrell, R. T. (ed.) *The Vikings*. London and Chichester: Phillimore, 95–115.

Nicolaisen, W. F. H. 1995 Pictish place names. In Nicoll, E. H. (ed.) *A Pictish Panorama*. Balgavies, Angus: Pinkfoot Press, 11–13.

Nieke, M. R. 1983 Settlement patterns in the first millennium AD: a case study on the island of Islay. In Chapman, J. C. and Mytum, H. C. (eds) *Settlement in North Britain 1000 BC–AD 1000*. Oxford: British Archaeological Reports British Series **118**, 299–325.

Nieke, M. R. 1988 Literacy and power: the introduction and use of writing in Early Historic Scotland. In Gledhill, J., Bender, B. and Larsen, M. T. (eds) *State and Society. The Emergence and Development of Social Hierarchy and Political Centralisation*. London: Unwin Hyman, 237–252. (One World Archaeology **4**.)

Noddle, B. 1974 Appendix D: Report on the animal bones found at Dun Mor Vaul. In MacKie, E. W. (ed.) *Dun Mor Vaul. An Iron Age Broch on Tiree*. Glasgow: University of Glasgow Press, 187–198.

Noddle, B. 1983 Appendix 4: Animal bone from Knap of Howar. In Ritchie, A. Excavation of a Neolithic farmstead at Knap of Howar, Papa Westray, Orkney. *Proceedings of the Society of Antiquaries of Scotland*, **113**, 92–100.

Noddle, B. 1986 Animal Bones. In Whittle, A., Keith-Lucas, M., Milles, A., Noddle, B., Rees, S. and Romans, J. C. C. (eds) *Scord of Brouster. An Early Agricultural Settlement on Shetland*. Oxford: Oxford University Committee for Archaeology Monograph **9**, 132.

Norris, R. 1988 Megalithic observatories in Britain, real or imagined. In Ruggles, C. L. N. (ed.) *Records in Stone*. Cambridge: Cambridge University Press, 262–276.

Nye, S. 1993 Botanical report. In Casey, P. J. and Davies, J. L. (eds) *Excavations at Segontium (Caernarfon) Roman Fort, 1975–1979*. York: Council for British Archaeology Research Report **90**, 82–96.

O'Connell, M. 1987 Early cereal-type pollen records from Connemara, western Ireland and their possible significance. *Pollen et Spores* **29**, 207–224.

O'Connor, T. P. 1991 Science, evidential archaeology and the new scholasticism. *Scottish Archaeological Review* **8**, 1–7.

O Còrrain, D. 1972 *Ireland before the Normans*. Dublin: Gill and Macmillan.

O'Nuallain, S. 1972 A Neolithic house at Ballyglass, County Mayo. *Journal of the Royal Society of Antiquaries of Ireland* **102**, 49–57.

Osborne, P. J. 1972 Insect faunas of Late Devensian and Flandrian Age from Church Stretton, Shropshire. *Philosophical Transactions of the Royal Society of London* **B263**, 327–367.

Osborne, P. J. 1974 An insect assemblage of Early Flandrian Age from Lea Marston, Warwickshire and its bearing on the contemporary climate and ecology. *Quaternary Research* **4**, 471–486.

Osborne, P. J. 1980 The Late Devensian–Flandrian transition depicted by serial insect faunas from West Bromwich, Staffordshire, England. *Boreas* **9**, 139–147.

O'Sullivan, P. E. 1976 Pollen analysis and radiocarbon dating of a core from Loch Pityoulish, eastern Highlands of Scotland. *Journal of Biogeography* **3**, 293–302.

Owen, O. A. 1992 Eildon Hill North. In Rideout, J. S., Owen, J. A. and Halpin, E. *Hillforts of Southern Scotland*. Edinburgh: AOC (Scotland) Monograph Series **1**, 21–71.

Owen, O. A. 1993 Tuquoy, Westray, Orkney: a challenge for the future? In Batey, C. E., Jesch, J. and Morris, C. D. (eds) *The Viking Age in Caithness, Orkney and the North Atlantic*. Edinburgh: Edinburgh University Press, 318–339.

Owen, O. A. and Dalland, M. forthcoming Scar, Sanday, Orkney: the rescue of a Viking boat burial. In Ambrosiani, B. and Clarke, H. (eds) Stockholm: *Proceedings of the Twelfth Viking Congress*.

Owen, J. A., Lyszkowski, R. M., Proctor, R. and Taylor, S. 1992 *Agabus wasastjernae* (Dytiscidae) Sahlberg new to Scotland. *The Coleopterist* **1** pt 2, 2–3.

Pálsson, H. and Edwards, P. 1978 *Egil's Saga*. Harmondsworth: Allen Lane.

Pálsson, H. and Edwards, P. 1981 *Orkneyinga Saga*. Harmondsworth: Allen Lane.

Parker Pearson, M. 1993 *English Heritage Book of Bronze Age Britain*. London: Batsford.

Parker Pearson, M. forthcoming *Between Land and Sea: Excavations at Dun Vulan, South Uist*. Sheffield: Sheffield Academic Press.

Parker Pearson, M. and Webster, J. 1994 *Bornish Mound 2: Viking Age Settlement, 1994 Excavations. Interim Report*. Sheffield: Department of Archaeology and Prehistory University of Sheffield.

Paterson, H. M. L. and Lacaille, A. D. 1936 Banchory microliths. *Proceedings of the Society of Antiquaries of Scotland* **70**, 419–434.

Peacock, J. D., Harkness, D. D., Housley, R. A., Little, J. A. and Paul, M. A. 1989 Radiocarbon ages for a glaciomarine bed associated with the maximum of the Loch Lomond Readvance in west Benderloch, Argyll. *Scottish Journal of Geology* **25**, 69–79.

Pears, N. V. 1968 Postglacial tree-lines of the Cairngorm Mountains, Scotland. *Transactions and Proceedings of the Botanical Society of Edinburgh* **40**, 361–394.

Pears, N. V. 1970 Postglacial tree-lines in the Cairngorm Mountains, Scotland: some modifications based on radiocarbon dating. *Transactions of the Botanical Society of Edinburgh* **40**, 536–544.

Peglar, S. M. 1993 The mid-Holocene *Ulmus* decline at Diss Mere, Norfolk, UK: a year-by-year pollen stratigraphy from annual laminations. *The Holocene* **3**, 1–13.

Peltenburg, E. J. 1982 Excavations at Balloch Hill, Argyll. *Proceedings of the Society of Antiquaries of Scotland* **112**, 142–214.

Pennington, W. 1974 *The History of British Vegetation*, 2nd edition. London: English Universities Press.

Pennington, W., Haworth, E. Y., Bonny, A. P. and Lishman, J. P. 1972 Lake sediments in northern Scotland. *Philosophical Transactions of the Royal Society of London* **B264**, 191–294.

Piggott, C. M. 1948 Excavations at Hownam Rings, Roxburghshire, 1948. *Proceedings of the Society of Antiquaries of Scotland* **82**, 45–67.

Piggott, C. M. 1953 Milton Loch crannog I: a native house of the 2nd century A.D. in Kirkcudbrightshire. *Proceedings of the Society of Antiquaries of Scotland* **87**, 134–152.

Piggott, S. 1954 *The Neolithic Cultures of the British Isles*. Cambridge: Cambridge University Press.

Piggott, S. 1956 Excavations in passage-graves and ring cairns of the Clava group, 1952–3. *Proceedings of the Society of Antiquaries of Scotland* **88**, 173–207.

Piggott, S. 1958 Native economies and the Roman occupation of North Britain. In Richmond I. A. (ed.) *Roman and Native in North Britain*. Edinburgh: Thomas Nelson and Sons, 1–27.

Piggott, S. (ed.) 1962 *The Prehistoric Peoples of Scotland*. London: Routledge and Kegan Paul.

Piggott, S. 1966 A scheme for the Scottish Iron Age. In Rivet, A. L. F. (ed.) *The Iron Age in Northern Britain*. Edinburgh: Edinburgh University Press, 1–15.

Piggott, S. 1972 Excavation of the Dalladies long barrow, Fettercairn, Kincardineshire. *Proceedings of the Society of Antiquaries of Scotland* **104**, 23–47.

Piggott, S. 1981 Early prehistory. In Piggott, S (ed.) *The Agrarian History of England and Wales*, Vol. 1. Cambridge: Cambridge University Press, 3–57.

Piggott, S. 1982 *Scotland before History*. Edinburgh: Edinburgh University Press. Revised edition; first published 1958.

Pitts, L. F. and St Joseph, J. K. 1985 *Inchtuthil. The Roman Legionary Fortress*. London: Society for the Promotion of Roman Studies, Britannia Monograph **6**.

Platt, M. I. 1934 Report on the animal remains. In Callander, J. G. (ed.) A long stalled chamber cairn or mausoleum near Midhowe, Rousay, Orkney. *Proceedings of the Society of Antiquaries of Scotland*, **68**, 348–350.

Platt, M. I. 1956 The animal bones. In Hamilton, J. R. C. (ed.) *Excavations at Jarlshof, Shetland*. Edinburgh: HMSO, 213–215.

Pollard, A. 1990 Down through the ages: a review of the Obanian cave deposits. *Scottish Archaeological Review* **7**, 58–74.

Pollard, T. and Humphreys, P. 1993 Bay of Sannick (Canisbay parish). Possible Mesolithic lithics. *Discovery and Excavation in Scotland 1993*, 42.

Pollard, T. and Morrison, A. (eds) 1996 *The Early Prehistory of Scotland*. Edinburgh: Edinburgh University Press.

Pons, A., Guiot, J., de Beaulieu, J. L. and Reille, M. 1992 Recent contributions to the climatology of the last glacial-interglacial cycle based on French pollen sequences. *Quaternary Science Reviews* **11**, 439–448.

Prestt, I., Cooke, A. S. and Corbett, K. F. 1974 British amphibians and reptiles. In Hawksworth, D. L. (ed.) *The Changing Flora and Fauna of Britain*. London: Academic Press, 229–254.

Price, R. J. 1983 *Scotland's Environment during the Last 30,000 Years*. Edinburgh: Scottish Academic Press.

Price, T. D. and Brown, J. A. (eds) 1985 *Prehistoric Hunter–Gatherers: The Emergence of Cultural Complexity*. London: Academic Press.

Proudfoot, E. V. W. (ed.) 1989 *Our Vanishing Heritage: Forestry and Archaeology*. Edinburgh: Council for Scottish Archaeology Occasional Paper **2**.

Proudfoot, E. V. W. 1995 Archaeology and early Christianity in Scotland. In Nicoll, E. H. (ed.) *A Pictish Panorama*. Balgavies, Angus: Pinkfoot Press, 27–30.

Pryor, F. M. M. 1991 *English Heritage Book of Flag Fen: Prehistoric Fenland Centre*. London: Batsford.

Rackham, D. J. 1989 Domestic and wild mammals. In Morris, C. D. (ed.) *The Birsay Bay Project* **1**. Durham: Department of Archaeology University of Durham Monograph **1**, 232–248.

Rackham, O. 1980 *Ancient Woodland*. London: Edward Arnold.

Raftery, B. 1990 *Trackways Through Time. Archaeological Investigations on Irish Bog Roads, 1985–1989*. Dublin: Headline Publishing.

Ralston, I. B. M. 1979 The Iron Age: Northern Britain. In Megaw, J. V. S. and Simpson, D. D. A. (eds) *Introduction to British Prehistory from the Arrival of* Homo Sapiens *to the Claudian Invasion*. Leicester: Leicester University Press, 446–501.

Ralston, I. B. M. 1980 *Sands of Forvie, Slains Parish, Grampian Region: Interim Report*. Aberdeen: Department of Geography University of Aberdeen.

Ralston, I. B. M. 1982 A timber hall at Balbridie Farm and the Neolithic in the North-East. *Aberdeen University Review* **168**, 238–249.

Ralston, I. B. M. 1984 Notes on the archaeology of Kincardine and Deeside District. *The Deeside Field*, **18**, 73–83.

Ralston, I. B. M. 1996 Recent work on the Iron Age settlement record in Scotland. In

Champion, T. C. and Collis, J. R. (eds), *The Iron Age in Britain and Ireland: Recent Trends.* Sheffield: Sheffield Academic Press, 133–153.

Ralston, I. B. M. and Inglis, J. C. 1984 *Foul Hordes: The Picts in the North-East and Their Background.* Aberdeen: University Anthropological Museum.

Ralston, I. B. M. and Smith, J. S. 1983 High altitude settlement on Ben Griam Beg, Sutherland. *Proceedings of the Society of Antiquaries of Scotland* **113**, 636–638.

RCAHMS (Royal Commission on the Ancient and Historical Monuments of Scotland) 1982 *Argyll 4: Iona.* Edinburgh: HMSO.

RCAHMS (Royal Commission on the Ancient and Historical Monuments of Scotland) 1984 *Argyll: An Inventory of the Monuments, 5: Islay, Jura, Colonsay and Oronsay.* Edinburgh: HMSO.

RCAHMS (Royal Commission on the Ancient and Historical Monuments of Scotland) 1988 *Argyll: An Inventory of the Monuments, 6: Mid Argyll and Cowal.* Edinburgh: HMSO.

RCAHMS (Royal Commission on the Ancient and Historical Monuments of Scotland) 1990 *North-East Perth: An Archaeological Landscape.* Edinburgh: HMSO.

RCAHMS (Royal Commission on the Ancient and Historical Monuments of Scotland) 1993a *Waternish, Skye and Lochalsh District, Highland Region. An Archaeological Survey.* Edinburgh: RCAHMS.

RCAHMS (Royal Commission on the Ancient and Historical Monuments of Scotland) 1993b *Strath of Kildonan. An Archaeological Survey.* Edinburgh: RCAHMS.

RCAHMS (Royal Commission on the Ancient and Historical Monuments of Scotland) 1994 *South-East Perth: An Archaeological Landscape.* Edinburgh: HMSO.

Reed, N. 1976 The Scottish campaigns of Septimius Severus. *Proceedings of the Society of Antiquaries of Scotland* **107**, 92–102.

Rees, S. E. 1979 *Agricultural Implements in Prehistoric and Roman Britain.* Oxford: British Archaeological Reports British Series **69** (2 vols).

Rees, S. E. 1981 Agricultural tools: function and use. In Mercer, R. J. (ed.) *Farming Practice in British Prehistory.* Edinburgh: Edinburgh University Press, 66–84.

Renfrew, A. C. 1979 *Investigations in Orkney.* London: Reports of the Research Committee of the Society Antiquaries of London **38**.

Renfrew, A. C. (ed.) 1985 *The Prehistory of Orkney BC 4000–1000 AD.* Edinburgh: Edinburgh University Press.

Renfrew, A. C. (ed.) 1990 *The Prehistory of Orkney BC 4000–1000 AD.* Edinburgh: Edinburgh University Press (revised edn).

Reynolds, D. M. 1982 Aspects of later prehistoric timber construction in south-east Scotland. In Harding, D. W. (ed.) *Later Prehistoric Settlement in South-East Scotland.* Edinburgh: Department of Archaeology University of Edinburgh Occasional Paper **8**, 44–56.

Reynolds, N. M. 1980 Dark Age timber halls and the background to excavations at Balbridie. *Scottish Archaeological Forum* **10**, 41–60.

Richards, C. 1992 Barnhouse and Maeshowe. *Current Archaeology* **131**, 444–448.

Richmond, I. A. and McIntyre, J. 1939 The Agricolan fort at Fendoch. *Proceedings of the Society of Antiquaries of Scotland* **73**, 110–154.

Rickman, G. 1980 *The Corn Supply of Ancient Rome.* Oxford: Oxford University Press.

Rideout, J. S. (with Owen, O. A.) 1992 Discussion. In Rideout, J. S., Owen, O. A., and Halpin, E. (eds) *Hillforts of Southern Scotland.* Edinburgh: AOC (Scotland) Ltd Monograph **1**, 139–144.

Ringrose, P. S. 1989 Recent fault movement and palaeoseismicity in western Scotland. *Tectonophysics* **163**, 305–315.

Ritchie, A. 1977 Excavation of Pictish and Viking-Age farmsteads at Buckquoy, Orkney. *Proceedings of the Society of Antiquaries of Scotland* **108**, 174–227.

Ritchie, A. 1983 Excavation of a Neolithic farmstead at Knap of Howar, Papa Westray, Orkney. *Proceedings of the Society of Antiquaries of Scotland* **113**, 40–121.

Ritchie, A. 1985 The first settlers. In A. C. Renfrew (ed.) *The Prehistory of Orkney BC 4000–1000 AD.* Edinburgh: Edinburgh University Press. 36–53.

Ritchie, A. 1993 *Viking Scotland.* London: Batsford/Historic Scotland.

Ritchie, A., Ritchie, J. N. G., Whittington, G. and Soulsby, J. 1974 A prehistoric field-boundary from the Black Crofts, North Connel, Argyll. *Glasgow Archaeological Journal* **3**, 66–70.

Ritchie, J. 1920 *The Influence of Man on Animal Life in Scotland.* Cambridge: Cambridge University Press.

Ritchie, J. N. G. 1981 Excavations at Machrins, Colonsay. *Proceedings of the Society of Antiquaries of Scotland* **111**, 263–281.

Ritchie, J. N. G. 1982 Archaeology and astronomy: an archaeological view. In Heggie, D. (ed.) *Archaeoastronomy in the Old World.* Cambridge: Cambridge University Press, 25–44.

Ritchie, J. N. G. 1990 Ritual monuments. In Renfrew, A. C. (ed.) *The Prehistory of Orkney BC 4000–1000 AD* (revised edition). Edinburgh: Edinburgh University Press, 118–130.

Ritchie, J. N. G. and Ritchie, A. 1981 *Scotland: Archaeology and Early History.* London: Thames and Hudson.

Ritchie, J. N. G. and Ritchie, A. 1991 *Scotland: Archaeology and Early History* (revised edition). Edinburgh: Edinburgh University Press.

Ritchie J. N. G. and Welfare H. 1983 Excavations at Ardnave, Islay. *Proceedings of the Society of Antiquaries of Scotland* **113**, 302–366.

Ritchie, W. 1979 Machair development and chronology in the Uists and adjacent islands. In Boyd, J. M. (ed.) *The Natural Environment of the Outer Hebrides.* Edinburgh: Royal Society of Edinburgh, 107–122. (Proceedings of the Royal Society of Edinburgh **77B**.)

Ritchie, W. 1985 Inter-tidal and sub-tidal organic deposits and sea level changes in the Uists, Outer Hebrides. *Scottish Journal of Geology* **21**, 171–176.

Rivet, A. L. F. (ed.) 1966 *The Iron Age in Northern Britain.* Edinburgh: Edinburgh University Press.

Rivet, A. L. F. and Smith, C. 1979 *The Place-names of Roman Britain.* London: Batsford.

Robertson, A. S. 1975 *Birrens (Blatobulgium).* Edinburgh: T. and A. Constable.

Robertson-Rintoul, M. S. E. 1986 A quantitative soil-stratigraphic approach to the correlation and dating of post-glacial river terraces in Glen Feshie, western Cairngorms. *Earth Surface Processes and Landforms* **11**, 605–617.

Robinson, D. E. 1987 Investigations into the Aukhorn peat mounds, Keiss, Caithness: pollen, plant macrofossil and charcoal analyses. *New Phytologist* **106**, 185–200.

Robinson, D. E. and Dickson, J. H. 1988 Vegetational history and land use: a radiocarbon-dated pollen diagram from Machrie Moor, Arran, Scotland. *New Phytologist* **109**, 223–251.

Roesdahl, E. 1992 *The Vikings.* Harmondsworth: Allen Lane.

Romans, J. C. C., Durno, S. E. and Robertson, L. 1973 A brown forest soil from Angus. *Journal of Soil Science* **24**, 125–128.

Romans, J. C. C. and Robertson, L. 1975 Soils and archaeology in Scotland in Evans, J. G., Limbrey, S. and Cleere, H. (eds) *The Effect of Man on the Landscape: The Highland Zone.* London: Council for British Archaeology Research Report **11**, 37–39.

Romans, J. C. C. and Robertson, L. 1983a An account of the soils at North Mains. In Barclay, G. J. Sites of the third millennium bc to the first millennium ad at North Mains, Strathallan, Perthshire, *Proceedings of the Society of Antiquaries of Scotland* **113**, 122–281.

Romans, J. C. C. and Robertson, L. 1983b The environment of North Britain: soils. In Chapman, J. C. and Mytum, H. C. (eds) *Settlement in North Britain. 1000 BC–AD 1000.* Oxford: British Archaeological Reports British Series **118**, 55–80.

Rose, J., Lowe, J. J. and Switsur, R. 1988 A radiocarbon date on plant detritus beneath till from the type area of the Loch Lomond Readvance. *Scottish Journal of Geology* **24**, 113–124.

Rowley-Conwy, P. 1985 The origin of agriculture in Denmark: a review of some theories. *Journal of Danish Archaeology* **4**, 188–195.

Ruddiman, W. F. and McIntyre, A. 1981 The North Atlantic Ocean during the last glaciation. *Palaeogeography, Palaeoecology, Palaeoclimatology* **35**, 145–214.

Ruggles, C. L. N. 1984 *Megalithic Astronomy: A New Archaeological and Statistical Study.* Oxford: British Archaeological Reports British Series **123**.

Russell, N. J., Bonsall, J. C. and Sutherland, D. G. 1995 The exploitation of marine molluscs in the mesolithic of western Scotland: evidence from Ulva Cave, Inner Hebrides. In

Fischer, A. (ed.) *Proceedings of the Man, Sea and the Mesolithic Conference, Horsholm.* Oxford: Oxbow Monograph **53**, 273–288.

Sadler, J. P. 1991 *Archaeological and Palaeoecological Implications of Palaeoentomological Studies in Orkney and Iceland.* Unpublished PhD thesis, University of Sheffield.

Samson, R. (ed.) 1992 *Social Approaches to Viking Studies.* Glasgow: Cruithne Press.

Samuelsson, G. 1910 Scottish peat mosses: a contribution to the knowledge of the Late-Quaternary vegetation and climate of northwestern Europe. *Bulletin of the Geological Institute, University of Uppsala* **10**, 197–260.

Saville, A. 1993 Bifaces of Lower Palaeolithic type from Scotland. *Lithics* 14, 1–7.

Saville, A. 1994a Exploitation of lithic resources for stone tools in earlier prehistoric Scotland. In Ashton, N. and David, A. (eds) *Stories in Stone.* London: Lithic Studies Society, 57–70.

Saville, A. 1994b *The Den of Boddam Project: Excavation and Survey on the Buchan Ridge Gravels, Grampian Region, in 1993.* Edinburgh: National Museums of Scotland.

Saville, A. 1996 Lacaille, microliths and the Mesolithic of Orkney. In Pollard, T. and Morrison, A. (eds) *The Early Prehistory of Scotland.* Edinburgh: Edinburgh University Press, 213–224.

Saville, A. and Hallén, Y. 1994 The 'Obanian Iron Age': human remains from the Oban cave sites, Argyll, Scotland. *Antiquity* **68**, 715–723.

Saville, A. and Miket, R. 1994 An Corran, Staffin, Skye. *Discovery and Excavation in Scotland 1994,* 40–41.

Sawyer, P. H. 1971 *The Age of the Vikings,* 2nd edition. London: Edward Arnold.

Scott, J. G. 1989 The hall and motte at Courthill, Dalry, Ayrshire. *Proceedings of the Society of Antiquaries of Scotland* **119**, 271–278.

Scott, J. G. 1992 Mortuary structures and megaliths. In Sharples, N. M. and Sheridan, A. (eds) *Vessels for the Ancestors.* Edinburgh: Edinburgh University Press, 104–119.

Scott, L. 1951 The colonisation of Scotland in the second millennium BC. *Proceedings of the Prehistoric Society* **17**, 16–82.

Scull, C. 1991 Post Roman Phase I at Yeavering: a re-construction. *Medieval Archaeology* **35**, 51–63.

Selkirk, A. 1992 Doughnuts and bananas, the Leuchars cropmark project. *Current Archaeology* **131**, 472–474.

Seller, T. J. 1986 Animal bone material. In Hunter, J. R. (ed.) *Rescue Excavations at the Brough of Birsay 1974–1982.* Edinburgh: Society of Antiquaries of Scotland Monograph Series **4**, 208–216.

Serebryanny, L. and Orlov, A. 1993 Debris in tills and moraines as a source of glaciological and palaeoenvironmental information; methodology and applications in the Caucasus. *The Holocene* **3**, 63–69.

Serjeantson, D. 1988 Archaeological and ethnographic evidence for seabird exploitation in Scotland. *Archaeozoologia* **2**, 209–288.

Serjeantson, D. 1990 The introduction of mammals to the Outer Hebrides and the role of boats in stock management. *Anthropozoologica,* **13**, 7–18.

Sharples, N. M. 1985 Individual and community: the changing role of megaliths in the Orcadian Neolithic. *Proceedings of the Prehistoric Society* **51**, 59–76.

Sharples, N. M. 1992a Aspects of regionalisation in the Scottish Neolithic. In Sharples, N. M. and Sheridan, A. (eds) *Vessels for the Ancestors.* Edinburgh: Edinburgh University Press, 322–331.

Sharples, N. M. 1992b Warfare in the Iron Age of Wessex. *Scottish Archaeological Review* **8**, 79–89.

Shepherd, I. A. G. 1976 Preliminary results from the beaker settlement at Rosinish, Benbecula. In Burgess, C. B. and Miket, R. (eds) *Settlement and Economy in the Third and Second Millennia BC.* Oxford: British Archaeological Report British Series **33**, 209–220.

Shepherd, I. A. G. 1981 The archaeology of the Rosinish machair. In Ranwell, D. S. (ed.) *Sand Dune Machair, 3.* Cambridge: Natural Environment Research Council, 24–29.

Shepherd, I. A. G. 1986 *Powerful Pots: Beakers in North-East Prehistory.* Aberdeen: Anthropological Museum, University of Aberdeen.

Shepherd, I. A. G. 1987 The early peoples. In Omand, D. (ed.) *The Grampian Book*. Golspie: The Northern Times, 119–130.

Shepherd, I. A. G. 1993 The Picts in Moray. In Sellar, W. D. H. (ed.) *Moray: Province and People*. Edinburgh: Scottish Society for Northern Studies, 75–90.

Shepherd, I. A. G. and Tuckwell, A. N. 1977 Traces of beaker-period cultivation at Rosinish, Benebecula. *Proceedings of the Society of Antiquaries of Scotland* **108**, 108–113.

Sheridan, A. 1992 Scottish stone axeheads: some new work and recent discoveries. In Sharples, N. M. and Sheridan, A. (eds) *Vessels for the Ancestors*. Edinburgh: Edinburgh University Press: 194–212.

Simmons, I. G. 1969 Evidence for vegetation changes associated with Mesolithic man in Britain. In Ucko, P. J. and Dimbleby, G. W. (eds) *The Domestication and Exploitation of Plants and Animals*. London, Duckworth, 111–119.

Simmons, I. G., Dimbleby, G. W. and Grigson, C. 1981 The Mesolithic. In Simmons, I. G. and Tooley, M. J. (eds) *The Environment in British Prehistory*. London: Duckworth, 82–124.

Simpson, D. D. A. 1976 The later Neolithic and Beaker settlement site at Northton, Isle of Harris. In Burgess, C. B. and Miket, R. (eds) *Settlement and Economy in the Third and Second Millennia BC*. Oxford: British Archaeological Reports British Series **33**, 221–231.

Sirks, B. 1991 *Food for Rome: The Legal Structure of the Transportation and Processing of Supplies for the Imperial Distributions in Rome and Constantinople*. Amsterdam: J. C. Gieben.

Sissons, J. B. 1967 *The Evolution of Scotland's Scenery*. Edinburgh: Oliver and Boyd.

Sissons, J. B. 1974 Late-glacial marine erosion in Scotland. *Boreas* **3**, 41–48.

Sissons, J. B. 1983 Shorelines and isostasy in Scotland. In Smith, D. E. and Dawson, A. G. (eds) *Shorelines and Isostasy*. London: Academic Press 209–226.

Sissons, J. B. and Brooks, C. L. 1971 Dating of early postglacial land and sea level changes in the Western Forth Valley. *Nature* **234**, 124–127.

Sloan, D. 1989 Shell and settlement: European implications of oyster exploitation. In Clutton-Brock, J. (ed.) *The Walking Larder: Patterns of Domestication, Pastoralism and Predation*. London: Unwin Hyman, 316–325.

Small, A. 1966 Excavations at Underhoull, Unst, Shetland. *Proceedings of the Society of Antiquaries of Scotland* **98**, 225–248.

Small, A. 1968 The distribution of settlement in Shetland and Faroe in Viking times. *Saga Book* **17**, 144–155.

Smith, A. G. 1970 The influence of Mesolithic and Neolithic man on British vegetation. In Walker, D. and West, R. G., (eds) *Studies in the Vegetational History of the British Isles*. London: Cambridge University Press, 81–96.

Smith, A. G. 1984 Newferry and the Boreal–Atlantic transition. *New Phytologist* **98**, 35–55.

Smith, A. G. 1985 Problems of inertia and threshold related to postglacial habitat changes. *Philosophical Transactions of the Royal Society of London* **B161**, 331–342.

Smith, A. G. and Cloutman, E. W. 1988 Reconstruction of vegetation history in three dimensions at Waun-Fignen-Felen, an upland site in South Wales. *Philosophical Transactions of the Royal Society of London* **B322**, 159–219.

Smith, A. G. (with Grigson, C., Hillman, G. and Tooley, M. J.) 1981 The Neolithic. In Simmons, I. G. and Tooley, M. J. (eds) *The Environment in British Prehistory*. London: Duckworth, 125–209.

Smith, A. N. 1993 Lothian: Ratho Quarry (Ratho parish). *Discovery and Excavation in Scotland 1993*, 59–61.

Smith, B. B. (ed.) 1994 *Howe. Four Millennia of Orkney Prehistory*. Edinburgh: Society of Antiquaries of Scotland Mongraph Series, **9**.

Smith, C. (with Hodgson, G. W. I., Armitage, P., Clutton-Brock, J., Dickson, C., Holden, A. and Smith, B. B.) 1994 Animal bone report. In Smith, B. B. (ed.) *Howe. Four Millennia of Orkney Prehistory*. Edinburgh: Society of Antiquaries of Scotland Monograph Series **9**, 139–153.

Smith, D. E., Firth, C. R., Turbayne, S. C. and Brooks, C.L. 1992 Holocene relative sea level changes and shoreline displacement in the Dornoch Firth area, Scotland. *Proceedings of the Geologists' Association* **103**, 237–257.

Smith, D. N. 1996 Thatch, turves and floor deposits: a survey of Coleoptera in materials from abandoned Hebridean blackhouses and the implications for their visibility in the archaeological record. *Journal of Archaeological Science* **23**, 161–174.

Smith, I. M. 1991 Sprouston, Roxburghshire: an early Anglian centre of the eastern Tweed basin. *Proceedings of the Society of Antiquaries of Scotland* **121**, 261–294.

Smyth, A. P. 1977 *Scandinavian Kings in the British Isles, 850–880*. Oxford: Oxford University Press.

Smyth, A. P. 1979 *Scandinavian York and Dublin II*. Dubin and New Jersey: Humanities Press.

Smyth, A. P. 1984 *Warlords and Holy Men. Scotland AD 80–1000*. London: Edward Arnold. (The New History of Scotland, **1**.)

Solem, T. 1986 Age, origin and development of blanket mires in Sor-Trondelag, Central Norway. *Boreas* **15**, 101–115.

Sommer, S. 1984 *The Military Vici of Roman Britain*. Oxford: British Archaeological Reports British Series **129**.

Soulsby, J. A. 1976 Palaeoenvironmental interpretation of a buried soil at Achnacree, Argyll. *Transactions of the Institute of British Geographers* **New Ser 1**, 279–283.

Spearman, M. R. 1988 Early Scottish towns; their origins and economy. In Driscoll, S. T. and Nieke, M. R. (eds) *Power and Politics in Early Medieval Britain and Ireland*. Edinburgh: Edinburgh University Press, 96–110.

Spearman, M. R. 1990 The Helmsdale Bowls: a re-assessment. *Proceedings of the Society of Antiquaries of Scotland* **120**, 63–77.

SSSES 1951 *Scientific Survey of South-Eastern Scotland*. Edinburgh: British Association.

Stevenson, A. C. and Birks, H. J. B. 1995 Heaths and moorland: long-term ecological changes, and interactions with climate and people. In Thompson, D. B. A., Hester, A. J. and Usher, M. B. (eds) *Heaths and Moorland: Cultural Landscapes*. Edinburgh: HMSO, 224–239.

Stevenson, J. B. 1975 Survival and discovery. In Evans, J. G., Limbrey, S. and Cleere, H. (eds) *The Effect of Man on the Landscape: The Highland Zone*. London: Council for British Archaeology Research Report **11**, 104–108.

Stevenson, J. B. 1984 The excavation of a hut circle at Cùl a'Bhaile, Jura. *Proceedings of the Society of Antiquaries of Scotland* **114**, 127–160.

Stevenson, J. B. 1991 Pitcarmicks and fermtouns. *Current Archaeology* **127**, 288–291.

Stewart, D. A., Walker, A. and Dickson, J. H. 1984 Pollen diagrams from Dubh Lochan, near Loch Lomond. *New Phytologist* **98**, 531–549.

Stewart, M. E. C. 1962 The excavation of two circular enclosures at Dalnaglar, Perthshire. *Proceedings of the Society of Antiquaries of Scotland* **95**, 134–158.

Stoneman, R. E., Barber, K. E. and Maddy, D. 1993 Present and past ecology of *Sphagnum imbricatum* and its significance in raised peat – climate modelling. *Quaternary Newsletter* **70**, 14–22.

Stuart, A. J. 1982 *Pleistocene Vertebrates in the British Isles*. London: Longman.

Stuart, A. J. 1995 Insularity and Quaternary vertebrate faunas in Britain and Ireland. In Preece, R. C. (ed.) *Island Britain: A Quaternary Perspective*. London: Geological Society of London Special Publication **96**, 111–126.

Stuart, J. 1822 Observations upon the various accounts of the progress of the Roman arms in Scotland, and of the scene of the great battle between Agricola and Galgacus. *Transactions of the Society of Antiquaries of Scotland* **2**, 289–313.

Stuiver, M. and Reimer, P. J. 1993 Extended ^{14}C data base and revised CALIB 3.0 ^{14}C age calibration program. *Radiocarbon* **35**, 215–230.

Sturlodottir, S. A. and Turner, J. 1985 The elm decline at Pawlaw mire: an anthropogenic interpretation. *New Phytologist* **28**, 1022–1030.

Sugden, D. E. 1968 The selectivity of glacial erosion in the Cairngorm Mountains, Scotland. *Transactions of the Institute of British Geographers* **45**, 79–92.

Sutcliffe, A. J. and Kowalski, K. 1976 Pleistocene vertebrates of the British Isles. *Bulletin of the British Museum Natural History (Geology)* **27**, 33–147.

Sutherland, D. G. 1984 The Quaternary deposits and landforms of Scotland and the neighbouring shelves: a review. *Quaternary Science Reviews* **3**, 157–254.

Sutherland, D. G. 1991 Late Devensian glacial deposits and glaciation in Scotland and the adjacent offshore region. In J. Ehlers, J., Gibbard P. L. and Rose, J. (eds) *Glacial Deposits in Great Britain and Ireland.* Rotterdam: A. A. Balkema, 53–60.

Sveinbjarnardóttir, G. 1991 Shielings in Iceland. An archaeological and historical survey. *Acta Archaeologica* **61**, 73–96.

Switsur, V. R. and Mellars, P. A. 1987 Radio-carbon dating of the shell middens. In Mellars, P. A. (ed.) *Excavations on Oronsay.* Edinburgh: Edinburgh University Press, 139–152.

Tallantire, P. A. 1992 The alder [*Alnus glutinosa* (L.) Gaertn.] problem in the British Isles: a third approach to its palaeohistory. *New Phytologist* **122**, 717–731.

Taute, W. 1968 *Die Stielspitzen-Gruppen im nördlichen Mitteleuropa: ein Beitrag zur Kenntnis der späten Altsteinzeit.* Fundamenta Reihe A, Band **5**. Cologne: Bölau.

Taylor, D. B. 1990 *Circular Homesteads in North-West Perthshire.* Dundee: Abertay Historical Society Publications, **29**.

Ten Hove, H. A. 1968 The *Ulmus* fall at the transition Atlanticum-Subboreal. *Palaeogeography, Palaeoclimatology, Palaeoecology* **5**, 359–369.

Terry, J. 1991 Lintshie Gutter: unenclosed platform settlement. In Terry, J. and Banks, I. (eds) *Lintshie Gutter: Unenclosed Platform Settlement. Stoneyburn: Burial Cairns.* Glasgow: Department of Archaeology University of Glasgow. (Archaeology Projects Glasgow, Report **22**.)

Thom, A. 1967 *Megalithic Sites in Britain.* Oxford: Oxford University Press.

Thom, A. 1971 *Megalithic Lunar Observatories.* Oxford: Oxford University Press.

Thomas, A. C. 1967 An Early Christian cemetery and chapel at Ardwall Island, Kirkcudbright. *Medieval Archaeology* **11**, 127–188.

Thomas, A. C. 1981a *Christianity in Roman Britain to A.D. 500.* London: Batsford.

Thomas, A. C. 1981b *A Provisional List of Imported Pottery in Post-Roman Western Britain and Ireland.* Redruth: Institute of Cornish Studies Special Report **7**.

Thomas, A. C. 1990 *Gallici nautae de Galliarum provincias*: a 6th–7th century trade with Gaul, reconsidered. *Medieval Archaeology* **34**, 1–26.

Thomas, G. D. 1988 Excavations at the Roman civil settlement at Inveresk, 1976–77. *Proceedings of Society of Antiquaries of Scotland* **118**, 139–176.

Thomas, J. 1987 Relations of production and social change in the Neolithic of north-west Europe. *Man* **22**, 405–430.

Thomas, J. 1988 Neolithic explanations revisited: the Mesolithic-Neolithic transition in Britain and South Scandinavia. *Proceedings of the Prehistoric Society* **54**, 59–66.

Thomas, J. 1990 Silent running: the ills of environmental archaeology. *Scottish Archaeological Review* **7**, 2–7.

Thomas, J. 1991 *Rethinking the Neolithic.* Cambridge: Cambridge University Press.

Thompson, R. Battarbee, R. W., O'Sullivan, P. E. and Oldfield, F. 1975 Magnetic susceptibility of lake sediments. *Limnology and Oceanography* **20**, 687–698.

Thoms, L. (ed.) 1980 *Settlements in Scotland 1000 BC–AD 1000.* Edinburgh: Edinburgh University Press. (Scottish Archaeological Forum **10**.)

Tipping, R. 1991 Climatic change in Scotland during the Devensian Late Glacial: the palynological record. In Barton, N., Roberts, A. J. and Roe, D. A. (eds) *The Late Glacial in North-West Europe: Human Adaptation and Environmental Change at the End of the Pleistocene.* London: Council for British Archaeology Research Report **77**, 7–21.

Tipping, R. 1992 The determination of cause in the generation of major prehistoric valley fills in the Cheviot Hills, Anglo-Scottish border. In Needham, S. and Macklin, M. G. (eds) *Alluvial Archaeology in Britain*, Oxford: Oxbow Monographs **27**, 111–121.

Tipping, R. 1994a The form and fate of Scotland's woodlands. *Proceedings of the Society of Antiquaries of Scotland* **124**, 1–54.

Tipping, R. 1994b 'Ritual' floral tributes in the Scottish Bronze Age – palynological evidence. *Journal of Archaeological Science* **21**, 133–139.

Tipping, R. 1994c Fluvial chronology and valley floor evolution of the upper Bowmont Valley, Borders Region, Scotland. *Earth Surface Processes and Landforms* **19**, 641–657.

Tipping, R. 1995a Holocene evolution of a lowland Scottish landscape: Kirkpatrick Fleming. Part II, regional vegetation and land-use change. *The Holocene* **5**, 83–96.

Tipping, R. 1995b Holocene landscape change at Carn Dubh, near Pitlochry, Perthshire, Scotland. *Journal of Quaternary Science* **10**, 59–75.

Tipping, R. 1996 Microscopic charcoal records, inferred human activity and climate change in the Mesolithic of northernmost Scotland. In Pollard, T. and Morrison, A. (eds) *The Early Prehistory of Scotland*. Edinburgh: Edinburgh University Press, 39–61.

Tipping, R., Carter, S. and Johnston, D. 1994 Soil pollen and soil micromorphological analyses of old ground surfaces on Biggar Common, Borders Region, Scotland. *Journal of Archaeological Science* **21**, 387–401.

Tipping, R., Edmonds, M. and Sheridan, A. 1993 Palaeoenvironmental investigations directly associated with a neolithic axe 'quarry' on Beinn Lawers, near Killin, Perthshire, Scotland. *New Phytologist* **123**, 585–597.

Tipping, R. and Halliday, S. P. 1994 The age of alluvial fan deposition at a site in the Southern Uplands of Scotland. *Earth Surface Processes and Landforms* **19**, 333–348.

Topping, P. 1989 Early cultivation in Northumberland and the Borders. *Proceedings of the Prehistoric Society* **55**, 161–179.

Triscott, J. 1982 Excavations at Dryburn Bridge, East Lothian. In Harding, D. W. (ed.) *Later Prehistoric Settlement in South-East Scotland.* Edinburgh: Department of Archaeology University of Edinburgh Occasional Paper **8,** 117–124.

Turner, J. 1965 A contribution to the history of forest clearance. *Proceedings of the Royal Society* **B161**, 343–354.

Turner, J. 1975 The evidence for land use by prehistoric farming communities: the use of three-dimensional pollen diagrams. In Evans, J. G., Limbrey, S. and Cleere, H. (eds) *The Effect of Man on the Landscape: The Highland Zone*, London: Council for British Archaeology, Research Report **11**, 86–95.

Turner, J. 1981 The Iron Age. In Simmons, I. G. and Tooley, M. J. (eds) *The Environment in British Prehistory*. London: Duckworth, 250–281.

Turner, J. 1983 Some pollen evidence for the environment of northern Britain 1000 B.C. to A.D. 1000. In Chapman, J. C., Mytum, H. C. (eds) *Settlement in North Britain 1000 BC–AD 1000*. Oxford: British Archaeological Reports British Series **118**, 3–27.

Turner, W. 1872 On human and animal bones and flints from a cave at Oban, Argyleshire. Edinburgh: *Report of the British Association 1871*, 160–161.

Turner, W. 1895 On human and animal remains found in caves at Oban, Argyllshire. *Proceedings of the Society of Antiquaries of Scotland*, **29**, 41–438.

van der Veen, M. 1992 *Crop Husbandry Regimes: An Archaeobotanical Study of Farming in Northern England 1000BC–AD500*. Sheffield: Sheffield Archaeological Monographs **3**.

Vasari, Y. and Vasari, A. 1968 Late- and Post-glacial macrophytic vegetation in the lochs of northern Scotland. *Acta Botanica Fennica* **80**, 1–20.

Vermeersch, P. M. and van Peer, P. (eds) 1990 *Contributions to the Mesolithic in Europe*. Leuven: Leuven University Press.

Wainwright, F. T. (ed.) 1962 *The Northern Isles*. Edinburgh: Thomas Nelson and Sons.

Wainwright, G. J. 1969 A review of henge monuments in the light of recent research. *Proceedings of the Prehistoric Society* **35**, 112–133.

Walker, M. J. C. 1984a Pollen analysis and Quaternary research in Scotland. *Quaternary Science Reviews* **3**, 369–404.

Walker, M. J. C. 1984b A pollen diagram from St Kilda, Outer Hebrides, Scotland. *New Phytologist* **97**, 99–113.

Walker, M. J. C. and Lowe, J. J. 1977 Postglacial environmental history of Rannoch Moor, Scotland. I. Three pollen diagrams from the Kingshouse area. *Journal of Biogeography* **4**, 333–351.

Walker, R. E. 1973 Roman veterinary medicine. In Toynbee, J. M. C. (ed.) *Animals in Roman Life and Art*. London: Thames and Hudson, 301–343, 404–414.

Walsh, M. 1992 *Palaeoenvironmental Investigations at Logie, Fife.* Unpublished MSc thesis, University of St Andrews.

Wasylikowa, K. 1986 Plant macrofossils preserved in prehistoric settlements compared with anthropogenic indicators in pollen diagrams. In Behre, K.-E. (ed.) *Anthropogenic Indicators in Pollen Diagrams.* Rotterdam: A. A. Balkema, 173–185.

Watkins, T. F. 1980 Excavation of a settlement and souterrain at Newmill, near Bankfoot, Perthshire. *Proceedings of the Society of Antiquaires of Scotland* **110**, 165–208.

Watkins, T. F. and Shepherd, I. A. G. 1980 A beaker burial at Newmill, near Bankfoot, Perthshire. *Proceedings of the Society of Antiquaries of Scotland* **110**, 32–43.

Watts, W. A. 1988 Europe. In Huntley, B. and Webb, T. III. (eds) *Vegetation History.* Dordrecht: Kluwer Academic Publishers, 155–192.

Webb III, T. 1986 Is vegetation in equilibrium with climate? *Vegetatio* **67**, 75–91.

Weber, B. 1994 Iron Age combs: analyses of raw material. In Ambrosiani, B. and Clarke, H. (eds) *Developments Around the Baltic and the North Sea in the Viking Age.* Stockholm: Birka Studies **3**, 190–194.

Welander, R. D. E., Batey, C. and Cowie, T. G. 1987 A Viking burial from Kneep, Uig, Isle of Lewis. *Proceedings of the Society of Antiquaries of Scotland* **117**, 149–174.

Welfare, H. and Swan, V. 1995 *Roman Camps in England: The Field Archaeology.* London: RCHM.

Wheeler, A. 1974 Changes in the freshwater fish fauna of Britain. In Hawksworth D. L. (ed.) *The Changing Flora and Fauna of Britain.* London: Academic Press, 157–178.

Wheeler, A. 1977 The fish bones from Buckquoy, Orkney. In Ritchie, A. (ed.) Excavation of Pictish and Viking-Age farmsteads at Buckquoy, Orkney. *Proceedings of the Society of Antiquaries of Scotland* **108**, 211–214.

Wheeler, A. 1979 The fish bones. In Renfrew, A. C. (ed.) *Investigations in Orkney.* London: Society of Antiquaries of London Research Report **38**, 144–149.

Wheeler, A. 1983 Appendix 6: Fish remains from Knap of Howar, Orkney. In Ritchie, A. (ed.) Excavation of a Neolithic farmstead at Knap of Howar, Papa Westray, Orkney. *Proceedings of the Society of Antiquaries of Scotland*, **113**, 103–105.

Whitaker, I. 1986 The survival of feral reindeer in northern Scotland. *Archives of Natural History* **13**, 11–18.

Whittington, G. 1975 Placenames and the settlement pattern of dark-age Scotland. *Proceedings of the Society of Antiquaries of Scotland*, **106**, 99–110.

Whittington, G. 1978 A sub-peat dyke at Shurton Hill, Mainland, Shetland. *Proceedings of the Society of Antiquaries of Scotland* **109**, 30–35.

Whittington, G. 1979 The archaeologist and the environment. *Scottish Archaeological Forum* **9**, 82–85.

Whittington, G. 1983 A palynological investigation of a second millennium BC bank-system in the Black Moss of Achnacree. *Journal of Archaeological Science* **10**, 283–291.

Whittington, G. 1984 Report on the pollen analysis. In Stevenson, J. B. (ed.) The excavation of a hut circle at Cùl a'Bhaile, Jura. *Proceedings of the Society of Antiquaries of Scotland* **114**, 147–157.

Whittington, G. 1993 Palynological investigations at two Bronze Age burial sites in Fife. *Proceedings of the Society of Antiquaries of Scotland* **123**, 211–213.

Whittington, G. and Edwards, K. J. 1993 *Ubi solitudinem faciunt pacem appellant*: the Romans in Scotland, a palaeoenvironmental contribution. *Britannia* **24**, 13–25.

Whittington, G. and Edwards, K. J. 1994 Palynology as a predictive tool in archaeology. *Proceedings of the Society of Antiquaries of Scotland* **124**, 55–65.

Whittington, G. and Edwards, K. J. 1995 A Scottish broad: historical, stratigraphic and numerical studies associated with polleniferous deposits at Kilconquhar Loch. In Butlin, R. and Roberts, N. (eds) *Human Impact and Adaptation: Ecological Relations in Historical Times.* Oxford: Blackwell, 68–87.

Whittington, G., Edwards, K. J. and Cundill, P. R. 1990 *Palaeoenvironmental Investigations at Black Loch, in the Ochil Hills of Fife, Scotland.* Aberdeen: Department of Geography University of Aberdeen O'Dell Memorial Monograph **22**.

Whittington, G., Edwards, K. J. and Caseldine, C. J. 1991a Late- and post-glacial pollen-analytical and environmental data from a near-coastal site in north-east Fife, Scotland. *Review of Palaeobotany and Palynology*, **68**, 65–85.

Whittington, G., Edwards, K. J. and Cundill, P. R. 1991b Late- and post-glacial vegetational change at Black Loch, Fife, eastern Scotland – a multiple core approach. *New Phytologist* **118**, 147–166.

Whittington, G., Edwards, K. J. and Cundill, P. R. 1991c Palaeoecological investigations of multiple elm declines at a site in north Fife, Scotland. *Journal of Biogeography* **18**, 71–87.

Whittington, G., Fallick, A. E. and Edwards, K. J. 1996 Stable oxygen isotope and pollen records from eastern Scotland and a consideration of Lateglacial and early Holocene climate change for Europe. *Journal of Quaternary Science* **11**, 327–340.

Whittington, G. and McManus, J. forthcoming Dark Age agricultural practices and environmental change: evidence from Tentsuir Sands, Fife, eastern Scotland. In Mills, C. M. and Coles, G. (eds) *On the Edge: Settlement in Marginal Areas*.

Whittington, G. and Ritchie, W. 1988 *Flandrian Environmental Evolution on North-East Benbecula and Southern Grimsay, Outer Hebrides, Scotland*. Aberdeen: Department of Geography University of Aberdeen O'Dell Memorial Monograph **21**.

Whittle, A. 1978 Resources and population in the British Neolithic. *Antiquity* **52**, 34–41.

Whittle, A., Keith-Lucas, M., Milles, A., Noddle, B., Rees, S. and Romans, J. 1986 *Scord of Brouster. An Early Agricultural Settlement on Shetland*. Oxford: Oxford University Committee for Archaeology Monograph **9**.

Wickham-Jones, C. R. 1986 The procurement and use of stone for flaked tools in prehistoric Scotland. *Proceedings of the Society of Antiquaries of Scotland* **116**, 1–10.

Wickham-Jones, C. R. 1990 *Rhum: Mesolithic and Later Sites at Kinloch. Excavations 1984–1986*. Edinburgh: Society of Antiquaries of Scotland Monograph Series **7**.

Wickham-Jones, C. R. 1994 *Scotland's First Settlers*. London: Batsford.

Wickham-Jones, C. R. and Collins, G. H. 1978 The sources of flint and chert in northern Britain. *Proceedings of the Society of Antiquaries of Scotland* **109**, 7–21.

Wickham-Jones, C. R. and Firth, C. 1990 Mesolithic survey. *Discovery and Excavation in Scotland 1990*, 22.

Wickham-Jones, C. R. and Macinnes, L. J. (eds) 1992 *All Natural Things. Archaeology and the Green Debate*. Oxford: Oxbow Monograph Series **21**.

Wilkins, D. A. 1984 The Flandrian woods of Lewis (Scotland). *Journal of Ecology* **72**, 251–258.

Williams Thorpe, O. and Thorpe, R. S. 1984 The distribution and sources of archaeological pitchstone in Britain. *Journal of Archaeological Science* **11**, 1–34.

Wilson, B. 1995 On the curious distortions behind the charge of scientism against environmental archaeology. *Scottish Archaeological Review* **9/10**, 67–70.

Wilson, D. 1851 *The Archaeology and Prehistoric Annals of Scotland*. Edinburgh: Sutherland and Knox.

Wilson, D. M. 1976 The Scandinavians in England. In Wilson, D. M. (ed.) *The Archaeology of Anglo-Saxon England*. Cambridge: Cambridge University Press, 393–403.

Woodman, P. C. 1988 Comment on Myers. *Scottish Archaeological Review* **5**, 34–35.

Woodman, P. C. 1989 A review of the Scottish Mesolithic: a plea for normality. *Proceedings of the Society of Antiquaries of Scotland* **119**, 1–32.

Woodman, P. C. and Monaghan, N. 1993 From mice to mammoths; dating Ireland's earliest faunas. *Archaeology Ireland* **7**, 31–33.

Wordsworth, J. 1985 The excavation of a Mesolithic horizon at 13–24 Castle Street, Inverness. *Proceedings of the Society of Antiquaries of Scotland* **115**, 89–103.

Wymer, J. J. 1988 Palaeolithic archaeology and the British Quaternary sequence. *Quaternary Science Reviews* **7**, 79–97.

Yalden, D. W. 1982 When did the mammal fauna of the British Isles arrive? *Mammal Review* **12**, 157.

Zutter, C. M. 1992 Icelandic plant and land-use patterns: archaeobotanical analysis of the Svalbar midden (6706–60), Northeastern Iceland. In Morris, C. D. and Rackham, D. J. (eds) *Norse and Later Settlement and Subsistence in the North Atlantic*. Glasgow: Department of Archaeology University of Glasgow Occasional Paper Series **1**, 139–148.

Zvelebil, M. (ed.) 1986 *Hunters in Transition*. Cambridge: Cambridge University Press.

Zvelebil, M. 1989a Economic intensification and Postglacial hunter-gatherers in north temperate Europe. In Bonsall, J. C. (ed.) *The Mesolithic in Europe*. Edinburgh: John Donald, 80–88.

Zvelebil, M. 1989b On the transition to farming in Europe, or what was spreading with the Neolithic: a reply to Ammerman (1989). *Antiquity* **63**, 379–383.

Zvelebil, M. 1992 Fear of flying, or how to save your own paradigm. *Antiquity* **66**, 811–814.

Zvelebil, M. 1994 Plant use in the Mesolithic and its role in the transition to farming. *Proceedings of the Prehistoric Society* **60**, 35–74.

Zvelebil, M. and Rowley-Conwy, P. A. 1986 Foragers and farmers in Atlantic Europe. In Zvelebil, M. (ed.) *Hunters in Transition*. Cambridge: Cambridge University Press, 167–188.

A Guide to the Literature since 1996

The following selection of the literature that has been published since 1996 is presented in the main by subject, according to our chapter headings. Brief comments are added when the contents are not self-evident from the titles. Where contributions have a particular relevance to other chapters, we have also indicated this, as far as possible, by annotation. Key publications of multi-period sites are grouped separately. A further section includes entries consisting of important regional studies of various types. We start by listing some general surveys, including edited collections pertaining to several topics, either uniquely of Scottish material, or where that material is considered in wider focus. We end with a brief consideration of newer developments, including matters such as genetic research not considered in the original publication. Throughout, articles within these publications are normally not listed separately.

During the 1990s, two general series on Scottish archaeology were published by Historic Scotland. Later volumes in the Batsford series (the first of which appeared in 1994) are signalled below. A differently-targeted set of eight well-illustrated booklets, under the general editorship of Gordon Barclay, appeared in 1998–1999.

GENERAL

Armit, I. 1998 *Scotland's Hidden History*. Stroud: Tempus.
Barclay, G. J. (ed.) 1998 Finlayson, B. *Wild Harvesters*; Barclay, G. *Farmers, Temples and Tombs*; Hingley, R. *Settlement and Sacrifice*; Maxwell, G. S. *A Gathering of Eagles*. Edinburgh: Historic Scotland/Canongate/Birlinn.
Prehistoric and Roman.
Barclay, G. J. (ed.) 1999 Carver, M. *Surviving in Symbols*; Campbell, E. *Saints and Sea-kings*; Lowe, C. *Angels, Fools and Tyrants*; Owen, O. *The Sea Road*. Edinburgh: Historic Scotland/Canongate/Birlinn.
Early Historic to the Norse.
Coles, B. J. 1998 Doggerland: a speculative survey. *Proceedings of the Prehistoric Society* **64**, 45–81.
Changing configurations of land and sea-level between Britain and Continental Europe have implications for Palaeolithic and Mesolithic colonization.
Charman, D. 2002 *Peatlands and Environmental Change*. Chichester: Wiley.
Wide-ranging text with many Scottish examples.
Edwards, K. J. and Sadler, J. P. (eds) 1999 *Holocene Environments of Prehistoric Britain*. = *Journal of Quaternary Science* **14** (*Quaternary Proceedings* **7**). Chichester: John Wiley.
Environmental and archaeological contributions on a wide range of topics.
Frodsham, P., Topping, P. and Cowley, D. (eds) 2000 *'We were always chasing time': Papers Presented to Keith Blood*. = *Northern Archaeology* **18/19**.
A range of papers, many based on survey.

Hunter, J. R. and Ralston, I. (eds) 1999 *The Archaeology of Britain: an Introduction from the Upper Palaeolithic to the Industrial Revolution*. London: Routledge.

Thomas, J. 1999 *Understanding the Neolithic*. London: Routledge.
  Revised edition of *Rethinking the Neolithic*, taking greater account of regional variation.

Ruggles, C. 1999 *Astronomy in Prehistoric Britain and Ireland*. London: Yale University Press.
  Archaeoastronomy, but also best practice in methodology, considered in detail.

Wickham-Jones, C. 2001 *The Landscape of Scotland: a Hidden History*. Stroud: Tempus.

REGIONAL STUDIES

Branigan, K. and Foster, P. (eds) 2000 *Barra to Berneray: Archaeological Survey and Excavation in the Southern Isles of the Outer Hebrides*. Sheffield: Sheffield Academic Press.
  From the Neolithic to the recent past.

Merritt, J. W., Connell, E. R. and Bridgland, D. R. (eds) 2000 *The Quaternary of the Banffshire Coast & Buchan. Field Guide*. London: Quaternary Research Association.
  Short accounts of current archaeological and environmental projects.

Ritchie, J. N. G. (ed.) 1997 *The Archaeology of Argyll*. Edinburgh: Edinburgh University Press.
  Papers take stock after full RCAHMS inventory of the county.

RCAHMS (Royal Commission on the Ancient and Historical Monuments of Scotland) 1997 *Eastern Dumfriesshire*. Edinburgh: HMSO.

Tipping, R. M. (ed.) 1999 *The Quaternary of Dumfries and Galloway. Field Guide*. London: Quaternary Research Association.
  As Merritt *et al*. 2002 *supra*.

Turner, V. 1998 *Ancient Shetland*. London: Batsford/Historic Scotland.

MULTI-PERIOD SITES/CULTURAL LANDSCAPES

Barber, J. (ed.) 1997 *The Archaeological Investigation of a Prehistoric Landscape: Excavations on Arran 1978–81*. Edinburgh: Scottish Trust for Archaeological Research Monograph **2**.

McCullagh, R. P. J. and Tipping, R. (eds) 1998 *The Lairg Project 1988–1996. The Evolution of an Archaeological Landscape in Northern Scotland*. Edinburgh: Scottish Trust for Archaeological Research Monograph **3**.
  Study of later prehistory, strong on environment and landscape.

Owen, O. and Lowe, C. 1999 *Kebister: the four-thousand-year-old Story of one Shetland Township*. Edinburgh: Society of Antiquaries of Scotland Monograph Series **14**.
  A comprehensive archaeological study of the site and its wider environmental context.

Speak, S. and Burgess, C. 1999 Meldon Bridge: a centre of the third millennium BC in Peeblesshire. *Proceedings of the Society of Antiquaries of Scotland* **129**, 1–118.

2: CLIMATE CHANGE

Anderson, D. E. 1998 A reconstruction of Holocene climatic changes from peat bogs in northwest Scotland. *Boreas* **27**, 208–224.
  Climate change inferred from peat characteristics.

Anderson, D. E., Binney, H. A. and Smith, M. A. 1998 Evidence for abrupt climatic change in northern Scotland between 3900 and 3500 calendar years BP. *The Holocene* **8**, 97–103.

Chambers, F. M., Barber, K. E., Maddy, D. and Brew, J. 1997 A 5500-year proxy-climate and vegetation record from blanket mire at Talla Moss, Borders, Scotland. *The Holocene* **7**, 391–399.
  Climatic signals detected by advanced statistical analyses of palaeoecological records.

Huntley, B. 1999 Climatic change and reconstruction. *Journal of Quaternary Science* **14**, 513–520. (*Quaternary Proceedings* **7**.)
Review of Holocene climate change, including spatial and temporal patterns, simulations and biotic responses.

Lowe, J. J., Birks, H. H., Brooks, S. J., Coope, G. R., Harkness, D. D., Mayle, F. E., Sheldrick, C., Turney, C. S. M. and Walker, M. J. C. 1999 The chronology of palaeoenvironmental changes during the Last Glacial-Holocene transition: towards an event stratigraphy for the British Isles. *Journal of the Geological Society of London* **156**, 397–410.
Multi-proxy examination of chronological parallelism of climatic, biotic and sedimentary indicators.

Oliver, M. A., Webster, R., Edwards, K. J. and Whittington, G. 1997 Multivariate, autocorrelation and spectral analysis of a pollen profile from Scotland and evidence of periodicity. *Review of Palaeobotany and Palynology* **96**, 121–144.
As Chambers *et al.* 1997 *supra*.

3: GEOMORPHOLOGY AND LANDSCAPE CHANGE

Ashmore, P., Brayshay, B. A., Edwards, K. J., Gilbertson, D. D., Grattan, J. P., Kent, M., Pratt, K. E. and Weaver, R. E. 2000 Allochthonous and autochthonous mire deposits, slope instability and palaeoenvironmental investigations in the Borve Valley, Barra, Outer Hebrides, Scotland. *The Holocene* **10**, 97–108.
Age-depth anomalies reflect episodic impacts of eroded hillslope materials (*c.* 3000 to 1750 BP).

Ballantyne, C. K. and Whittington, G. 1999 Late Holocene floodplain incision and alluvial fan formation in the Central Grampian Highlands, Scotland: chronology, environment and implications. *Journal of Quaternary Science* **14**, 651–671.
Environmental changes are attributed to a few extreme climate events rather than direct human interference or long-term climatic change.

Ballantyne, C. K., McCarroll, D., Nesje, A., Dahl, S. O. and Stone, J. O. 1998 The last ice sheet in North-West Scotland: reconstruction and implications. *Quaternary Science Reviews* **17**, 1149–1184.
Three-dimensional reconstruction of ice cover at the last glacial maximum, extending to the Hebrides.

Curry, A. M. 2000 Holocene reworking of drift-mantled hillslopes in the Scottish Highlands. *Journal of Quaternary Science* **15**, 529–541.
Repeated episodes of hillslope instability over 6000 years are primarily due to extreme climate events (rainstorms), with limited evidence of anthropogenic activity.

McEwen, L. J. 1997 Geomorphological change and fluvial landscape evolution during the Holocene. In Gordon, J. E. (ed.), *Reflections on the Ice Age in Scotland,* 116–129. Glasgow: Scottish Association of Geography Teachers and Scottish Natural Heritage.
Synthesis of changes in fluvial systems, with reference to climate, extreme climatic events, deforestation and landuse.

Smith, D. E., Cullingford, R. A. and Firth, C. R. 2000 Patterns of isostatic land uplift during the Holocene: evidence from mainland Scotland. *The Holocene* **10**, 87–103.
Synthesis: also considers sea-level change and the ages of raised shoreline and estuarine deposits.

Smith, D. E., Firth, C. R., Brooks, C. L., Robinson, M. and Collins, P. E. F. 1999 Relative sea-level rise during the Main Postglacial Transgression in NE Scotland, UK. *Transactions of the Royal Society of Edinburgh: Earth Sciences* **90**, 1–27.
Indicates the longevity and diachroneity of Holocene sea-level changes; deposits are ascribed to the Second Storegga Slide tsunami.

de la Vega Leinart, A. C., Keen, D. H., Jones, R. L., Wells, J. M. and Smith, D. E. 2000 Mid-Holocene environmental changes in the Bay of Skaill, Mainland Orkney, Scotland: an

integrated geomorphological, sedimentological and stratigraphic study. *Journal of Quaternary Science* **15,** 509–528.
Coastal change and dune development indicates the influence of nearby Neolithic activity (Skara Brae): environmental effects on prehistoric agriculture are noted.

4: SOILS AND THEIR EVOLUTION

Davidson, D. and Smout, C. 1996 Soil change in Scotland: the legacy of past land improvement processes. In Taylor, A. G., Gordon, J. E. and Usher, M. B (eds) *Soil Sustainability in Scotland,* 44–54. Edinburgh: HMSO.
Stresses the positive aspects of soil improvement through manuring, stone clearance and drainage.
Edwards, K. J. and Whittington, G. 2001 Lake sediments, erosion and landscape change during the Holocene in Britain and Ireland. *Catena* **42,** 143–173.
Accelerated sediment accumulation can be related to anthropogenicically-related soil erosion, especially from the Neolithic onwards.
Gilbertson, D. D., Schwenninger, J.-L., Kemp, R. A. and Rhodes, E. J. 1999 Sand-drift and soil formation along an exposed North Atlantic coastline: 14,000 years of diverse geomorphological, climatic and human impacts. *Journal of Archaeological Science* **26,** 439–469.
A chronological exploration of natural and human influences on dune and soil formation.
Simpson, I. A., Dockrill, S. J., Bull, I. D. and Evershed, R. P. 1998 Early anthropogenic soil formation at Tofts Ness, Sanday, Orkney. *Journal of Archaeological Science* **25,** 729–746.
A range of techniques address soil enhancement in the Bronze Age. Manuring allowed arable activity in a highly marginal environment.

5: VEGETATION CHANGE

Many archaeobotanical accounts are to be found within excavation monographs and appendices to articles.

Dickson, C. and Dickson, J. H. 2000 *Plants & People in Ancient Scotland.* Stroud: Tempus.
An invaluable synthesis on plant remains from habitation sites of all periods.
Edwards, K. J., Mulder, Y., Lomax, T. A., Whittington, G. and Hirons, K. R. 2000 Human-environment interactions in prehistoric landscapes: the example of the Outer Hebrides. In Hooke, D. (ed.) *Landscape, the Richest Historical Record.* Society for Landscape Studies Supplementary Series **1,** 13–32.
Changing vegetational landscapes are examined; emphasis upon persistence of woodland and arguments for natural versus human agency in environmental change.
Edwards, K. J. and Whittington, G. 1998 Landscape and environment in prehistoric West Mainland, Shetland. *Landscape History* **20,** 5–17.
Emphasis upon vegetation history and the interplay between pastoral and arable activities.
Edwards, K. J. and Whittington, G. 2000 Multiple charcoal profiles in a Scottish lake: taphonomy, fire ecology, human impact and inference. *Palaeogeography, Palaeoclimatology, Palaeoecology* **164,** 67–86.
Raises questions about the nature of evidence for fire history; a direct fire-vegetation relationship is unproven.
Ritchie, W., Whittington, G. and Edwards, K. J. 2001 Holocene changes in the physiography and vegetation of the Atlantic littoral of the Uists, Outer Hebrides, Scotland. *Transactions of the Royal Society of Edinburgh: Earth Sciences* **92,** 121–136.
Sea-level, landform and vegetational changes are examined.
Tipping, R. and Milburn, P. 2000 The mid-Holocene charcoal fall in southern Scotland: spatial and temporal variability. *Palaeogeography, Palaeoclimatology, Palaeoecology* **164,** 193–209.

Charcoal falls may reflect late Mesolithic climatic dryness and natural fire incidence rather than intentional burning.

Wells, J. M., Mighall, T. M., Smith, D. E. and Dawson, A. G. 1999 Brighouse Bay, southwest Scotland: Holocene vegetational history and human impact at a small coastal valley mire. In Andrews, P. and Banham, P. (eds), *Late Cenozoic Environments and Hominid Evolution: a Tribute to Bill Bishop,* 217–233. London: Geological Society.

Evidence for human impact on the vegetation from Mesolithic times onward, with landscape instability and erosion from the Neolithic to Iron Age periods.

6: FAUNAL CHANGE

Many specialist accounts are to be found within excavation monographs and appendices to articles.

Barrett, J. H., Nicholson, R. A. and Cerón-Cerrrasco, R. 1999 Archaeo-ichthyological evidence for long-term socio-economic trends in northern Scotland: 3500 BC to AD 1500. *Journal of Archaeological Science* **26**, 353–388.

Long-term patterns of marine resource exploitation are examined.

Brooks, S. J., Mayle, F. E. and Lowe, J. J. 1997 Chironomid-based Lateglacial reconstruction for southeast Scotland. *Journal of Quaternary Science* **12**, 161–167.

Chironomid (non-biting midge) records show rapid warming at the start of the Holocene.

Gonzales, S., Kitchener, A. C. and Lister, A. M. 2000 Survival of the Irish elk into the Holocene. *Nature* **405**, 753–754.

Radiocarbon dates, including one for an example from the River Cree, Galloway, indicate that giant deer survived beyond their previously supposed extinction date in the Lateglacial period.

McCormick, F. 1998 Calf slaughter as a response to marginality. In Mills, C. M. and Coles, G. (eds) *Life on the Edge: Human Settlement and Marginality,* 49–51. Oxford: Oxbow Monograph **100**.

Evidence for fodder shortage is here considered as an explanation for this phenomenon on Atlantic Scottish sites.

Sadler, J. P. and Jones, J. C. 1997 Chironomids as indicators of Holocene environmental change in the British Isles. *Quaternary Proceedings* **5**, 219–232.

Yalden, D. 1999 *The History of British Mammals.* London: Poyser.

7: THE MESOLITHIC

Finlayson, B. 1999 Understanding the initial colonization of Scotland. *Antiquity* **73**, 879–884.

Macklin, M. G., Bonsall, C., Davies, F. M. and Robinson, M. R. 2000 Human-environment interactions during the Holocene: new data and interpretations from the Oban area, Argyll, Scotland. *The Holocene* **10**, 109–121.

Mithen, S. 2000 Mesolithic sedentism on Oronsay: chronological evidence from adjacent islands in the southern Hebrides. *Antiquity* **74**, 298–304.

Mithen, S. (ed.) 2001 *Hunter-gatherer Landscape Archaeology: The Southern Hebrides Mesolithic Project 1988–1998.* Cambridge: McDonald Institute Monographs. 2 vols. Multidisciplinary investigations into the archaeology and environments of sites in Islay and Colonsay.

Mithen, S., Finlay, N., Carruthers, W., Carter, S. and Ashmore, P. 2001 Plant use in the Mesolithic: evidence from Staosnaig, Isle of Colonsay, Scotland. *Journal of Archaeological Science* **28**, 223–234.

Consideration of an exceptional pit containing great quantities of lithic material and charred hazelnut shells.

Richards, M. P. and Mellars, P. A. 1998 Stable isotopes and the seasonality of the Oronsay middens. *Antiquity* **72**, 178–184.
Analysis of human bones shows marine resources provided most protein, supporting the possibility of year-round occupation.

Richards, M. P. and Sheridan, J. A. 2000 New AMS dates on human bone from Mesolithic Oronsay. *Antiquity* **74**, 313–315.
Late Mesolithic occupation overlaps with early Neolithic dates.

Saville, A. 1998 Studying the Mesolithic period in Scotland: a bibliographic gazetteer. In Ashton, N., Healy, F. and Pettitt, P. (eds), *Stone Age Archaeology: Essays in Honour of John Wymer,* 211–224. Oxford: Oxbow Monograph **102**/Lithic Studies Society Occasional Paper **6**.
Extensive bibliography and commentary.

Young, R. (ed.) 2000 *Mesolithic Lifeways: Current Research from Britain and Ireland.* Leicester: Leicester University. (Leicester Archaeology Monographs **7**.)
Several essays treat Scottish topics.

8: THE NEOLITHIC

Barclay, G. J. 2001 'Metropolitan' and 'Parochial'/'Core' and 'Periphery': a historiography of the Neolithic of Scotland. *Proceedings of the Prehistoric Society* **67**, 1–16.
Includes an assessment of archetypal landscapes interpreted after the English tradition.

Barclay, G. J and Maxwell, G. S. 1998 *The Cleaven Dyke and Littleour: Monuments in the Neolithic of Tayside.* Edinburgh: Society of Antiquaries of Scotland Monograph **13**.
Investigation of a cursus monument/bank barrow and its wider context; includes a range of environmental studies.

Barclay, G. J., Brophy, K. and Macgregor, G. 2002 A Neolithic building at Claish Farm, near Callander, Stirling Council, Scotland, UK. *Antiquity* **76**, 13–14.
The discovery of a second Balbridie-type structure, undermining a tendency to dismiss Balbridie as exceptional.

Bradley, R. J. 1998 *The Significance of Monuments.* London: Routledge.
Considers the origins and development of monument building in the Neolithic of Europe.

Ruggles, C. and Barclay, G. J. 2000 Cosmology, calendars and society in Neolithic Orkney: a rejoinder to Euan MacKie. *Antiquity* **74**, 62–74.
The arguments against precise astronomy and the 'megalithic yard'.

Ritchie, A. (ed.) 2000 *Neolithic Orkney in its European Context.* Cambridge: McDonald Institute Monographs.
Contains numerous papers on Orkney in its wider context; while the emphasis is on the Neolithic, useful material pertaining to the Mesolithic and the Bronze Age is included.

Saville, A. 1999 A cache of flint axeheads and other flint artefacts from Auchenhoan, near Campbeltown, Kintyre, Scotland. *Proceedings of the Prehistoric Society* **65**, 83–123.
Imports from County Antrim, Northern Ireland.

9: THE BRONZE AGE

Barrett, J. C. and Gourlay, R. B. 1999 An early metal assemblage from Dail na Caraidh, Inverness-shire, and its context. *Proceedings of the Society of Antiquaries of Scotland* **129**, 161–187.
Important Early Bronze Age metalwork hoard considered in its landscape setting.

Bradley, R. 2000 *The Good Stones. A New Investigation of the Clava Cairns.* Edinburgh: Society of Antiquaries of Scotland Monograph **17**.
Recent excavations and survey establish the cairns as Early Bronze Age in date; includes soil and other environmental studies.

Bunting, M. J. and Tipping, R. 2001 Anthropogenic pollen assemblages from a Bronze Age cemetery at Linga Fiold, West Mainland, Orkney. *Journal of Archaeological Science* **28**, 487–500.
  Pollen analyses are used to reconstruct aspects of activity in and around the site.
Cowie, T. G., Hall, M., O'Connor, B. and Tipping, R. 1996 The late Bronze Age hoard from Corrymuckloch, near Amulree, Perthshire: an interim report. *Tayside and Fife Archaeological Journal* **2**, 60–69.
  Sheds sideways light on high society.
Hunter, F. 2000 Excavation of an Early Bronze Age cemetery and other sites at West Water Reservoir, West Linton, Scottish Borders. *Proceedings of the Society of Antiquaries of Scotland* **130**, 115–182.
  Remarkable find of lead; pollen evidence for floral tributes.
Needham, S. 1996 Chronology and periodisation in the British Bronze Age. In Randsborg, K. (ed.) *Absolute Chronology: Archaeological Europe 2500–500* BC, 121–140. Copenhagen: *Acta Archaeologica Supplementa* **1**.
Needham, S., Bronk Ramsey, C., Coombs, D., Cartwright, C. and Pettitt, P. 1997 An independent chronology for British Bronze Age metalwork: The results of the Oxford radiocarbon accelerator programme. *Archaeological Journal* **154**, 55–107.
Ralston, I. B. M. and Sabine, K. A. 2000 *Excavations of Second and First Millennia* BC *Remains on the Sands of Forvie, Slains, Aberdeenshire*. Aberdeen: Department of Geography and Environment, University of Aberdeen. = O'Dell Memorial Monograph **28**.
  Settlement, agricultural and funerary evidence juxtaposed.
Rohl, B. and Needham, S. 1998 *The Circulation of Metal in the British Bronze Age: the Application of Lead Isotope Analysis*. London: British Museum. (*British Museum Occasional Paper* **102**.)
Sheridan, J. A. 1999 Drinking, driving, death and display: Scottish Bronze Age artefact studies since Coles. In Harding, A. F. (ed.), *Experiment and Design: Archaeological Studies in Honour of John Coles*, 49–59. Oxford: Oxbow.
  Developments since the 1960s.

10: THE IRON AGE

Armit, I. 1997 *Celtic Scotland*. London: Batsford/Historic Scotland.
  Overview of the pre-Roman Iron Age.
Armit, I. 1999 Life after Hownam: the Iron Age in south-east Scotland. In Bevan, B. (ed.) *Northern Exposure: Interpretative Devolution and the Iron Ages of Britain*, 65–79. Leicester: Leicester University. (Leicester Archaeology Monographs **4**.)
Gwilt, A. and Haselgrove, C. (eds) 1997 *Reconstructing Iron Age Societies: New Approaches to the British Iron Age*. Oxford: Oxbow Monograph **71**.
  Several papers treat cultural and environmental aspects, extending into the Roman Iron Age.
Harding, D. W. 2000 *The Hebridean Iron Age: Twenty Years Research*. Edinburgh: University of Edinburgh Department of Archaeology Occasional Paper **20**.
  Overview of long-term research project focused on west Lewis.
Haselgrove, C. and McCullagh, R. (eds) 2000 *An Iron Age Coastal Community in East Lothian: the Excavation of Two Later Prehistoric Enclosure Complexes at Fisher's Road, Port Seton, 1994–5*. Edinburgh: Scottish Trust for Archaeological Research Monograph **6**.
Parker Pearson, M. and Sharples, N. M. 1999 *Between Land and Sea: Excavations at Dun Vulan, South Uist*. Sheffield: Sheffield Academic Press.
  Examination of a coastal broch, including wider environmental and cultural contexts.
Nicholson, R. A. and Dockrill, S. J. (eds) 1998 *Old Scatness Broch, Shetland: Retrospect and Prospect*. Bradford: Department of Archaeological Sciences, University of Bradford = NABO Monograph **2**.

Preliminary studies on multi-period site; environmental reports and wider contexts are included.

Sharples, N. 1998 Scalloway: a broch, Late Iron Age settlement and Medieval cemetery in Shetland. Oxford: Oxbow Monograph **82**.
Excavations and environmental studies.

11: THE ROMAN PRESENCE

Armit, I. 1999 The abandonment of souterrains: evolution, catastrophe or dislocation? *Proceedings of the Society of Antiquaries of Scotland* **129**, 577–596.
An attempt to relate the abandonment of souterrains to the departure of the Romans.

Bishop, M. (ed.) 2002 *Roman Inveresk: Past, Present and Future*. Duns: Armatura.
Useful collation of many years work in and around the most important Roman civil settlement.

Dumayne-Peaty, L. 1998 Human impact on the environment during the Iron Age and Romano-British times: palynological evidence from three sites near the Antonine Wall, Great Britain. *Journal of Archaeological Science* **25**, 203–214.
A consideration of the minimal impact of the Roman presence on forest clearance and agriculture.

Erdrich, M., Giannotta, K. M. and Hanson, W. S. 2000 Traprain Law: native and Roman on the northern frontier. *Proceedings of the Society of Antiquaries of Scotland* **130**, 441–456.
Assessment of the nature of Roman contacts with this important native site.

Hunter, F. 1997 Iron age hoarding in Scotland and northern England. In Gwilt, A. and Haselgrove, C. (eds) *Reconstructing Iron Age Societies: New Approaches to the British Iron Age*, 108133. Oxford: Oxbow.
A consideration of the role of Roman material culture in indigenous ritual practices.

Hunter, F. 2001 Roman and native in Scotland: new approaches. *Journal of Roman Archaeology* **14**, 289–309.
Analysis of the distribution, nature and use of Roman material culture found on non-Roman sites.

Hutcheson, A. R. J. 1997 Ironwork hoards in northern Britain. In Meadows, K., Lemke, C. and Heron, J. (eds), *TRAC 96: Proceedings of the Sixth Annual Theoretical Roman Archaeology Conference Sheffield 1996*, 65–72. Oxbow: Oxford.
Analysis which demonstrates the absence of transmission of technological information between Rome and the indigenous Iron Age population.

12: THE EARLY HISTORIC PERIOD

Campbell, E. 2001 'Were the Scots Irish?' *Antiquity* **75**, 285–292.
Inward migration across the North Channel re-assessed.

Crone, A. B. 2000 *The History of a Scottish Lowland Crannog: Excavations at Buiston, Ayrshire*. Edinburgh: Scottish Trust for Archaeological Research Monograph **4**.
Re-excavation and re-assessment of a key first millennium AD site.

Driscoll, S. and Yeoman, P. 1997 *Excavations within Edinburgh Castle*. Edinburgh: Society of Antiquaries of Scotland Monograph **12**.
The earlier occupations of the Castle Rock investigated.

Henry, D. (ed.) 1997 *The Worm, the Germ and the Thorn*. Balneaves, Angus: Pinkfoot Press.
Papers on the Picts.

Hill, P. 1997 *Whithorn*. Stroud: Sutton.
Important early Christian centre in Galloway excavated and assessed.

Lane, A. and Campbell, E. 2000 *Dunadd: an early Dalriadic Capital*. Oxford: Oxbow.
Excavation and re-assessment of this key nucleated fort in Argyll.

Perry, D. 2000 *Castle Park, Dunbar: Two Thousand Years on a Fortified Headland*. Edinburgh: Society of Antiquaries of Scotland Monograph **16**.
Excavations revealed Iron Age, British and Anglian use.
Ritchie, A. 1997 *Iona*. London: Batsford/Historic Scotland.

13: THE EARLY NORSE

Barrett, J. H. 1997 Fish trade in Norse Orkney and Caithness: a zooarchaeological approach. *Antiquity* **71**, 616–638.
Although slightly later in emphasis, this paper considers economic factors of the Scandinavian colonisation.
Barrett, J. H., Beukens, R. P. and Brothwell, D. R. 2000 Radiocarbon dating and marine reservoir correction of Viking Age Christian burials from Orkney. *Antiquity* **74**, 537–543.
Determinations from two chapel sites indicate the possible adoption of Christianity at a relatively precocious date.
Barrett, J., Beukens, R., Simpson, I., Ashmore, P., Poaps, S. and Huntley, J. 2000 What was the Viking Age and when did it happen? A view from Orkney. *Norwegian Archaeological Review* **33**, 1–39.
An investigation of 'core' and 'periphery' in the Norse world with special reference to Orkney.
Buteux, S. (ed.) 1997 *Settlements at Skaill, Deerness, Orkney*. Oxford: British Archaeological Reports, British Series **260**.
Final report of excavations at an important site which spans the Iron Age/Viking periods.
Graham-Campbell, J. and Batey, C. E. 1998 *Vikings in Scotland*. Edinburgh: Edinburgh University Press.
Synthesis of published and unpublished material from throughout Scotland.
Owen, O. and Dalland, M. (eds) 1999 *Scar. A Viking Boat Burial on Sanday, Orkney*. East Linton: Tuckwell Press/Historic Scotland.
Sharples, N. and Parker Pearson, M. 1999 Norse settlement in the Outer Hebrides. *Norwegian Archaeological Review* **32**, 41–61.
New evidence from the machair lands, mainly of South Uist, and a consideration of settlement continuity from the Bronze Age.

NEWER DEPARTURES

A number of innovations have materialised, or become more prominent, since we wrote the initial and final chapters in 1996. These include the application of new techniques, or modifications of existing methods (some of which, like OSL dating, are included in works cited above). Of likely major future significance are genetic approaches. Other means of identifying new arrivals in the population are also being actively researched. New approaches are also being taken to the examination of the erosion of the archaeological record. The following references are indicative of some of the more recent cross-disciplinary lines of enquiry that are being applied to Scottish material.

Davidson, D. A., Grieve, I. C., Tyler, A. N., Barclay, G. J. and Maxwell, G. S. 1998 Archaeological sites: assessment of erosion risk. *Journal of Archaeological Science* **25**, 857–860.
An evaluation of soil erosion rates at a crop mark site in Perthshire.
Evison, M. P. 1999 Perspectives on the Holocene in Britain: human DNA. *Journal of Quaternary Science* **14**, 615–623 (*Quaternary Proceedings* **7**).

Summary of gene-sequence research and perspectives at the British scale. The hunter-gatherer/agricultural transition and indigenous versus diffusionist arguments figure prominently.

Helgason, A., Hickey, E., Goodacre, S., Bosnes, V., Stefánsson, K., Ward, R. and Sykes, B. 2001 mtDNA and the islands of the North Atlantic: estimating the proportions of Norse and Gaelic ancestry. *American Journal of Human Genetics* **68**, 723–737.

Matrilinear ancestry analysed by mtDNA sequences, using Icelandic, Orcadian and Hebridean populations. The majority of Iceland's original female settlers were from Scotland and Ireland. Viking women were also the ancestors of some modern Scottish islanders.

Hewitt, G. M. 1999 Post-glacial re-colonization of European biota. *Biological Journal of the Linnean Society* **68**, 87–112.

Biota other than humans can also be investigated *via* their genetic composition. The paper is a European-scale overview of the spread of selected plants and animals.

Hoaen, A. and Coles, G. 2000 A preliminary investigation into the use of fungal spores as anthropogenic indicators on Shetland. In Nicholson, R. A. and O'Connor, T. P. (eds) *People as an Agent of Environmental Change*, 30–36. Oxford: Oxbow.

The identification of fungal spores that can be correlated with herbivore dung.

Jones, M. 2001 *The Molecule Hunt: Archaeology and the Search for Ancient DNA*. Harmondsworth: Penguin Allen Lane.

An overview of approaches and findings associated with genetic material derived from plants, animals and humans.

Mills, C. M. and Coles, G. (eds) 1998 *Life on the Edge: Human Settlement and Marginality*. Oxford: Oxbow Books. (Oxbow Monograph **100**.)

Environmental limits to human activity and the interrelationship of environmental, economic and social systems explored.

Simpson, I. A., van Bergen, P. F., Ellmmah, M., Roberts, D. J. and Evershed, R. P. 1999 Lipid biomarkers of manuring practice in relict anthropogenic soils. *The Holocene* **9**, 223–229.

A test of the extent to which free soil lipids reflect known manuring practices, suggesting that in this regard historic documentation forms only a partial record.

Sommerville, A. A., Sanderson, D. C. W., Hanson, J. D. and Housley, R. A. 2001 Luminescence dating of aeolian sands from archaeological sites in Northern Britain: a preliminary study. *Quaternary Science Reviews* **20**, 913–919.

Preliminary tests include determinations from Tofts Ness.

Sykes, B. 2001 *The Seven Daughters of Eve*. London: Bantam Press.

Wider treatment of human genetics, *cf.* Evison *supra*.

Whittington, G. and Edwards, K. J. 1999 Landscape scale soil pollen analysis. *Journal of Quaternary Science* **14**, 595–604 (*Quaternary Proceedings* **7**).

Points the way towards exploiting more fully the complex pollen records contained within soils, using *inter alia* multiple profiles from Jura and Shetland.

Index

Aberlemno Kirkyard 236
Achany Glen 173
Achnacree 54–5, 152
Ackergill 222
Aerial photography 5, 129, 153, 189, 236
Agriculture 53, 60, 141, 148, 188, 191,
    207–8, 231–3, 253–4
  beginnings 64–72
  establishment 72–4
  land capability for (LCA) 51–2
  potential 32–3
  production 6
  recession 148
  regimes 6
Airthrey 84
Alcohol 143
Alder (*Alnus glutinosa*) 15, 17, 67, 73, 251
Allt na Fearna Mor, Lairg 222
Alnus glutinosa see Alder
An Corran 111, 117, 119, 121
An Sithean 18, 54–5, 57, 152
Angles 219, 222
Angus 175, 178, 191
Animal bones 89, 189, 260
Animal products 233
Animal traction 164
Animals 83, 164, 205
Ant (*Formica lemannii*) 107
Antonine Wall 108, 183, 196, 197, 201, 203,
    206
Arable farming 189–91
Archaeological record, survival and
    detection 4–5
Archaeological subdivisions 8
Arctic fox (*Alopex lagopus*) 86
Ard marks *see* Cultivation
Ardnamurchan Peninsula 26
Ardnave 84, 99, 152
Ardwall Island 222
Arkle 25
Arran 72, 158
Arran pitchstone 122
Artefacts 30–3, 71, 91
  Bronze Age 153
  Mesolithic 114–17, 120
  Neolithic 140

Ash (*Fraxinus excelsior*) 16, 211
Assynt 86
Astronomy 139–40
Atlantic period 17
Atlantic Scotland 218
Aurochs (*Bos primigenius*) 87
Avena see Oats

Badger (*Meles meles*) 99
Bágh Siar 113
Balbirnie 130
Balbridie 130, 131, 143, 146, 147, 264
Baleshare 85, 92–8
Balfarg 130, 134, 146, 147
Ballevulin 111, 114
Balloch Hill 130, 147
Ballyglass, Ireland 146
Balnabroich 152, 159, 160
Balneaves 130, 134, 137
Balnuaran of Clava 133
Bank barrows 134
Bank vole (*Clethrionomys glareolus*) 90
Barley (*Hordeum vulgare*) 76, 143, 189, 204,
    212, 232, 251
Barmekin of Echt 130, 147
Barnhouse 130, 145, 149
Barr Hill 85
Barra 76, 78–9
Bay of Sannick 111, 115
Beaquoy 152, 159
Beaver (*Castor fiber*) 87
Beetle (Coleoptera) 105–8
Beetle (*Laemostenus* sp.) 108
Benbecula 41
Bennybeg 130, 137
Bernicia 219
Bertha 196, 203
Bettyhill 111
Betula see Birch
Betula nana see Dwarfbirch
Biological databases 14
Biostratigraphical evidence 14
Birch (*Betula*) 15, 16, 65, 211
Birch bark beetle (*Scolytus ratzeburgi*)
    107

Bird fragments 92–6, 231
Birrens 85
Birsay 85, 97, 102–3
Black grouse (*Lyrus tetrix*) 101
Black henbane (*Hyoscyamus niger*) 143
Black Loch 43, 59, 68, 69, 73–7, 173, 192, 200, 209
Blair Drummond Moss 152, 156, 157
Blairhall 130, 137
Blind colydiid (*Aglenus brunneus*) 108
Bloak Moss 173, 192, 200, 209
Blue whale (*Balaenoptera musculus*) 90
Boar (*Sus scrofa*) 87, 102
Boddam Den 130, 140
Boghead 54–5, 58, 130, 143
Bolsay Farm 111, 115, 117
Bolton Fell Moss 200, 209
Bone assemblages 88, 91, 102
Boonies 173
Boreal phase 13
Bos taurus see Cattle
Boysack Mills 222, 236, 238
Bracken (*Pteridium aquilinum*) 76
Braeroddach Loch 42, 43, 55, 59, 60, 68, 73, 75, 111, 123, 152
Bridgend 111, 114
Brighouse Bay 20, 105, 106
Britons (Strathclyde) 218, 222
Broad-leaved pinehole borer (*Xyleborus dispar*) 106
Broch architecture 183, 184
Broch towers 185, 225
Brochs 170, 171, 215
Bronze Age 5, 8, 32, 43, 59, 75, 151–68
    archaeological monuments 151
    artefactual evidence 153
    burials 154–5
    environmental dimensions 75–9, 165–7
    lowland zone landscapes 151–7
    settlement and economy 157–62
    settlement record 153
    subsistence economy 162–7
Broomend of Crichie 130
Brough of Birsay 222, 232, 239, 249, 250, 252
Broughty Ferry 40
Brown bear (*Ursus arctos*) 86, 87
Brown Caterthun 130, 147
Broxmouth 85, 173, 176, 177, 182, 214, 218
Bu 173, 184, 186
Buchan 25, 140
Buckquoy 85, 92–8, 222, 226, 227, 249
Buildings 145, 160, 206, 228
Buiston 108, 222, 226, 265

Burghead 54–5, 221, 222
Burial cairns 160
Burial sites 132, 235–6
Burials, Bronze Age 154–5
Burnfoothill Moss 68, 75
Burnswark 173, 187

Cairngorms 21
Cairnholy 130, 131
Cairnmore 172, 173
Cairnpapple 222
Cairns 158, 159, 160
Caisteal nan Gillean 111, 118
Caithness 25, 187, 248, 253
Caledonian Fold Belt 25
Caledonian forest 210
Callanish 68, 72, 75, 81
Candidula intersecta 104
Carabid (*Odacantha melanura*) 20
Carboniferous 25
Carboniferous sedimentary rocks 26
Carding Mill Bay 88, 89, 111, 119
Carlungie 222, 227
Carn Dubh 68, 75
Carpow 200
Carved stone balls 137, 140
Carwinning Hill 130, 147
Cas chrom 142
Castle Fraser 130
Castle Hill 54–5
Castle Rock, Edinburgh 222, 224
Castlesteads 173, 190
Cat 231
Catchment studies 59
Catherinefield Farm 152
Catpund 152
Cattle 91, 148, 173, 176, 205, 232, 252
Causewayed enclosures 147
Cave deposits 19
Cellular buildings 225–6
Cemeteries 235
Centralization 221–2
Cereals 61, 72, 127, 143, 148, 164, 189–91, 207, 208
    adaptation 143
    drink 143
    impressions on pottery 143
    see also Barley, Oats, Wheat
Cernuella virgata 104
Cerylon histeroides 106
Chambered cairns 129, 131
Charcoal 71
Chariots 189
Chesters 173
Cheviot Hills 73, 124

Chew Green 196, 199, 203
Chicken (*Gallus gallus*) 101
Chironomidae (midges) 20, 107
Christianity 233–9
Cill Donain 41
Cladh Hallan 41, 262
Clambid (*Calyptomerus dubius*) 108
Clatchard Craig 222, 224
Clatteringshaws Loch 16
Clava series of monuments 133, 136
Cleaven Dyke 54–5, 58, 130, 133–5, 264
Clettraval 186
Clientage 222
Climate change 11–22, 64
    and faunal evidence 19–20
    and peat development 17–19, 21
    and sand movement 21–2
    and vegetational change 14–27
    Holocene 13–14
    long-term 14–20
    postglacial 17
    revised view 13–14
    short-term 20–2
    studies 262
    traditional view 13
    vegetational evidence of 20–1
Climatic implications and geographical
        location 11–13
Climatic modelling 14
Climatic Optimum 16
Clyde Estuary 39
Clyde region 135
Clydesdale 219
Cnip 85, 97–8, 155, 156, 173, 186, 222,
        226
Coalfish (*Pollachius virens*) 90
Cochlicella acuta 104
Cod (*Gadus morhua*) 88
Coileagean an Udail (Udal) 85, 92–8, 222,
        227
Coleoptera (beetles) 20, 105–8, 260, 263
Coll 247
Colonization, Mesolithic 113–17
Colonsay 242
Common dolphin (*Delphinus delphis*) 88
Common porpoise (*Phocaena phocaena*) 88
Common rorqual (*Balaenoptera physalus*)
        88
Common seal (*Phoca vitulina*) 88
Common shrew (*Sorex araneus*) 88
Conger eel (*Conger conger*) 91
Coppicing 70
Corkwing wrasse (*Crenilabrus melops*)
        99
Cormorant (*P. carbo*) 101
Corrimony 130, 132

Corstorphine 84
Corylus avellana see Hazel
Cossonine weevil (*Eremotes ater*) 107
Counties, prior to local government
        reorganization 8, 10
Courthill, Dalry 222
Cramond 196, 200
Crane (*Grus* sp.) 100
Crannogs 5, 170, 226
Creag na Caillich 130, 140
Creag nan Uamh 84, 86
Creich 38
Crichton Mains 173, 183
Crofting 148
Crop plants 232
Cropmarks 5, 202, 210, 213, 238
Crops 189–91
Crosskirk 85, 100–1, 173
Crowberry (*Empetrum nigrum*) 64
Croy Hill 197
Cuillin Hills 26
Cùl a'Bhaile 54–5, 152, 158, 163
Cultivation 60, 148
    ard marks 142, 163
    impact of 59
    in relation to soils 143
    Neolithic 141–4
    potato 148
    ridging 142
    spade marks 142
    systems 191
    terraces 224
Cultural framework 6–8
    impact of calibration 7–8
Cultural groups 221
Cultural mix 219
Cursus monuments 134, 135, 137
Cyperaceae *see* Sedges

Dail na Caraidh 152
Dalladies 54–5, 130, 143
Dallican Water 68, 71
Dalnaglar 18, 54–5, 58
Dalriada 221
Dalrulzion 173, 176
Deglaciation 27
Deil's Dyke 222
Deroceras agreste 104
Devonian 25
Dimlington Stadial 27
Districts 8
Documentary evidence 217, 221
Dod, The Roxburgh 218, 222
Dog 88, 231
Doon Hill 222, 226, 228

Douglasmuir 130, 137, 138, 173, 175
Drem 173
Drift deposits 27, 31
Drift geology 47
Druimvargie 111, 119
Drumturn Burn 167
Dryburn Bridge 172, 173, 177
Drystone roundhouses 184
Duck (*Anas platyrrhynchos*) 101
Dumbarton Rock 222
Dun Carloway 172, 173, 184
Dun Mor Vaul 85
Dunadd 222, 224
Dunbar 222, 228
Dundurn 222, 224, 225
Dunollie 222
Duns 170, 171, 181, 187, 225
Dupplin 222
Dupplin Cross 237
Durness Limestone 83
Durrington Walls, Wiltshire 139
Dwarf birch (*Betula nana*) 64
Dwarf willow (*Salix herbacea*) 64
Dytiscid (*Agabus wasastjernae*) 107

Early Historic Period 217–39
    centralization 221, 222
    continuities from pre-Roman period
        218
    economy and manufacture 230–3
    legacy of Rome 217–218
    material culture 221
    protohistory 218–22
    settlement evidence 222–30
    warfare 231
Early Norse Period 241–54
Earn valley 139
East Lothian 175, 178, 180, 228
Easter Kinnaird 222
Ecclesiastical sites 236–7
Edinburgh *see* Castle Rock
Edin's Hall 214, 215
Eel (*Anguilla anguilla*) 88
Eider (*Somateria mollissima*) 91
Eildon Hill North 172, 173, 192, 200,
    214
Eilean Dhomnuill, N Uist 130, 144,
    146
Eilean Olabhat 222, 230
Elginhaugh 199
Elk (*Alces alces*) 87, 102
Elm (*Ulmus*) 6, 16, 66, 72–4, 141, 211
Emmer (*Triticum dicoccum*) *see* Wheat
Empetrum nigrum see Crowberry
Enclosures 170

Neolithic 144–9
Environment 1–8
Environmental analysis 259
Environmental change 2, 3
    chronologies 6
Environmental conditions 2
Ertebolle society 124
Ettrick Association 49
Excavation reports 261
Excavations assigned to archaeological
        periods 258
External contacts 238–9

Fair Isle 242
Farming
    activities 76
    communities 127
    economy 123–4
    Neolithic 141, 148
    patterns 165
    settlement 32
Father lasher (*Myoxocephalus scorpio*)
        99
Faunal assemblages 143
Faunal change 83–108
Faunal evidence and climate change
        19–20
Fendoch 199, 203
Ferns 64
Fertilizers 59
Field mouse (*Apodemus sylvaticus*) 99
Field systems 142, 158–60, 167, 190, 191,
        213, 245
    boundaries 142
Fieldwork 188
Fife 40, 74–7
Filipendula ulmaria see Meadowsweet
Fish species 97–8
Fishing 188, 231
Flanders Moss 200, 209
Flandrian 26
Flandrian peat beds 40
Flax (*Linum usitatissimum*) 143, 233,
        251
Flounder (*Plathichthys flesus*) 91
Forest clearance 209, 213
Forest utilization model 74
Forteviot 222, 225
Forth-Clyde lowlands 16
Forth Estuary 39
Forth Valley 38, 39, 156
Fortifications 179
Forts 202, 203, 208, 214, 224–5, 230
    nucleated 224
    see also Hillforts

Forvie 21, 152
Fossil clifflines 36
Fowling 188
Fox (*Vulpes vulpes*) 99
Fozy Moss 200
Fraxinus excelsior see Ash
Freswick Links 85, 103, 111, 115
Friarton 111, 121
Fulmar (*Fulmarus glacialis*) 90

Gaelic influences 248
Gask Ridge 201
Gaulcross 222
Geographical location and climatic
    implications 11–13
Geographical units 8
Geology 23–6
Geometry 139–40
Geomorphology 23–44
    and human activity 41–4
    changes during the Holocene 34–7
Germans 217
Giant deer (*Megaloceros giganteus*) 87
Glacial erosion 27, 32
Glaciation 26–30
    and early human occupance 30–3
    and sea-level change 33–4
Glaciofluvial deposits 27, 30
Glacio-isostatic processes 35
Gleann Mor 111, 117
Glen Coe 36
Glen Etive 36
Glen Feshie 37
Glenbatrick 111, 115
Glenlochar 202, 203
Gleying 45–6
Goat, *see also* Sheep
Gododdin 219
Goose (*Anser anser*) 101
Goshawk (*Accipiter gentilis*) 99
Grain 204, 207
Grain weevil (*Sitophilus granarius*) 107
Gramineae *see* Grasses
Grampians 25, 27
Granite mountains 26
Grasses (Gramineae, Poaceae) 64
Grassland *see* Pasture
Grave goods 247
Great auk (*Pinguinus impennis*) 90
Great Glen Fault 24, 25
Green Castle, Portknockie 222, 225
Green Knowe 152, 158
Grey seal (*Halichoerus grypus*) 88
Grooved Ware 134, 143, 144
Ground beetle 106

Grübenhäuser 228, 229
Guillemot (*Uria aalge*) 88
Gurness 173, 185, 186, 226

Haddock (*Melanogrammus aeglefinus*)
    90
Hadrian's Wall 195, 200, 209
Halibut (*Hippoglossus hippoglossus*) 91
Hallow Hill 222
Hallucinogen 143
Hare (*Lepus* sp.) 99
Harris 41, 42
Hawthorn (*Crataegus*) 211
Hazel (*Corylus avellana*) 15, 16, 66, 67, 70,
    73, 211, 251
Hazelnut shells 120
Heather (*Calluna vulgaris*) 71, 76, 81
Hebridean Craton 25
Hebrides 26, 27, 30, 110, 185, 189
Hedgehog (*Erinaceus europaeus*) 90
Hekla 166
Helicella itala 104
Helix aspersa 104
Helmsdale 218, 222
Henges 133, 135, 136, 137
Hengiform enclosures 133, 136
Herald Hill 130, 132
Heron (*Ardea cinerea*) 101
Highland Boundary Fault 24, 25, 201
Highlands 23, 25, 32
Hillforts 170, 171, 174, 176, 180–3, 185,
    187, 192
    see also Forts
Historic period 61
Hoards 147, 218
    coins 218
Hoddom 222, 229
Holocene 2, 13–14, 26, 30, 33
    geomorphological changes 34–7
    sea-level changes 37–41
    soil evolution 56–8
    vegetational development 64–79
    vertebrate fauna 86–103
Holywood 130, 134
Hooded crow (*Corvus corone*) 90
Hordeum see Barley
Hordeum vulgare var. nudum 232
Hordeum vulgare var. vulgare 232
Hornish Point 85, 92–6
Horses 189, 221, 231
House mouse (*Mus musculus*) 99
Housefly (*Musca domestica*) 108
Houses 131, 144–9, 176, 185, 225–226
Howe 85, 92–8, 101, 173, 184, 186, 222,
    226, 227

Hownam 178
Hownam Law 173, 174
Hownam Rings 173, 176, 177
Human activity 57, 84
    and geomorphological response 41–4
Human–environmental interactions 2
Human impact 58–61, 83
Human occupance 30–3
Human settlement 110
Hummocky moraines 30, 32
Hunter-gatherers 122–3, 127, 142
Hunting 112, 188
Hut circles 158–60, 167, 170, 176
Hut Knowe 173, 191

Ice Age 19, 26
Imports, high quality 218
Inchnadamph 111, 113
Inchtuthil 85, 199
Insect species 105–8, 263
Inveresk 85, 196, 213
Inverness 111, 115
Iona 54–5, 97–8, 222, 242
Iron Age 8, 43, 169–93
    architecture and social change
        183–7
    Atlantic regions 183–7
    Eastern Scotland 175–83
    economic development and change
        192–3
    economy and environment 188–93
    economy and society 193
    environment 75–81
    landform and province 169–70
    Provinces 170
    regionalism 170, 188–9
    settlements 170–2, 175–82
    society and change 174
    South-West Scotland 187–8
    subdivision 171
Iron nails 206
Isbister 6, 84, 99
Islay 18, 91
Isle of Mull 70

Jarlshof 85, 101, 152, 159
Jet necklaces 154
Juniper (Juniperus communis) 64
Jura 35, 36, 39, 158

Kenneth Macalpin 221
Kilmartin Valley 152, 153
Kilphedir 54–5, 57, 173, 188

Kinalty 130, 137
King Lists 218
Kingdoms 239
Kings Cave 85
Kinloch 42, 70, 115, 116
Kinloch Farm, Fife 130, 147
Kinloch, Rhum 68, 75, 84, 86, 111, 117,
    130
Kintyre 72
Kirkbuddo 54, 55, 58, 209, 210
Knap of Howar 84, 91–8, 130, 143–5,
    148

Labour requirements 6
Laigh of Moray 189
Lairg 54–7, 60, 152, 158, 161, 162, 173,
    188, 222
Lamprophyre 35
Land
    communilization 149
    division 142
    fencing 143
    holdings 219, 222
    stone clearance from 142
Land capability
    analysis 49
    classes 51, 52
    for agriculture (LCA) 51–2
Land resources 49–53
Land snail fauna 104
Land use 61, 140–4
Landscape 23–44
    Bronze Age 151–7
    division 189
    geology and relief units 23–6
    physical characteristics 23–34
    study of 265
    subdivisions 190
    Tertiary evolution 26
Langdale, Cumbria 140
Languages 219
Late Devensian 27, 28, 30
Late Neolithic agricultural recession
    141
Lateglacial period 33, 63
    ice sheets 2
    vertebrate fauna 86
Lathrisk 222
Leaching 45–6, 56
Leadketty 130, 147
Leckie 218, 222
Lemming (Dicrostonyx torquatus) 86
Lewisian Gneiss 25, 33, 57
Liddle Farm 54, 55, 152, 159
Lime (Tilia) 16

Ling (*Calluna vulgaris*) *see* Heather
Ling (*Molva molva*) 99
Links of Noltland 84, 130, 142, 148, 149
Lintshie Gutter 152, 158
Lismore Fields, Derbyshire 131, 146
Literacy 234
Lithics 70
Loanhead of Daviot 138
Local government reorganization 8
Loch a'Bhogaidh 68, 75
Loch an t-Sìl 68, 70, 71
Loch Bharabhat 68, 75
Loch Cill an Aonghais 68, 73
Loch Cuithir 42
Loch Doon 68, 81, 120
Loch Druidibeg 107
Loch Etive 38
Loch Glashan 222
Loch Hourn 29
Loch Lomond 38, 39, 68, 200, 209
Loch Lomond Readvance 30, 36
Loch Lomond Stadial 27, 32–5
Loch Maree 65
Loch Meodal 68, 73
Loch na Berie 222, 226
Loch Nell 152, 164
Loch of Brunatwatt 68, 71
Loch of Huxter 172, 173
Loch of Park 42
Loch Quoich 29
Loch Sionascaig 56, 68, 75
Loch Tay 226
Lochan na Cartach 75, 76, 78, 79
Longhorn (*Arhopalus rusticus*) 107
Lower Palaeozoic 49
Luce Bay 21
Luib 32
Lunan Valley 189, 238
Lundin Links 222
Lundin Tower, Fife 20
Lussa Bay 111, 114
Lussa Wood 111, 114, 115
Lynx (*Felis lynx*) 86, 87

Machair 161, 164, 189, 262
Machrie Moor 72, 75, 130, 143
McNaughton's Fort 173
Maes Howe 130, 134, 139
Main Lateglacial Shoreline 33–4, 39
Main Postglacial Shoreline 38
Main Postglacial Transgression 38–40
Mammals 19–20, 85
Manure 148
Manx shearwater (*Puffinus puffinus*)
    101

Material culture 221
Meadowsweet (*Filipendula ulmaria*) 64
Mediterranean Basin 7
Megalithic sites 139–40
Meiklewood 84
Melasis buprestoides 106
Meldon Bridge 130, 147
Memsie 152
Mesolithic 4, 5, 8, 42, 64–73, 109–25,
    263
    artefacts 114–17, 120
    chronological framework 117–19
    colonization 113–17
    economics 119–22
    environment 64–72, 112–13
    nature of evidence 117–19
    radiocarbon dating 118
    settlement 39
    shell middens 110
    site types and archaeological perceptions
        117
    sites 40
    society 122–4
    state of research 112
    subsistence 119–21
    technology 121–2
    vertebrate fauna 112
Meso-neolithic transition 72–4, 123–4,
    263
Metalworking 188, 230
Metamorphic rocks 25
Midden deposits 110, 163
Midhowe 173, 185, 186
Midland Valley 23–5, 32
Military works 183
Millfield Farm 111, 114
Milton Loch 173, 187
Minch 87
Moine Thrust 24, 25
Mollusca 260
Monamore 54–5
Monasteries 222
Monboddo 222
Monuments
    Clava series of 133
    Neolithic 132–40
Moorlands 72, 111
Moray 222
Moray Firth 25, 41, 170, 175, 189, 221,
    225
Morton 40, 84, 89, 111, 115
Mousa 173
Mull 26, 70
Mull of Galloway 222
Mullins 199
Myrehead 152

Neolithic 5, 7, 8, 42, 57, 58, 59, 72–9, 123,
    127–49, 263
    agricultural recession 148
    artefacts 140
    astronomy, geometry, theocracy 139, 140
    burial and ceremonial structures 132
    cultivation 141–4
    economy 144–9
    enclosures 144–9
    environment 72–9, 141–4
    farming 141, 148
    houses 144–9
    land use 140–4
    monuments 132–40
    origins of concept 127
    problems of data 129
    recent trends 128–9
    regional variation 135–9
    resources 140
    settlement 139, 140, 141, 148–9
    society 132, 133, 139
    survival of evidence 131
Neolithic-Bronze Age transition 43
New Stone Age *see* Neolithic
Newmill 152, 173, 176
Newstead 85, 196, 199
Newton 84, 173, 190
Norse influence 225, 242, 264
    geography and chronology 241–4
    relationship between native and incomer
        249
    settlement sites 245–9
North Mains 54, 55, 58, 59, 68, 72, 111,
    130, 142, 222
North Pole 12
North Uist 242
North-West Highlands 24
Northern Isles 26, 247
Northton 41, 42, 84, 92–9, 130, 148
Northumbria 222
North-West Highlands 25, 27, 29, 32

Oak (*Quercus*) 15, 16, 65, 73, 211
Oats (*Avena* spp.) 76, 189, 233, 251
Oban 84, 110
Ochil Hills 192
Off-site records 256–7
On-site records 257–62
Orcadian broch tower settlements 185
Orkney 6, 40, 135, 139, 143, 145, 148, 159,
    183–5, 187, 245, 248, 250, 252
Oronsay 84, 88, 89, 110, 121
Oronsay middens 119, 120
Otoliths 119
Otter (*Lutra lutra*) 88, 148

Outer Hebrides 17, 21, 40, 41, 72, 81,
    87
Ox yoke 164
Oxychilus draparnaudi 104

Palaeolithic 109, 113, 124
Papa Stour 108
Pass of Ballater 152
Pastoralism 42
Pasture 143, 148
Pax romana 214, 218
Peat accumulation 57
Peat cutting 81
Peat development and climate change
    17–19, 21
Peat growth 5
Peat spread 79–81
Peoples 219–21
Pickletillem 59
Pictish period 102
Pictish wars 217
Pictland 235
Picts 218, 219, 221, 222, 264
    ships 221
    silver chains 220, 221
    symbol stones 220, 221
Pierowall 84
Pig 88, 148, 149, 205, 231
Pine (*Pinus sylvestris*) 21, 65, 67, 75
Pine marten (*Martes martes*) 88
Pinus sylvestris see Pine
Pit alignment systems 189
Pit defined enclosures 137
Pit names 219
Pitcarmick 222, 227
Pitcarmick houses 229–30
Pitkennedy 154
Pitnacree 130, 142
Place-names 219, 220, 245
Plant macrofossils 260
Plantago lanceolata see Ribwort plantain
Pleistocene 26, 27
Ploughs 142
Poaceae *see* Grasses
Podzolization 45–6, 56–8, 60
Polar Front 12–13, 21
Pollack 88
Pollen
    analysis 17, 59, 141, 209, 210, 260
    data 63
    diagrams 67, 69, 73, 77, 78, 80, 141, 209,
        210, 214
    preservation 64
    profiles 74
    spectra 71, 76

stratigraphical record   14
types   64
Ponies   189
Pool   222, 226, 250, 251, 253
Population   5–6
Potato   61
Pottery   134, 143, 206, 207, 249
Promontory forts   218
Pteropsida *see* Ferns
Puffin (*Fratercula arctica*)   102

Quanterness   84, 86, 99, 173, 184, 185
Quarrying   140
Quartzite   35
Quartzite mountains   25
Quaternary period   9, 26
Quercus see Oak
Querns   145
Quernstones   208

Rabbit (*Oryctolagus cuniculus*)   83
Radiocarbon dating   6–7, 41, 54, 71, 88,
113, 115, 118, 140, 169, 191, 208, 218,
233
Raigmore   130, 146
Rainfall   21
Rannoch Moor   18, 25, 33
Rat holes   263
Ratho   222
Rattray   152
Razorbill (*Alca torda*)   88, 101
Recumbent stone circles   136, 137
Red deer (*Cervus elephus*)   71, 87, 112, 148,
231, 252
Red grouse (*Lagopus lagopus*)   101
Red squirrel (*Sciurus vulgaris*)   88
Regions   8, 170, 183–9
Reindeer (*Rangifer tarandus*)   19, 86
Reineval   81
Religion   233–9
Resources, Neolithic   140
Restenneth   222
Rhoin Farm   72
Rhum   42, 70, 115, 116
Rhum bloodstone   122
Rhynie   222
Ribwort plantain (*Plantago lanceolata*)   76,
192
Ring-ditch houses   176
Ring-forts   225
Ring-groove houses   187
Rinyo   130, 145, 148
Rispain   173, 187
Ritual monuments   160
River Dee   121

Robertshaven   103
Rock art   144
Rocks   24, 25, 47
Roe deer (*Capreolus capreolus*)   87
Roman activity   180
Roman army   6
Roman camps   192
Roman Christianity   234, 239
Roman forts   200
Roman occupation   169, 179, 182, 183, 187,
192, 195–216, 264
campaign armies   203–4
chronology   195–8
diet   207
early Antonine period   196
economic demands   203–6
economic return from taxation   215
Flavian period   199, 202, 203
food requirements   212
grain supplies   204, 207
impact of demands   206–12
impact on local environment   212–16
influence of geography   198–203
legacy of   217–18
main manifestations   202
material possessions   206
military diet   205
military land confiscation   214–15
settlement patterns   214
timber supplies   208
Rosinish   41, 84, 152, 163, 164
Rotary quern   190
Roundhouses   162, 185, 186, 218
complex Atlantic   218
stone   218
Rousay   242
Rowan (*Sorbus aucuparia*)   211
Rumex see Sorrel

St Andrews   222, 230
St Kilda   90
Saithe   90, 103, 119
Salix see Willow
Salix herbacea see Dwarf Willow
Salmonid (*Salmo* sp.)   90
Sand movement amd climate change   21–2
Sanday   242, 246
Sandray   263
Sands of Forvie   21, 152
Sandstones   25
Saw-toothed grain beetle (*Oryzaephilus
surinamensis*)   107
Saxa Vord   68, 76, 79, 80
Scalloway   85, 102
Scandinavian influence   241–54

Scandinavian settlement 243
Scar 246
Scone 222, 230
Scord of Brouster 18, 54, 55, 57–9, 68, 73,
    99, 130, 142, 149
Scots 219
Scots pine *see* Pine
Scotstarvit 173, 176
Scotto-Pictish kingdom 239
Sculptor's Cave, Covesea 222
Sea birds 148
Sea-level change 33–4, 37–41
Sea scorpion (*Taurolus bubalis*) 99
Sea wrasse (*Labrus* sp.) 86
Seal 148
Seaweed 148
Sedge smut beetle (*Phalacrus caricis*) 106
Sedges (Cyperaceae) 64
Sediment deposition 42, 43
Senchus fer nAlban 219
Settlement
    Bronze Age 157–62
    Early Historic Period 222–30
    Iron Age 170–2, 175–82
    Mesolithic 39
    Neolithic 140, 141, 148–9
    Norse 245–9
    nucleated 141
    patterns 61, 214
    record 218
    Roman occupation 214
    Scandinavian 243
    Southern Uplands 189
Shag (*Phalocrocorax aristotelis*) 101
Sheep (or goat) 127, 148, 205
Sheep (*Ovis aries*) 91, 205
Shellfish 148, 231
Sheshader 152
Shetland 17, 18, 40, 58, 71, 79, 80, 187, 245,
    248
Shurton Hill 130, 142
Skaill 250, 251, 253
Skara Brae 84, 91, 107, 129, 130, 139, 143,
    145, 148, 149
Skate (*Raja batis*) 91
Skye 26, 32, 42
Smittons 120
Snails 104
Soil–climate–crop interactions 52
Soil–human interactions 61
Soil(s) 45–62, 260
    acidity 21
    associations 46, 48
    buried 53–6, 58
    change, evaluation 53–61
    classification 47

    conservation 148
    distribution 50
    droughtiness 52
    erosion 41–3, 60, 210
    evolution, Holocene 56–8
    formation 45–6
    groups 48–9
    human impact on 58–61
    landscape 49
    maintenance 60
    present-day 46–9
    resource 45–53
    series 46
    sources of past evidence 53–6
    survey maps 46, 47
    Survey of Scotland 58
    types 47, 48
    variability 60
Sorisdale 247
Sorrel (*Rumex*) 64, 76
Souterrains 170, 218
South Uist 41, 70, 71, 81, 106
Southern Uplands 23, 26, 50
    settlement 189
Southern Uplands Fault 24, 26
Spades *see* Cultivation
Spelt wheat (*Triticum spelta*) 207
Sphagnum 18, 76
Sprouston 222
Starr 68, 81, 111, 120
States, emergence of 239
Stewartry of Kirkcudbright 16, 187
Stock-raising 231–3
Stone circles 139
Stone implements 140
    Arran pitchstone 140
    chert 140
    flint 140
    of jadeite 140
    quarries 140
    quartz 140
    Rhum bloodstone 140
Storegga Slide 40
Strageath 54–5, 58
Strathallan 58
Strathearn 38, 39
Strathmore 58, 218
Streng Moss 200, 209
Structural provinces 23–5
Sturgeon (*Acipenser sturio*) 90
Subsistence 53, 127
    Bronze Age 162–7
    Mesolithic 119–21
Suisgill 130, 142, 152
Swallow (*Hirundo rustica*) 88
Symbolism 234

Tankardstown, Ireland 146
Tap O'Noth 222
Tay Estuary 39, 40, 170
Tay Firth 175
Tay Valley 39, 139
Tentsmuir Forest 41, 152, 222
Tephra effect 21
Tephrochronology 166
Terracing of hillslopes 233
Tertiary period 26
Tertiary Volcanic Province 26, 33
Thanages 219, 222
Theocracy 139–40
Thrush (*Turdus* sp.) 90
Tilia see Lime
Tills 27, 30
Timber 158, 178, 208, 212, 226–30, 264
Tofts Ness 54, 55, 59, 152, 173, 184
Tormore 54–5, 57, 152, 158
Torridonian sandstone 57
Torrs 173
Torrs Warren 41
Transhumance 144, 149
Traprain Law 32, 173, 179, 180, 182, 200, 218, 222
'Treb dykes' 63
Tree-rings 166
Tree species 15–17, 64, 211
Trelystan, Wales 149
Triticum see Wheat
Tulloch Wood 152, 163
Tuquoy 108
Turbot (*Scophthalmus maximus*) 90
Tweed Valley 30
Tyne-Forth Province 170

Udal *see* Coileagean
Ulmus see Elm
Ulva Cave 110, 111, 119
Unenclosed centres 215, 230
Unst 242

Vatersay 113
Vegetation change 63–82, 141
  and climate change 14–27
  Holocene 64–79
  sources of evidence 63–4
Venerable Bede 147
Vertebrate fauna 83–103
  Holocene 86–103
  Lateglacial 86
  Mesolithic 112
  nature of the evidence 83–6

Vikings 221, 241, 249
Vitrified forts 178
Volcanic dome 180
Volcanic eruptions 166
Volcanic events 5
Volcanic tephra 263
Volcanoes 21
Vole (*Microtus* sp.) 86, 88
Votadini 219
Votadinian houses 176, 179, 191, 192

Wag of Forse 222, 227
Walton Moss 200, 209
Warfare 221
Water bodies 5
Waternish 152
Weasel (*Mustela nivalis*) 88
Weathering 45–6
Wemyss 84
Wessex 129, 133
Western Isles 129, 135, 143, 146, 185, 187, 248
Westray 242
Wetland sites 265
Whale 148
Whale bones 90
Wheat (*Triticum dicoccum* = Emmer) 76, 113, 164, 189, 204, 233
Wheelhouses 170, 185, 186, 226
White Meldon 173
White Moss, Shapinsay 152
White-tailed eagle (*Haliaetus albicilla*) 99
Whitekirk 222
Whithorn 222
Whiting 88
Wild cat (*Felis sylvestris*) 91
Wild resources 148, 149, 231–3
Willow (*Salix* sp.) 211, 251
Windermere Interstadial 27
Wolf (*Canis lupus*) 87
Wooden artefacts 226
Woodland 64–72, 191–2, 210, 212, 214
  management 143
  rate of expansion 67
  reduction from late Neolithic times 74–9, 141
  regeneration 72–4, 141
  time-transgressive nature of spread 65
Workshops 229
Wormy Hillock 130
Wrack (*Fucus* sp.) 251

Yorkshire 129

Index compiled by Geoffrey C. Jones